VOLUME IX (in preparation)

CHEMICAL EXPERIMENTATION UNDER EXTREME CONDITIONS

Edited by Bryant W. Rossiter

VOLUME X

APPLICATIONS OF BIOCHEMICAL SYSTEMS IN ORGANIC CHEMISTRY, in Two Parts

Edited by J. Bryan Jones, Charles J. Sih, and D. Perlman

VOLUME XI

CONTEMPORARY LIQUID CHROMATOGRAPHY

R. P. W. Scott

VOLUME XII

SEPARATION AND PURIFICATION, Third Edition

Edited by Edmond S. Perry and Arnold Weissberger

VOLUME XIII

LABORATORY ENGINEERING AND MANIPULATIONS, Third Edition

Edited by Edmond S. Perry and Arnold Weissberger

VOLUME XIV

THIN-LAYER CHROMATOGRAPHY, Second Edition

Justus G. Kirchner

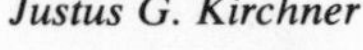

TECHNIQUES OF CHEMISTRY

ARNOLD WEISSBERGER, *Editor*

VOLUME XII

SEPARATION AND PURIFICATION

Third Edition

TECHNIQUES OF CHEMISTRY

VOLUME XII

SEPARATION AND PURIFICATION

Third Edition

Edited by
EDMOND S. PERRY AND ARNOLD WEISSBERGER
Research Laboratories
Eastman Kodak Company
Rochester New York

A WILEY-INTERSCIENCE PUBLICATION

JOHN WILEY & SONS

New York · Chichester · Brisbane · Toronto

Library of Congress Cataloging in Publication Data:

Main entry under title:

Techniques of chemistry.

Supersedes Technique of organic chemistry and its companion, Technique of inorganic chemistry.
Includes bibliographical references.
CONTENTS: v. 1. Physical methods of chemistry, edited by A. Weissberger and B. W. Rossiter (incorporating 4th completely rev. and augm. ed. of technique of organic chemistry) [etc.]
1. Chemistry—Manipulation—Collected works. I. Weissberger, Arnold, 1898— ed. II. Technique of organic chemistry. III. Technique of inorganic chemistry.

QD61.T4 542 77-114920
ISBN 0-471-026557(v. 12)

Printed in the United States of America

10 9 8 7 6 5 4 3 2 1

AUTHORS

CHARLES M. AMBLER

Sharples-Stokes Division, Pennwalt Corporation, Warminster, Pennsylvania

A. J. BARNARD, JR.

Research and Development, J. T. Baker Chemical Company, Phillipsburg, New Jersey

E. R. BAUMANN

Anson Marston Distinguished Professor of Engineering, Iowa State University, Ames, Iowa

PHYLLIS R. BROWN

Department of Chemistry, University of Rhode Island, Kingston, Rhode Island

FREDERICK W. KEITH, JR.

Sharples-Stokes Division, Pennwalt Corporation, Warminster, Pennsylvania

ANTÉ M. KRSTULOVIC

Department of Chemistry, University of Rhode Island, Kingston, Rhode Island

SHELDON W. MAY

School of Chemistry, Georgia Institute of Technology, Atlanta, Georgia

C. S. OULMAN

Professor of Civil Engineering, Iowa State University, Ames, Iowa

EDWARD G. SCHEIBEL

Suntech, Inc., Marcus Hook, Pennsylvania

LLOYD R. SNYDER

Clinical Chemistry Department, Technicon Instruments Corporation, Tarrytown, New York

MORRIS ZIEF

Research and Development, J. T. Baker Chemical Company, Phillipsburg, New Jersey

INTRODUCTION TO THE SERIES

Techniques of Chemistry is the successor to the Technique of Organic Chemistry Series and its companion—Technique of Inorganic Chemistry. Because many of the methods are employed in all branches of chemical science, the division into techniques for organic and inorganic chemistry has become increasingly artificial. Accordingly, the new series reflects the wider application of techniques, and the component volumes for the most part provide complete treatments of the methods covered. Volumes in which limited areas of application are discussed can be easily recognized by their titles.

Like its predecessors, the series is devoted to a comprehensive presentation of the respective techniques. The authors give the theoretical background for an understanding of the various methods and operations and describe the techniques and tools, their modifications, their merits and limitations, and their handling. It is hoped that the series will contribute to a better understanding and a more rational and effective application of the respective techniques.

It is my pleasure to acknowledge the collaboration of Dr. Edmond S. Perry in the editorial work on volumes of the preceding and the present series dealing with aspects of separation and purification:

Technique of Organic Chemistry, Vol. IV, *Distillation*, 2nd ed., Vol. XII, Kirchner, *Thin-Layer Chromatography*, Vol. XIII, Schupp, III, *Gas Chromatography*; Techniques of Chemistry, Vol. VII, Hwang and Kammermeyer, *Membranes in Separations*, Vol. XI, Scott, *Contemporary Liquid Chromatography*, and the present volume.

The authors and editors hope that readers will find the volumes useful and will communicate to them any criticisms and suggestions for improvements.

ARNOLD WEISSBERGER

Research Laboratories
Eastman Kodak Company
Rochester, New York

PREFACE

A collection of techniques useful for the isolation and purification of compounds appeared originally in 1950 as Volume III of the series Techniques of Organic Chemistry. In its second edition, published in 1956, the subjects of the first edition were divided into two treatises: *Separation and Purification*, Parts I and II; the latter bore the subtitle *Laboratory Engineering*. The second edition was also expanded to include a section on *diffusion methods*, comprising "Thermal Diffusion," "Barrier Separations," "Dialysis," "Electrodialysis," and "Zone Electrophoresis."

In the present edition, transferred to the more comprehensive series Techniques of Chemistry, the discussion of separation techniques and laboratory engineering is continued in two separate volumes: *Separation and Purification* and *Laboratory Engineering and Manipulations*.

For several reasons the diffusion methods have been deleted from the third edition. "Barrier Separations" has been expanded into Volume VII of the present series, *Membranes in Separations*, by Hwang and Kammermeyer. "Electrophoresis," likewise, will be treated in a separate volume, together with other electrokinetic methods. "Thermal Diffusion of Organic Liquids" was eliminated because little progress has been made in this technique since the previous edition.

We deeply regret the death of Lyman C. Craig. He had planned to write a fundamental and comprehensive treatise on countercurrent distribution and dialysis. The enlightenment that his unique mind and wisdom would have given is missed, as is his friendship. His renowned *Countercurrent Distribution* had a phenomenal effect on separation science. Although countercurrent distribution is now largely superseded by chromatography, the essentials of the method are retained and included in Chapter III, "Liquid-Liquid Extraction."

"Crystallization and Recrystallization" is to become part of a new volume. "Solvent Removal, Evaporation, and Drying" has been transferred to *Laboratory Engineering and Manipulations*.

To the remaining chapters of the second edition, which are all revised for the third edition, are added the following new chapters: "Purity," "Solvent Selection for Separation Processes," "Ion-Exchange Chromatography," and "Affinity Chromatography."

Other methods concerned with separation and purification are treated in separate volumes of the present or the parent series. They are in the series Techniques of Organic Chemistry, *Distillation* (Vol. IV), *Gas Chromatography* (Vol. V), and *Thin-Layer Chromatography* (Vol. XII); in the series Techniques of Inorganic Chemistry, the chapters "Gas Chromatography" (Vol. III) and "Ion-Exchange Techniques" (Vol. IV); and in the present series, "Liquid Chromatography" (Vol. XI).

EDMOND S. PERRY
ARNOLD WEISSBERGER

Research Laboratories
Eastman Kodak Company
Rochester, New York

CONTENTS

Chapter I
Purity: Concept and Reality 1
MORRIS ZIEF AND A. J. BARNARD, JR.

Chapter II
Solvent Selection for Separation Processes 25
LLOYD R. SNYDER

Chapter III
Liquid-Liquid Extraction 77
EDWARD G. SCHEIBEL

Chapter IV
Ion-Exchange Chromatography 197
PHYLLIS R. BROWN AND ANTÉ KRSTULOVIC

Chapter V
Affinity Chromatography 257
SHELDON W. MAY

Chapter VI
Centrifuging 295
CHARLES M. AMBLER AND FREDERICK W. KEITH, JR.

Chapter VII
Filtration As a Laboratory Tool 349
C. S. OULMAN AND E. R. BAUMANN

Name Index 423
Subject Index 431

TECHNIQUES OF CHEMISTRY

ARNOLD WEISSBERGER, *Editor*

VOLUME XII

SEPARATION AND PURIFICATION

Third Edition

Chapter I

PURITY: CONCEPT AND REALITY

Morris Zief
A. J. Barnard, Jr.

1 The Evolution of Purity 1

2 Concepts of Purity 2

3 Numerical Expression of Purity 4

4 Water and "Polywater" 6

5 High-Purity Substances 7

Needs for High Purity 7
High-Purity Substances as Analytical Standards 10
Ultrapurification 11
Handling and Contamination Control 14
Containment 16
Characterization 16

1 THE EVOLUTION OF PURITY

For chemicals, the evolution of quality standards has proceeded from the rule of *caveat emptor* to specification "by sample or standard" to definition by physical and chemical measurements. This final stage became possible only with the development of measurement as a science and with an understanding of chemical combination; reliable results from routine chemical and physical analysis were also necessary.

Some early efforts to evaluate the quality of materials were associated with metals and alloys, particularly gold and silver. Possibly the earliest reference to a "purity" test appears in the Babylonian Tell "el" Amarna tablets of the fourteenth century BC. One tablet is a letter from the king of Babylon informing the king of Egypt that "of twenty *mines* of gold sent to him, only five came from the furnace as real gold." The analytical technique used was cupellation, in which base metals are oxidized and sink to the bottom of the melt. It is intriguing that this early analytical report alludes to a fraud or at least a significant departure from expected purity.

The biblical reference to "trial by fire" is an allusion to a simple test for the purity of gold, namely, that only high-purity gold remains bright when fused in air. The technique for precious metals described in ancient literature as "degree of fineness" and in the Middle Ages as "touchstone" is based on examination of the color of the streak left when the sample is scratched with a rock (usually a dense flint). The simple touchstone method permits the detection of one part of either copper or silver in one thousand parts of gold.

No survey of the early period of purity can fail to mention the alchemist. In many ways the alchemist's quest for the *universal* substance parallels the modern chemist's demand for the *pure* substance. The recipes for the "philosopher's stone" contain directions for the purification of starting materials and ingredients. Failures in accomplishing the transmutation of base metals into precious metals were often ascribed to the presence of impurities, and their removal was attempted by repeated processing.

The historian F. J. Moore has observed that the alchemists should not be judged in twentieth century terms. They had no adequate theory on which to drape speculations. Assignment of meaning to chemical purity is difficult without some knowledge of the quantitative relations involved in chemical combination. Moore has stated that there was nothing in the knowledge of the alchemist's time from which one had the right to conclude that it was impossible to obtain gold from lead.

Science in many ways only refines old knowledge. The alchemical period contributed much of modern practical knowledge of equipment and elementary techniques for purification by precipitation, filtration, crystallization, distillation, and sublimation.

2 CONCEPTS OF PURITY

With the recognition that chemical compounds exist as molecules or ions of relatively invariant composition, chemical purity could be related to the degree of identity of such species. As Eyring [1] pointed out almost 30 years ago, a pure chemical compound is one in which a *single* molecular species is present. In equivalent terms, a chemical compound is pure if all molecules present are identical. The expression "singular molecular species," or its equivalent "identical molecules," is the operative expression, and the implications can be appreciated by simple examples.

Are various polymeric or associated forms of a molecule to be treated as "identical"? If not, "pure" water is a rarity. Do isotopic differences rule out molecular identity? If so, neither "high-purity" water nor standard reference material hydrocarbons are especially pure because one in about 5000 hydrogen atoms is twice as heavy as the others. A final example to be credited to Eyring [1] is illuminating: in the most microscopic definition of purity, it would be maintained that two molecules having different energies are not identical. The third law of thermodynamics holds that all molecules of a (crystalline) substance are identical in energy only at absolute zero. Consequently, a pure compound exists only at absolute zero!

The more microscopic the viewpoint in defining a single molecular species, the more difficult it is to realize or to approach the pure chemical compound. In some ways, purity is therefore a relative property of a sample or simply a way of viewing or idealizing a sample.

A further complication appears if the concept of purity is considered from an operational viewpoint. With molecules, only a limited number of those present in the sample can be examined for "identity" or "nonidentity." The certainty with which the purity can be established is consequently limited by the uncertainties involved in this "examination" and the random errors associated with that process.

In practical considerations (see below), purity is often defined in terms of the percentage of the major constituent, often called an assay value. Here the "pure" system is implicit in the assumption of a reference standard, in the application of the laws of constant and combining proportions and in the use of atomic weights.

Another approach to defining purity is to record some property of the major constituent that is influenced by the presence of impurities. The mere recording of the boiling point or the melting point value or range is a simple example. The measurement of more than one property is especially helpful in the preparation of a highly pure material through repeated fractionation. A pure compound is considered to be attained when further fractionation fails to produce fractions with different properties.

Helium presents a fascinating case of fractionation [2]. Below 2.186°K, liquid helium passes to a "superfluid" form that has no viscosity and posses-ses the ability to penetrate minute openings. By use of these properties for the actual fractionation and for the analytical deductions, helium has been prepared with only one mole of impurity in 1×10^{10} moles of the sample. This value corresponds to a purity expressed in mole percent of 99.99999999! Perhaps this is a unique and possibly limiting case.

Purity and impurity are two sides of the same coin. For real materials, the question is often not "how pure?" but "how impure?" Specific impurities are determined and usually expressed in terms of percentage or some other fractional part of the total sample. Even with a relatively pure material, there may be less concern about the exact expression of its purity than with the nature and content of impurities present. For example, a high iron content in a glass fiber wave guide was initially believed to increase the attenuation of light passed; now it is known that only the iron(II) content is critical [3].

Ambrose [4] of the British Standards Institution has presented another interesting example. The lower pyridine bases, including the three isomeric picolines, have a refractive index in the range 1.50 to 1.51. Consequently, in refractometric measurements of one of the picolines, the presence of another as a minor impurity will have a negligible effect on the value obtained. However pyridine bases readily absorb water, which has a much lower refractive index of 1.33. A small content of water, therefore, would introduce a serious error. In contrast, if a picoline is to be used as a standard in infrared spectrophotometry, the presence of an isomeric picoline introduces serious error, but the presence of water can be easily tolerated.

Unlike the proverbial case of blind men examining an elephant, information must be pooled from numerous directions to provide a broad-based assessment of the quality, purity, and usefulness of chemicals.

3 NUMERICAL EXPRESSIONS OF PURITY

The content of a chemical species in a material can be expressed in various ways, depending on the data available and the intended purpose.

For a major or minor component, content is usually expressed in percent weight by weight (or weight by volume for solutions). For trace components (10^{-2} to 10^{-4} %) or ultratrace components ($<10^{-4}$ %), other fractional parts are often preferable, notably parts per million (μg/g), parts per billion, and so on. Unfortunately, the European concept of a billion differs from that in the United States (10^{12} vs. 10^{9}). To avoid this difficulty, IUPAC commissions have proposed the use of parts per milliard (abbreviated ppM) to designate 10^{9}. It is good practice in articles and monographs to define usages early in the text and also in tables, by such expressions as "... parts per billion (i.e., ng/g)."

Korenman [5] has urged that the Sorensen operator be used to record and report p% values (i.e., −log%) for contents. The approach warrants more attention than it has received, and it is attractive, since for trace and ultratrace components, "order of magnitude" thinking is often appropriate. Some of these various modes of expression of content are highlighted in Table 1.1.

For metals, Melchior [6] has introduced the parameter "degree of purity" (i.e., Reinheitgrad), R, defined by the relation $R = -\log(100 - w)$, where w is the weight percent of the major element. For w values of 99.0, 99.5, 99.90, 99.95, . . . , %, the R values would be 0.0, 0.3, 1.0, 1.3, . . .

The concept of "purity by difference" finds use as a simple index to purity. The percentages established for particular impurities are added and the sum is subtracted from exactly 100 to give a purity percentage. A chemical with

Table 1.1 Expressions of Content (w/w) [5]

Constituent	*Percentage*	ppm (μg/g)	ppb (ng/g)	p% (= −log %)
Macro				
Major	1–100	—	—	0 to −2
Minor	0.01–1	100–1000	—	2–0
Micro				
Trace	≤0.01	≤100	—	≥2
Ultratrace	≤0.0001	≤1	≤1000	≥4
	0.0000001	0.001	1	7

a single impurity at a content of 100 ppm (i.e., 0.01 %) would be described as 99.99 % pure. This index has greater validity for high-purity metals and nonmetals because the impurities are probably present in their elemental state. In the case of compounds, the exact nature of the impurities is not always determined. For example, all trace sulfur species may be codetermined and expressed as sulfate. Often for elements and simple inorganic compounds, the values for spectrographically detected impurities are added and the sum subtracted from 100 to give the "spectrographic purity." The value is not intended to imply that impurities not detectable by spectrography are absent. Purchasers of "spectrographic grade" materials sometimes have overlooked this point and have been surprised to find significant contents for nonmetals and their oxygenated species [7].

In the use of the purity-by-difference concept, significant figures have sometimes carelessly been ignored. For example, the sum of 0.001 % and 0.0002 %, each value expressed to the number of significant figures allowed by the precision of the measurements, is 0.001 %, not 0.0012 %!

An impurity content of 1 ppm (μg/g) corresponds to a purity by difference of 99.9999 %. For simplicity, this percentage can be read as "six nines" and the material is termed 6N [8]. Similarly, an impurity content of 0.5 ppm corresponds to 99.99995 % purity by difference and would be expressed as either 6N5 or 6N+. (It is interesting to contrast this approach with that using p%; the impurity content for the two examples reduces to 4.0 and 4.3 p%, respectively.)

For the description of high-purity forms of elements, the purity-by-difference concept has been refined by some workers with the listing of two values, one prefixed by "m" to designate reference to metallic impurities and the other by "t" to indicate that total impurities are considered, including nonmetallic ones such as oxygen, carbon, and nitrogen [9]. In this system, for example, the expression m5N5 : t4N indicates the purity is 99.9995 % based on assessment of metallic impurities as 5 μg/g by spectrographic and spectrometric methods, but 99.99 % based on the assessment of both metallic impurities (5 μg/g) and nonmetallic ones (95 μg/g).

Purity values may also be reported in molar terms. Mole percentage values are often secured by methods of analysis based on colligative properties.

In materials science, notably semiconductor phenomena, the number of *atoms* of various impurities in a unit weight or volume may be of greater interest than the weight of the impurities. A convenient unit for this purpose is atoms per cubic centimeter.

4 WATER AND "POLYWATER"

Water has probably been subjected to more purification studies and by more techniques than any other substance. As noted in an earlier section, water can be considered "microscopically impure" even after extreme purification because a mixture of isotopes is present. The geonormal isotopic composition involves 99.98 and 0.02 atom percent for ^{1}H and ^{2}H, respectively, and 99.76, 0.04, and 0.2 atom percent for ^{16}O, ^{17}O, and ^{18}O, respectively. The isotopic composition of water from other planets may be different [10]. It is noteworthy that so-called heavy water, deuterium oxide, is now available even in drum quantities, with the price directly related to the atom percent of deuterium versus hydrogen present.

Purification of water by means of mixed-bed ion-exchange columns (strongly acidic cation exchangers containing sulfonic acid functionality plus strongly basic anion resins having pendant quaternary ammonium hydroxide groups) has largely supplanted distillation for the preparation of ion-free water. A single pass through the mixed bed yields water with a greater resistance than that obtained by three distillations in vitreous silica; 18 MΩ/cm at 25°C is routine in many laboratories (theoretical resistance at 25°C = 18.3 MΩ/cm). Membrane filters (0.22 μ) can now be supplemented with hollow fibers ($\sim$0.02 μ) to separate additional quantities of particulate matter [11]. Granular carbon columns or distillation from permanganate can be adopted to remove organic material. The employment of ion-exchange resins, carbon, distillation, and membrane filtration in sequence represents the state-of-the-art approach for the elimination of inorganic, organic, and particulate materials.

The passage of water through an octadecylsilane-bonded reverse phase column removes organic impurities, which would interfere in gradient elution high-performance liquid chromatography [12].

The isolation of "anomalous" water was reported by Deryaguin [13] in 1965 from condensation of water in minute, freshly drawn capillaries. Numerous publications in the following years considered this "anomalous" water to be a polymeric form of ordinary water, and thereby the term "polywater" was introduced. Even though both sodium and silicon were demonstrated to be present in the samples, the controversy continued. Finally, by 1972 most investigators had acknowledged that polywater was simply polycontaminated water! The polywater experience dramatically illustrates the necessity of isolating a substance in an analytically "pure" condition before making final judgments about the composition and physical properties of that substance.

5 HIGH-PURITY SUBSTANCES

The preparation and characterization of chemicals of extreme purity has evolved as a new interdisciplinary science. The introduction of new separation and purification techniques has attracted investigators to their use in isolating high-purity materials. The development of analytical techniques, such as gas-liquid chromatography (GLC), mass spectrometry (MS), combined GLC–MS, radioactive tracer methods, and neutron activation analysis (NAA), to detect and determine organic and inorganic trace impurities at or below the 0.1-ppm level, made possible the prompt assessment of purification efforts. The facile accumulation of analytical data by such techniques is reflected by the increasing number of papers devoted to the characterization of purified materials.

Trace elemental analysis has demonstrated that the purity attainable is affected by the methods of handling raw materials and intermediates as well as by the materials of construction for apparatus and storage containers. Four key parameters that merit consideration in the investigation of a highly pure material are separation and purification, handling, containment, and analysis [7, 14]. This volume considers in detail only the separation and purification parameters; this chapter, therefore, treats the other parameters briefly with reference to literature.

Needs for High Purity

The invention of the transistor stimulated enormous growth for the electronics industry in the 1950s. Because electrically active impurities have a profound effect on the yield, properties, performance, and stability of semiconductor devices, massive programs were undertaken to produce elemental germanium and silicon with extremely low impurity contents. As technology advanced, concerted efforts were required to obtain key dopants in ultrapure forms, including phosphorus tribomide and oxychloride and boron oxide and tribromide (see Table 1.2). The early emphasis on semiconductor devices and materials at the Cambridge Research Laboratories of the U.S. Air Force, coupled with two sponsored conferences and derived monographs [15, 16], stimulated interdsciplinary approaches for many aspects of high purity. In retrospect, it is apparent that the requirements of the expanding electronic device industry and the associated research and development catalyzed and supported many of the studies on the preparation and characterization of high-purity materials.

In the early 1970s research toward the use of glass fibers as wave guides for optical communication initiated research on the preparation of low-impurity inorganic oxides and salts [17]. The miniaturization of the vacuum tube by the transistor is being paralleled by the replacement of the heavy copper wire cable by a small bundle of glass fibers. Because the transmission of light along a fiber is impaired by absorption, notably by transition elements such as chromium, cobalt, copper, iron, manganese, and vanadium, the contents of those elements in the glass drawn to fiber must be extremely low. For long-distance "low-attenuation" applications, the individual transition elements should be present in the fiber at contents of 20 ng/g or less [18]. Preparation of the glass-making raw materials and the glass itself for low-attenuation purposes still presents a challenge [3].

Beginning in 1969 the lunar sample studies that formed part of the Apollo program required high-purity reagents to assure low blanks [19]. Acids (hydrochloric, nitric, hydrofluoric), oxides (phosphorus pentoxide), and fluxes (sodium carbonate) of extreme purity were needed for sample dissolution and adjustments. Additionally, some organic solvents were specially purified for extraction operations.

Developments in the biomedical and biochemical fields have also created needs for chemicals extremely low in impurity contents. Studies directed to essential trace elements and their discovery have required diet components "free" of the elements of interest. For example, calcium hydrogen phosphate with lead, cadmium, arsenic, and vanadium contents less than 100 ng/g have been required for some studies [20]. The sensitivity of enzyme systems to trace metals suggests that some earlier nutritional studies based on diets without close control of trace metal contents are suspect [21]. Recent literature indicates that a new era is at hand in which the effect of trace impurities on fundamental life processes will be elucidated.

Table 1.2 delineates some of the key applications for high-purity compounds. Although the major uses are for inorganic compounds, organic compounds of high purity are vitally important in research and analytical laboratories as standards, buffers, chelating agents, reagents, fluorescent probes, and so on.

Table 1.2 Some Uses for High-Purity Chemicals [22]

Analytical standards

1. Spectral methods: $CaCO_3$, KCl, Li_2CO_3, LiCl, Na_2CO_3, NaCl, and other metal salts and oxides
2. Organic elemental analysis
   - Carbon-hydrogen: anthracene, benzoic acid, biphenyl
   - Carbon-hydrogen and halogens: chloro-, fluoro-, and iodobenzoic acids
   - Nitrogen: $NH_4H_2PO_4$, 8-quinolinol, sulfamic acid, urea
   - Phosphorus: diphenylphosphinic acid
   - Sulfur: benzyl disulfide, sulfamic acid
3. Titrimetry: benzoic acid, $CaCO_3$, EDTA, Na_2CO_3, K biphthalate, NaCl, sulfamic acid, Tris buffer
4. Diagnostic laboratory: $CaCO_3$, cholesterol, creatinine, dextrose, glycine, KCl, KH_2PO_4, Li_2CO_3, magnesium gluconate dihydrate, $(NH_4)_2SO_4$, NaBr, NaCl, urea, uric acid
5. Fluorescence measurements: 9-aminoacridine, fluorescein and its esters, quinine sulfate dihydrate, rhodamine B
6. Gas chromatography: various organic compounds

Reagents

1. Acids: HCl, $HClO_4$, HF, HNO_3, HOAc, H_2SO_4, P_2O_5
2. Bases: Li_2CO_3, Na_2CO_3, NH_4OH
3. Fluxes: $CaCO_3$, Li_2CO_3, Na_2CO_3, P_2O_5
4. Buffers: HOAc, K biphthalate, KH_2PO_4, K citrate, Tris
5. Chelating agents: EDTA, 8-quinolinol

Biochemical and nutritional research

1. Trace element research: $CaHPO_4$, KH_2PO_4, KBr, KCl, NaCl
2. Diverse: cholesterol, EDTA, glycine, sugars, urea
3. Fluorescent probes: acridine orange, 1,8-ANS, fluorescein and its esters, 2,6-TNS, rhodamine B

Electronics and materials science

1. Semiconductor dopants: BBr_3, CrO_3, $POCl_3$, P_2O_5
2. Fiber optics wave guides: Al_2O_3, B_2O_3, BaF_2, $CaCO_3$, CaF_2, CeO_2, H_3BO_3, Li_2CO_3, Na_2CO_3, $POCl_3$, SiO_2, ZnO
3. Laser and photochemical studies: various dyes and aromatic hydrocarbons

High-Purity Substances as Analytical Standards

Compounds of relatively high purity have long been applied in chemical analysis and notably as standards for titrimetry. In the contemporary period, high-purity substances, both elements and compounds, are receiving ever-increasing attention as standards for physical methods of analysis and for procedures based on instrumental techniques. Some selected applications of high-purity compounds as analytical standards are listed in Table 1.2 [22].

Chalmers [23] and Beegly, Mears, and Michaelis [24] have considered the needs and availability of standards for analysis. In 1965 the International Union of Pure and Applied Chemistry (IUPAC) [25] advanced the following classification of standard materials:

A. *Reference or Atomic Weight Standard*: for example, atomic weight silver.

B. *Ultimate Standard*: a substance that can be purified to virtually atomic weight standard.

C. *Primary Standard*: a commercially available material of 100 ±0.02% purity.

D. *Working Standard*: a commercially available material of 100 ±0.05% purity.

E. *Secondary Standard*: a material of lower purity than can be standardized against a primary standard.

Today probably the majority of standards employed in physical or chemical analysis are of levels D and E. Materials of levels A and B are virtually unavailable, with the possible exception of metals. Indeed, until the past few years, materials of level C had only limited availability beyond those used as standards for titrimetry. The degree of difficulty in attaining even a level C standard is reflected by the purity (assay) of some representative standard reference materials introduced by the U.S. National Bureau of Standards in recent years that are listed in Table 1.3.

The most difficult purification problems are presented by organic compounds, especially those that are multifunctional or complex. The need for improved standards for diagnostic laboratories was identified some years ago, and the National Bureau of Standards has responded with the assistance of other organizations and firms. A 1962 survey of the variability of serum cholesterol values prompted work toward a cholesterol standard, which was introduced in 1967 [26]. The material corresponds to an IUPAC level E material (Table 1.3). The relatively low assay achieved demonstrates the difficulties in the purification and the characterization of complex organic compounds. Bilirubin has proved more intractable to purification (Table 1.3). In spite of the relatively low purity of the cholesterol and bilirubin standards, they have aided effectively in improving the reliability of the relevant diagnostic determinations.

Table 1.3 Purity (Assay) of Some NBS Standard Reference Materials[a]

	Percentage Purity	
Material	*Assay of Major Component*	*Based on Impurity Contents*[b]
Metals		
Gold	—	99.9997 (5N7)
Platinum	—	99.9991 (5N1)
Zinc	—	99.9997 (5N7)
Inorganic compounds		
Arsenic trioxide	99.98	
Potassium chloride	99.99	
Organic compounds		
Benzoic acid	99.98	
Bilirubin	99.0	
Cholesterol	99.4	

[a] U.S. Department of Commerce, National Bureau of Standards Special Publication 260, "Catalog of NBS Standard Reference Materials," 1975–1976 edition.

[b] Impurities include contained gases, of which oxygen is the main contaminant.

Ultrapurification

The early 1950s brought techniques that introduced a new era in the production of high-purity materials. In 1952 Pfann [27] reported zone-refining experiments that soon afforded ultrapure metals and eventually, by extension of the technique, the purification of diverse inorganic and organic compounds [28–30].

In the same year James and Martin [31] announced gas-liquid chromatography. The GLC approach virtually revolutionized the assessment of the purity of volatile organic compounds and some nonvolatile ones that could be converted to volatile derivatives. In the 1960s GLC was studied on the preparative scale in the laboratory; in the 1970s some industrial applications were reported [32–34]. It should be added that high-performance liquid chromatography (HPLC) is repeating the GLC story with initial analytical applications and, more recently, the purification of complex compounds, both inorganic and organic [35, 36].

In the 1950s, the HEPA (*h*igh-*e*fficiency *p*articulate *a*ir) filter, which was developed for nuclear energy programs in the 1940s, became generally available. This filter remains the *sine qua non* for clean room and clean air technology, including the exclusion of airborne particulates during ultrapurification processes and the transfer and subdivision of ultrapure materials [37, 38].

Ultrapurification focuses on the reduction of impurity content to a level lower than that commonly attained as a result of ordinary preparation and purification efforts. The term, therefore, is imprecise but useful [39]. Ultrapurification often requires that material (or an intermediate) be brought to a relatively high degree of purity and low impurity content by conventional methods, whereupon it is subjected to one or more stringently defined and controlled techniques. In other words, *pre*purification often precedes *ultra*purification. In some cases a classical technique such as crystallization, fractional distillation, or solvent extraction is sufficient where an appropriate starting material can be secured by prepurification. For ultrapurification, techniques are obviously favored that facilitate repeated partitioning of material between phases; for example, multipass fractional solidification and zone refining, multiplate fractional distillation, preparative scale HPLC, and preparative GLC. Our experience, and that of many other workers as well, is that ultrapurification often can be achieved in a few discrete operations. Extended schemes involving many transfers often fail because of contamination from air, containers, reagents, and so on. Zief and co-workers [14, 29, 40, 41] have published a number of monographs providing practical information on ultrapurification techniques and presenting practical examples. The following paragraphs sketch out a few considerations and approaches.

Frequently quite different processes can be applied for prepurification and ultrapurification. In fact, the chance of success is often improved when different physical and chemical phenomena are involved in the two stages. For example, if the boiling point and the freezing point of a thermally stable organic substance permit purification by means of both gas-liquid and solid-liquid phase equilibria, an excellent route to ultrapurity may be at hand. An early example from our laboratories is instructive [42]. Benzene was brought to a gas chromatographic assay of 99.9 area percent by fractional distillation of reagent grade material. The distillate was then subjected to two-stage fractional freezing at 5°C. The second state crystals, amounting to 72% of the distilled material, showed no resolvable gas chromatographic impurity peak (99.99 area percent). It is noteworthy that organic solvents of extreme purity prepared by multipass fractional freezing processes now appear to be of interest for materials science and for application in advanced laboratory studies.

The purification approaches for many organic solvents are dictated by the impurities present. For example, impurities that form an azeotrope with the main component are difficult to remove by conventional distillation. The problem of azeotropes can be circumvented by the use of preparative gas chromatography. Also, appropriate chemical treatments before fractional distillation may be a satisfactory solution to the azeotrope problem. For example, acetonitrile commonly contains traces of allyl alcohol and as a result exhibits an ultraviolet cutoff near 225 nm. Treatment of the solvent before distillation with potassium permanganate and sodium carbonate destroys the allyl alcohol. Consequently, a spectrophotometric-use acetonitrile can be obtained with an ultraviolet cutoff at 190 nm.

Since zone-refined materials that are solid at room temperature are usually isolated as large chunks, they are unsuited for many applications. Reduction of the crystalline material is sometimes difficult and must, of course, be conducted under contamination-free conditions. In favorable cases, subsequent sublimation can provide free-flowing fine crystals, possibly with added purification and greater uniformity [43].

Prepurification of an organic or inorganic compound by crystallization can result in the occlusion of solvent (mother liquors) within the resulting crystals. For example, sodium chloride crystallized from water contains occluded mother liquor. Heating of the crystals to 600°C is required to assure complete release of water, and thereby an assay of 99.995 % or greater. Up to 240 ppm of water has been detected in primary standard potassium dichromate [44]. In addition to heating, zone refining and sublimation are effective means for obtaining material free of occlusions.

When a lowering of impurity content is not readily achievable by crystallization or distillation, solvent extraction sometimes can be effective. For the removal of metal ions from compounds that can be dissolved in a substantially aqueous medium, the use of a chelating extractant can be valuable. For example, the use of dithizone, 8-quinolinol, or ammonium 1-pyrrolidinecarbodithioate allows the extractive separation of up to 23 elements into methyl isobutyl ketone or chloroform [45]. For the removal of anions and metals as anionic complexes, alkyl amines of molecular weight 351 to 393 can be applied; such amines have been termed "liquid ion exchangers" [46].

Preparative GLC was described by Kirkland as early as 1957 [32]. Industrial units are now producing material of 99.5 to 99.9 % purity at the rate of 5 to 15 metric tons/year [33]. However the technique has not lived up to its initial promise for the facile batch production of diverse ultrapure organic compounds. One of the difficulties has been the bleeding of the liquid phase into the isolate, requiring its further purification—for example, by fractional distillation [34]. High-performance liquid chromatography appears to offer many exciting possibilities for the purification of organic and inorganic compounds, both liquid and solid. Since 80 % or more of known chemical compounds are nonvolatile and only a limited number can easily be made volatile by derivatization, HPLC is favored over GLC. The scale-up of HPLC to the purification of multigram quantities has been initiated. One automated unit now available allows 500 ml of solution to be eluted per minute [47].

The high resolution of liquid chromatography is illustrated by the preparation of polystyrene standards having various molecular weights in the range 900 to 800,000. For these standards, the polydispersity (i.e., the ratio of the weight average and number average molecular weights) is reported to range from 1.0025 to 1.009 [48]. HPLC is being advanced by instrumentation and the availability of optimized supports, including highly uniform macropellicular materials and closely fractionated microparticles of silica gel or aluminum oxide [49].

For the removal of electroactive metal ions from some water-soluble salts, controlled potential electrolysis of their aqueous solution at the mercury cathode can be valuable. The technique has been applied to calcium nitrate to lower its transition metal content [50].

Purification of tank gases before their use in ultrapurification and some trace analysis is a routine requirement. Many trace impurities can be removed from inert gases by adsorption on 5 Å molecular sieves at cryogenic temperatures; carbon dioxide, ammonia, and hydrogen sulfide can be adsorbed at room temperature [51]. Scrubbing with a potassium hydroxide solution has long been applied to remove carbon dioxide. Combustion by passage over copper oxide serves for the destruction of hydrogen at 300°C and saturated hydrocarbons at 800°C.

Handling and Contamination Control

Although the control of temperature and humidity is now routine for many laboratories, the need for reducing airborne contamination is not commonly appreciated. The electronics, aerospace, and nuclear industries, followed by the pharmaceutical industry, were the first to adopt elaborate installations to remove airborne contaminants in the form of dusts, mists, and fumes.

Classes of clean environments can be defined by the number and size of the particulates present in the air. U.S. Federal Standard 209a was established in 1963 to define the concentration of particulates in sensitive work areas [52]. Classes were defined in terms of the maximum number of particles 0.5 and 5.0 μ in diameter present in a cubic foot of air. For class 100, the "cleanest" environment specified, no more than 100 particles 0.5 to 5.0 μ in diameter are permitted. No particles 5.0 μ or larger in diameter are permitted.

Class 100 conditions are attained by air filtration involving HEPA filters [53]. Two types of airflow have been applied to remove particulate matter. The first is a clean room containing HEPA filters spaced at intervals in the ceiling. The air is removed by grilles located at the sidewalls at floor level. Because of the turbulent airflow conditions existing within such a room, an airborne particle can pass a critical work area several times before exiting.

The second type is the laminar-flow clean room derived from Whitfield's concept of laminar patterns of a unidirectional airflow [54]. Air moves in one pass from a bank of HEPA filters in a laminar fashion to the work area and exits without turbulence. Where a HEPA filter bank is mounted over the entire ceiling, the entire room can virtually be maintained at a class 100 level.

In the 1960s the high cost of class 100 laminar flow clean rooms discouraged many applications. The introduction of clean air stations and modules made possible the upgrading of laboratories with a minimum of alterations and relatively small cost, and without loss of existing bench and floor space. Zief and Nesher [55] early reported what could be accomplished with a two-man laboratory for about $25,000.

Clean air modules mounted about 1 m above existing bench tops provide maximum usefulness. Dust control hoods and laminar-flow work stations are recommended for localized activities. Where noxious vapors are evolved, clean air fume hoods can be installed; by balancing the flow of air through the HEPA filters and the exhaust flow, no noxious vapors developed within the hood are allowed to enter the laboratory proper. Trace organic compounds in the gaseous state can be removed from laboratory air by its repeated passage through a bed of activated carbon that is either incorporated into the clean air module or is contained in a separate fan-driven unit.

As an initial step in reducing particulate contamination, all incoming air inlets from heating, ventilating, and air-conditioning ducts should be provided with a "85–95 %" pleated glass-fiber filter. These filters are 85 % efficient for the removal of particles 0.5 to 5 μ in diameter and 95 % efficient for particles greater than 5 μ. A filter is commonly placed behind an outer plastic grid inserted into a stainless steel rectangular frame, which is fastened to the original air inlet by stainless steel screws. The area of the filter should be considerably larger than that of the original air inlet, to ensure that the effective pressure at the air duct is not reduced significantly [56].

Humidity control can be an important factor in the control of particulate contamination. At low relative humidity (<40 %), airborne particles develop electrostatic charge and thereby present a serious contamination problem even in a class 10,000 area! Synthetic fibers collect charged particles more effectively than cotton; consequently nylon laboratory coats are preferred over cotton ones.

Laboratory walls are best coated with a resistant epoxy paint containing no pigments or tinting additives. When there is doubt regarding the elemental composition of paints or other materials of construction, a sample should be analyzed appropriately. It is best to coat the tops of laboratory benches with epoxy paint, and for further protection they can be covered with Teflon FEP or polyethylene sheeting.

Containment

In the isolation and storage of high-purity materials, contamination from the walls of containers rivals that from contact with laboratory air. Although the leaching of elements from walls by liquids is the primary difficulty, organic compounds can also present problems. For example, bis(2-ethylhexyl) phthalate used as a plasticizer for polyvinyl chloride enters blood stored in bags of that polymer and is localized in the lipoprotein fraction of the plasma [57].

The problem of contamination from various materials has been considered by a number of workers and from different viewpoints [40, 58–60]. As a source of contamination, key container materials can be rated as follows: polyfluorocarbon $<$ polyolefin $<$ vitreous silica $<$ platinum $<$ borosilicate glass. Experience in many laboratories attests to the advantages of Teflon FEP, polyethylene, and vitreous silica. When bottles, beakers, and separatory funnels of Teflon FEP became commercially available, they quickly became standard equipment wherever trace element contamination was a consideration.

Often the literature on the leaching of trace elements from glass and plastics is difficult to evaluate because only fragmentary data on essential parameters are reported. For an understanding of the leaching behavior of a glass, for example, its composition, the upper temperature of working, the manufacturer, and the nature and duration of all cleaning procedures are important.

The cleaning of laboratory ware must receive careful attention to eliminate trace element contamination. In fact, the validity of published studies is sometimes in doubt because of inattention to exhaustive cleaning of equipment. A good general-purpose cleaner for glass and many polymeric materials is warm concentrated nitric acid. For glass and polyethylene, a contact time with this acid of 5 to 6 hr at 60°C is appropriate, followed by copious washing with high-purity water and drying at room temperature in a laminar-flow, clean air station. Polypropylene should be treated with the warm acid for no longer than one hour. Metal ions can be removed from glassware by washing with a 1:1 mixture of concentrated sulfuric acid and concentrated nitric acid, then rinsing with high-purity water. Because chromium is adsorbed by borosilicate to the extent of 10 ng/cm^2 and can be removed only with difficulty, the use of dichromate–sulfuric acid cleaning solution is not recommended [61]. Often after preliminary cleaning of a container, the best leaching agent is the liquid or solution that is to be worked or stored in it.

Characterization

That a compound of high purity has indeed been obtained by one or more purification steps can be confirmed only by analysis. The characterization of high-purity substances continues to present a challenge. Four approaches have been delineated by Barnard [25]: (1) evaluate the major constituent ("assay") or the overall purity, (2) determine minor and trace constituents, (3) conduct various general tests, and (4) measure physical and general properties that can aid in defining quality. The extent to which these four goals can be realized clearly depends on the nature of the compound, the nature and extent of impurities present, and the state of the development of relevant analytical techniques. Table 1.4, updated from its most recent appearance [62], reflects thinking about the methods and techniques applicable to the *practical* analysis of both inorganic and organic compounds of high purity.

When purity greater or equal to 99.95 % must be assured, chemical assay methods are severely limited. For research samples and some standard materials, high-precision coulometry and isotope dilution mass spectrometry are feasible. Of greater practical utility are precision gravimetry and precision titrimetry. Precision gravimetry, early applied in the establishment of chemical atomic weights, is important in the assessment of high-purity inorganic compounds. For example, the gravimetric determination of chloride performed with meticulous attention to details provides agreement between replicates better than 1 part in 5000 [63].

Precision weight titrimetry can approach the reliability of gravimetry [64, 65] and is often overlooked. The technique is best reduced to a "difference" measurement. In the simplest case, to a weighed amount of the substance to be assessed is added a reagent of unambiguous purity in a weighed amount either slightly greater or less than that required for complete reaction. This small difference, which often corresponds to less than 0.1 % of the assay value, is then measured by a volume-based titration. The titrant is often a dilute solution of either the substance to be assessed or the comparison reagent. Buoyancy corrections are made for all weighings, and the end point is established by an instrumental technique. In acid-base titrations the second differences of pH measurements made after addition of portions of the titrant serve to establish the end point. The technique has been extended to complexometric titrations with photometric end point detection [64].

It may be noted that titrimetry and gravimetry in a nonprecision mode can approach reliabilities of 0.1 % and thereby provide assay values of 99.9+ % and the assurance that no blunder has occurred in purification processes. For this level of reliability, introduction of constant potential coulometers may facilitate the assay of inorganic and organic compounds presenting an electroactive ion or function.

Another direct evaluation of purity is the measurement of a property of the major constituent that undergoes a change proportional to the mole fraction of impurities. Precision cryoscopy represents a powerful tool; unfortunately

Table 1.4 Methods for the Practical Analysis of High-Purity Chemicals

Major Constituent and Assessment of Purity	*Minor and Trace Constituents*	*Physical Properties*	*General Tests*
Coulometry	Activation analysis	Boiling point	Acidity or alkalinity
Differential scanning calorimetry	Atomic absorption photometry (flame and furnace)	Chemical microscopy	Ash
Elemental organic analysis	Cold-vapor atomic absorption photometry	Density	Carbonizable substances
High-performance liquid chromatography	Emission spectrography	Differential thermal analysis	Clarity of solution
Nuclear magnetic resonance	Extraction-spectrophotometry	Electrical conductivity (low level)	Heavy metals
Phase-solubility analysis	Flame-emission photometry	Fluorescence quantum yield	Loss on drying
Precision gas chromatography	Fluorimetry	Melting and freezing points	Particulate matter
Precision gravimetry	Gas chromatography	Refrative index	pH of solution

Precision weight titrimetry (instrumental end point detection)	High-performance liquid chromatography	Specific rotation	Residue after evaporation
	Isotope dilution	Spectral curves and maxima (UV, visible, IR, fluorescence emission)	Thermogravimetric analysis
	Kinetic analysis	Transition temperatures	
	Mass spectrometry		
	Nuclear magnetic resonance		
	Polarography		
	Spot tests		
	Stripping analysis		
	Thin-layer chromatography		
	Titrimetry (electrometric or photometric end point detection)		
	Turbidimetry		
	Ultraviolet, visible, and infrared spectroscopy		
	X-Ray fluorescence		

large samples, extended equilibration times, and the need for precision thermometry make this approach unsuited to practical analysis [66]. Differential scanning calorimetry (DSC), however, has proved to be an acceptable refinement. Joy and co-workers [67] extended DSC to the high-purity region and concluded that for an *absolute* determination, the upper limit is 99.95 mole percent. Differences in impurity content as little as 0.005 mole percent can be detected, thereby allowing the *relative* purity of two or more lots of a thermally stable compound to be assigned up to 99.99 mole percent. A differential scanning calorimeter presents an analog recording of the energy involved in a thermal transition as a function of the scanning temperature. The mole impurity content is calculated by application of the Van't Hoff equation [68].

Lawrenson [69] has shown that nuclear magnetic resonance (NMR) can be used directly for purity measurements as well as for phase diagram measurements [70]. Possibilities exist for exploiting the improved detection of the solidus provided by NMR to advance the accuracy of the DSC method [71].

Phase-solubility analysis (PSA) allows the assessment of the purity of non-ionic organic compounds of high purity [72]. With use of resonable sample sizes and a microbalance for weighings, as little as 0.05 % by weight of an impurity can be determined [64].

Gas chromatography has been widely applied to the evaluation of the purity of thermally stable low- and high-boiling organic compounds. "GC assays" are highly dependent on instrumental conditions and blind to any impurities that are not detected or resolved. Thermally unstable compounds can be converted to stable derivatives—for example, by silylation or esterification. However the derivatization may be incomplete and some impurities may not form volatile derivatives. HPLC offers an advantage in that assay values can be established without regard to the thermal stability of the compound or impurities; here also, however, the values are dependent on instrument conditions.

In the initial study of an ultrapurification process it is often appropriate to apply more than one method of assay. Caution should be exercised when the analytical method is based on the same principles as the purification. For example, where multipass zone refining fails because of solid solution formation, DSC may fail to detect the impurity remaining [67].

Some of the techniques useful for the practical determination of minor and trace elements in high-purity substances are listed in Table 1.4. It is imperative in ultratrace analysis (<1 μg/g) to employ "clean air" techniques (see sections on ultrapurification and containment) and reagents of special or high purity [40].

For trace metals and a few nonmetals, survey dc-arc emission spectrography is probably the most valuable technique. The limits of detection, as in many survey techniques, can be improved by preconcentration techniques. For example, organic compounds can be ashed either wet or dry [64]. Metal impurities in inorganic compounds can be enriched by collection, extraction, or ion-exchange processes [73, 74]. For the trace determination of individual trace elements, atomic absorption spectroscopy finds ever-increasing application in its various forms (flame, furnace, cold vapor). Extraction with ammonium 1-pyrrolidinecarbodithioate into methyl isobutyl ketone has been applied to the atomic absorption determination of about 20 elements [75]. The organic extract is fed to the flame, providing enhancement of the signal over the use of an aqueous solution. Differential pulse polarography offers opportunities for the determination of a few trace metals in high-purity substances and can be combined with anodic stripping from a pendant mercury drop [76].

Since a high-purity compound provides a "clean" reaction environment, kinetic methods of trace analysis should be relevant to the determination of particular impurities in such substances. About 200 kinetic methods have been applied to the determination of about 34 elements [77]. The long-known catalytic effect of trace iodide on the arsenite reduction of cerium(IV) has been applied to the determination of trace iodide in high-purity alkali halides, acetic acid, and certain other chemicals down to contents of 0.005 ppm with sample sizes of 2 g or less [77, 78].

The determination of trace organic impurities remains a challenge. Often the impurities present must first be identified, and thin-layer and liquid chromatography techniques are often helpful in this effort.

Many general tests long used for the characterization of pharmaceutical and reagent-grade chemicals can be used in the evaluation of high-purity compounds (see Table 1.4). Often the tests need only simple modification, such as an increase in sample size, a change in the concentration of reagents, or more precise measurements. For organic liquids, the determination of the residue after evaporation (nonvolatile matter) is appropriate. Any significant residue can be examined by infrared and other techniques. To obtain reliable results below about 10 ppm, it is imperative that the evaporation be performed under "clean air" conditions with proper selection of sample size and weighing conditions. Values as small as 0.5 ppm can be reported [79].

Physical properties can be used to monitor the purification of a substance. For example, measurement of the absorption of 8-quinolinol at selective wavelengths in the 410–600 nm region allows the removal of trace metals in a multistage purification process to be followed [80].

References

1. H. Eyring, *Anal. Chem.*, **20**, 98 (1948).
2. M. A. Biondi, *Rev. Sci. Instr.*, **22**, 535 (1951).
3. A. D. Pearson, in *Applied Solid State Science*, Vol. 6, Academic Press, New York, 1976, p. 185.
4. D. Ambrose, *Nature*, **171**, 902 (1953).
5. I. Korenman, *Analytical Chemistry of Low Concentrations*, J. Schmorak, Transl., Israel Program for Scientific Translations, Jerusalem, 1968.
6. P. Melchior, *Metall*, **12**, 822 (1958).
7. M. Zief and F. W. Michelotti, *Clin. Chem.*, **17**, 833 (1971).
8. Light and Co., England, *Catalog of Ultrapure Elements*, 1961.
9. Alfa Products, Ventron Corp., Beverly, Mass., 1971–1973 catalog, p. 147.
10. R. H. Eastman, *General Chemistry: Experiment and Theory*, Holt, Rinehart and Winston, New York, 1970, p. 46.
11. Romicon Hollow Fibers, Romicon, Inc., Woburn, Mass.
12. R. E. Majors, *J. Assoc. Off. Anal. Chem.*, **60**, 186 (1977).
13. B. V. Deryaguin, M. V. Talaev, and N. N. Fedyakin, *Proc. Acad. Sci. USSR, Phys. Chem.*, **165**, 807 (1965).
14. M. Zief and R. M. Speights, Eds., *Ultrapurity: Methods and Techniques*, Dekker, New York, 1972.
15. A. F. Armington, M. S. Brooks, and B. Rubin, "Purification of Materials," *Ann. N.Y. Acad. Sci.*, **137**, Art. 1, 1–402 (January 20, 1966).
16. M. S. Brooks and J. K. Kennedy, *Ultrapurification of Semiconductor Materials*, Macmillan, New York, 1962.
17. A. D. Pearson, paper presented at the Tenth International Congress on Glass, Kyoto, Japan, July 1974.
18. A. D. Pearson and W. G. French, *Bell Lab. Record*, **50**, 103 (1972).
19. A. J. Barnard, Jr., and E. F. Joy, *Chemist* (New York), **47**, 243 (1970).
20. Dr. Klaus Schwarz, Veterans Hospital, Long Beach, Calif., personal communication, 1975.
21. K. Schwarz, "Elements Newly Identified as Essential for Animals," paper presented at the International Atomic Energy Agency Symposium on Nuclear Activation Techniques in Life Sciences, Bled, Yugoslavia, April 10, 1972.
22. A. J. Barnard, Jr., *Proceedings of the Sixth Materials Research Symposium, Institute for Materials Research*, U.S. National Bureau of Standards, Gaithersburg, Md., October 29–November 2, 1973, p. 320.
23. R. A. Chalmers, in *Comprehensive Analytical Chemistry*, Vol. III, G. Svehla, Ed., Elsevier, Amsterdam, 1975, p. 185.
24. H. F. Beeghly, T. W. Mears, and R. E. Michaelis, in *Treatise on Analytical Chemistry*, Part III, Vol. 3, I. M. Kolthoff, P. J. Elving, and F. H. Stross, Eds., Wiley, New York, 1976.
25. Report prepared by Analytical Standards Committee, *Analyst* (London), **90**, 251 (1965).
26. NBS Certificate of Analysis, Standard Reference Material 911, October 20, 1967.

27. W. G. Pfann, *Trans. AIME*, **194**, 747 (1952).
28. E. F. G. Herington, *Zone Melting of Organic Compounds*, Wiley, New York, 1963.
29. M. Zief and W. R. Wilcox, *Fractional Solidification*, Dekker, New York, 1967.
30. G. R. Atwood, in *Recent Developments in Separation Science*, Vol. 1, N. N. Li, Ed., Chemical Rubber Co., Cleveland, 1972, p. 1.
31. A. T. James and A. J. P. Martin, *Biochem. J.*, **50**, 679 (1952).
32. J. J. Kirkland, in *Gas Chromatography*, J. Coates, H. J. Noebels, and I. S. Fagerson, Eds., Academic Press, New York, 1958, p. 203.
33. F. W. Karasek, *Res. Devel.*, **27** (9), 30 (1976).
34. J. R. Gruden and M. Zief, in Ref. 14, p. 131.
35. C. T. Enos, G. L. Geoffroy, and T. H. Risby, *J. Chromatogr. Sci.*, **15**, 83 (1977).
36. J. N. Little, R. L. Cotter, J. A. Prendergast, and P. D. McDonald, *J. Chromatogr.*, **126**, 439 (1976).
37. J. A. Paulhamus, in Ref. 14, p. 263.
38. *Contamination Control Handbook*, NASA-CR-61264, NASA–George C. Marshall Space Flight Center, Alabama, February 1969.
39. P. Jannke, J. K. Kennedy, and G. H. Moates, in Ref. 29, p. 463.
40. M. Zief and J. W. Mitchell, *Contamination Control in Trace Element Analysis*, Wiley, New York, 1976.
41. M. Zief, Ed., *Purification of Inorganic and Organic Materials*, Dekker, New York, 1969.
42. Ref. 40, p. 123.
43. E. C. Kuehner and R. T. Leslie, in *Encyclopedia of Industrial Chemical Analysis*, Vol. 3, F. D. Snell and C. L. Hilton, Eds., Wiley-Interscience, New York, 1966, p. 572.
44. J. Knoeck and H. Diehl, *Talanta*, **16**, 181 (1969).
45. H. Malissa and E. Schöffmann, *Mikrochim. Acta*, **1**, 187 (1955).
46. T. Susuki and T. Sotobayashi, *Bunseki Kagaku*, **12**, 910 (1963).
47. Preparative Liquid Chromatograph LC-500, Waters Associates, Milford, Mass.
48. J. L. Waters, *J. Chromatogr.*, **55**, 213 (1971).
49. F. M. Rabel, *Am. Lab.*, **7** (5), 53 (1975).
50. M. Zief and J. Horvath, *Lab. Prac.*, **23** (4), 175 (1974).
51. C. K. Hersh, *Molecular Sieves*, Reinhold, New York, 1961.
52. "Clean Room and Work Station Requirements, Controlled Environment," Federal Standard 209a, General Services Administration Business Service Centers, August 10, 1966.
53. H. Gilbert and J. H. Palmer, "High Efficiency Particulate Air Filter Units," TID-7023, U.S. Atomic Energy Commission, Washington, D.C., August 1961.
54. W. J. Whitfield, "A New Approach to Clean Room Design," Sandia Corp., Report SC-4673 (RR), Office of Technical Services, Department of Commerce, Washington, D.C., March 1962.
55. M. Zief and A. Nesher, *Clin. Chem.*, **18**, 446 (1972).
56. Ref. 40, p. 50.
57. R. J. Rubin, technical paper presented at the Regional Technical Conference of the Society of Plastics Engineers, Palisades Section, March 20–22, 1973, p. 81.
58. D. E. Robertson, in Ref. 14, p. 207.

59. C. C. Patterson and D. M. Settle, *Proceedings of the Seventh Materials Research Symposium*, U.S. National Bureau of Standards, Gaithersburg, Md., October 7–11, 1974, p. 321.
60. E. C. Kuehner and D. H. Freeman, in Ref. 41, p. 297.
61. R. E. Thiers, in *Methods of Biochemical Analysis*, Vol. 5, D. Glick, Ed., Wiley-Interscience, New York, 1957, p. 274.
62. A. J. Barnard, Jr., in Ref. 14, p. 413.
63. K. Little, *Talanta*, **18**, 927 (1971).
64. A. J. Barnard, Jr., E. F. Joy, K. Little, and J. D. Brooks, *Talanta*, **17**, 785 (1970).
65. A. J. Barnard, Jr., E. F. Joy, and F. W. Michelotti, *Clin. Chem.*, **17**, 841 (1971).
66. F. W. Schwab and E. Wickers, *Temperature, Its Measurement and Control in Science and Industry*, Reinhold, New York, 1941, pp. 246–264.
67. E. F. Joy, J. D. Bonn, and A. J. Barnard, Jr., *Thermochim. Acta*, **2**, 67 (1971).
68. G. L. Driscoll, I. N. Duling, and F. Magnotta, in *Analytical Calorimetry*, R. S. Porter and J. F. Johnson, Eds., Plenum Press, New York, 1968, pp. 271–278.
69. I. J. Lawrenson, *Chem. Ind.* (London), **1972**, 172.
70. E. F. G. Herington and I. J. Lawrenson, *J. Appl. Chem.*, **19**, 337 (1969).
71. P. D. Garn and J. J. Houser, personal communication.
72. W. J. Mader, *Crit. Rev. Anal. Chem.*, **1**, 193 (1970).
73. E. F. Joy, N. A. Kershner, and A. J. Barnard, Jr., *Spex Speaker*, **16** (3), 1 (1971).
74. N. A. Kershner, E. F. Joy, and A. J. Barnard, Jr., *Appl. Spectrosc.*, **25**, 542 (1971).
75. J. W. Robinson, P. F. Lott, and A. J. Barnard, Jr., in *Chelates in Analytical Chemistry*, Vol. 4, H. A. Flaschka and A. J. Barnard, Jr., Eds., Dekker, New York, 1972, pp. 233–275.
76. I. Shain, in *Treatise on Analytical Chemistry*, Part I, Vol. 4, I. M. Kolthoff and P. J. Elving, Eds., Wiley-Interscience, New York, 1963, pp. 2533–2568.
77. K. B. Yatsimirskii, *Kinetic Methods of Analysis*, Pergamon Press, Oxford, 1966.
78. K. Little, A. J. Barnard, Jr., and J. D. Brooks, J. T. Baker Chemical Co., Phillipsburg, N.J., unpublished work.
79. B. H. Campbell and L. G. Hallquist, J. T. Baker Chemical Co., Phillipsburg, N.J., unpublished work.
80. K. Eckschlager, P. Stopka, and J. Veprek-Siska, *Chem. Prum.*, **12**, 667 (1967).

Chapter **II**

SOLVENT SELECTION FOR SEPARATION PROCESSES

Lloyd R. Snyder

---

1 Introduction 26

Peripheral Properties of the Solvent 27
Boiling Point 32
Viscosity 32
Solvent Properties Affecting Detection 33
Other Properties 34

2 Factors Affecting Solubility and Separation 36

Intermolecular Interactions 38
Dispersion Interactions 38
Dipole Interactions 39
Hydrogen Bonding 41
Covalent Bonding 47
Other Interactions 47
The "Polarity" of Solvents and Solutes 47
Hydrophobic Interactions 49
Solvent Selectivity and the Failure of (2.3a) 50

3 Solvent Classification Schemes 51

Solvent (and Solute) Polarity Scales 51
The Hildebrand Solubility Parameter δ 51
Values of δ for Pure Compounds 52
The Rohrschneider Polarity Scale 54
Solvent (and Solute) Selectivity Parameters 59
Subdividing the Solubility Parameter δ 59
Subdividing the Polarity Parameter P' 60

4 General Approach to Solvent Selection 62

Maximizing Solubility for a Given Solute 62
Maximizing the Separation of Two Solutes 63

5 Application in Differential Separation Methods 66

Solvent Extraction (Leaching) 66
Liquid-Liquid Partition 66
Liquid-Liquid Chromatography 67
Fractional Crystallization 67
Extractive Distillation 68
Solvent-Adsorption Partition and Liquid-Solid Chromatography 68
Gas Chromatography 71
Dialysis, Ultracentrifugation, and Thermal Diffusion 72
Acknowledgment 72
Symbols 73

1 INTRODUCTION

The general importance of separation methods to the chemist needs little elaboration. Separation is essential in many analytical schemes, in the purification of synthetic products, and in the isolation of natural products from plant, animal, or mineral sources. Most laboratory separations involve one or more solvents that play a basic role in the separation process. Table 2.1, by way of example, provides a partial listing of some common solvent-based separation methods. Although in some cases a solvent is already part of the starting sample, more often the solvent(s) must be added during the separation process.

Table 2.1 Some Common Separation Processes Requiring a Solvent

Solvent extraction (leaching)
Liquid-liquid partition
Fractional crystallization[a]
Extractive distillation
Solvent-adsorption partition
Liquid chromatography
Gas-liquid chromatography
Dialysis, ultrafiltration
Thermal diffusion[a]

[a] Can be used without added solvent in some cases.

The selection of a specific solvent or solvent mixture* for use in a given separation is one of the more complex and poorly understood operations required of the chemist. Many factors and solvent properties should be considered, apart from those bearing directly on the ability of the solvent to affect separation. Unfortunately the basic physical chemistry involved in such separations is not well developed, and what is known is not commonly used by most workers. As a result, the chemist has been forced to rely mainly on chemical intuition, making use of simple acid-base or complexation equilibria, a rough understanding of the properties of "polar" (hydrophylic) versus "nonpolar" (hydrophobic) molecules, or such qualitative concepts as hydrogen bonding.

In this chapter I attempt to develop a simple yet reliable approach for selecting the right solvent for a given application. My main emphasis is on solvent properties that directly affect the separation of the sample, rather than on peripheral considerations such as safety, economics, and compatibility of the solvent with operations that precede or follow separation. However in the balance of this introductory section we look briefly at these "peripheral" solvent properties.

This chapter emphasizes laboratory-scale separation schemes, at the expense of commercial-scale applications. One reason is that in commercial operations economic optimization is paramount, and this requires a rather precise description of how separation varies with separation conditions. The present approach, on the other hand, avoids quantitative calculations where possible, because complex equations are rarely used by bench workers in selecting solvents, and because these equations lack broad reliability.

"Peripheral" Properties of the Solvent

The "peripheral" properties of pure solvents include those that are of interest in choosing an appropriate solvent but often do not directly affect separation *per se*. Many of these peripheral properties can be found in Riddick and Bunger's *Organic Solvents* [1] or—for higher-boiling hydrocarbon solvents—in the tabulation of the American Petroleum Institute [2]. Often only the boiling point or density of the solvent is of interest, and this information is available in more general handbooks that the average chemist has in his personal library (e.g., Ref. 3). Table 2.2 provides an abbreviated listing of such solvent properties for a number of common solvents (arranged roughly according to solvent "polarity").

* A "solvent" as discussed in this chapter can be either a pure compound or a mixture of pure solvents. When a distinction is important, we use "pure solvent" to mean a single pure compound (e.g., pure benzene, pure chloroform, etc.).

Table 2.2 Peripheral Properties of Some Common Solvents [1, 6]

Solvent		*Boiling Point*					*Solubility*[e] *(weight percent)*		*UV Cutoff*
Number	*Name*	(°C)	n^a	η^b	ε^c	d^d	*Water/ Solvent*	*Solvent/ Water*	(nm)
1	FC-78[f]	50	1.267	0.4	1.88	1.70			190
	FC-75	102	1.276	0.8	1.86	1.76			190
	FC-43	174	1.291	2.6	1.9	1.81			190
2	Isooctane[g]	99	1.389	0.47	1.94	0.69	0.011	0.0002	190
3	*n*-Pentane	36	1.355	0.22	1.84	0.61	0.010	0.0038	190
4	*n*-Hexane	69	1.372	0.30	1.88	0.65	0.010	0.0009	190
5	*n*-Heptane	98	1.385	0.40	1.92	0.68	0.010	0.0003	190
6	*n*-Octane	126	1.395	0.52	1.95	0.70	0.010	0.0001	190
7	Cyclohexane	81	1.423	0.90	2.02	0.77	0.012	0.0055	190
8	Cyclopentane	49	1.404	0.42	1.97	0.74	0.014	0.0156	
9	Carbon disulfide	46	1.624	0.34	2.64	1.26			380
10	Carbon tetrachloride	77	1.457	0.90	2.24	1.58	0.008	0.08	265
11	2-Chloropropane	36	1.375	0.30	9.82	0.86			230
12	Ethyl sulfide	92	1.44	0.42	5.7[h]	0.83			290
13	Ethyl bromide	38	1.421	0.38	9.39	1.45		0.7	
14	Butyl ether	142	1.397	0.64	3.08	0.76	0.19	0.03	220
15	Triethylamine	89	1.398	0.36	2.4[h]	0.72			275
16	Isopropyl ether	68	1.365	0.38	3.9[h]	0.72	0.62	1.2	220
17	Toluene	110	1.494	0.55	2.4[h]	0.86	0.046	0.054	285
18	*p*-Xylene	138	1.493	0.60	2.27	0.86			290

19	Chlorobenzene	132	1.521	0.75	5.6[h]	1.10			
20	Bromobenzene	156	1.557	1.04	5.4[h]	1.49			
20a	Perchloroethylene	121	1.505	0.84	2.3[h]	1.61	0.02		
21	Phenyl ether	258	1.580	3.3	3.7	1.07			
22	Phenetole	170	1.505	1.14	4.2	0.96			
23	Ethyl ether	35	1.350	0.24	4.3	0.71	1.3	6.9	220
24	Benzene	80	1.498	0.60	2.3[h]	0.87	0.058	0.178	280
25	1-Octanol	195	1.427		10.3		3.9	0.06	210
26	Fluorobenzene	85		0.55	5.4[h]	1.02			
27	Benzyl ether	288	1.538	4.5		1.04			
28	Methylene chloride	40	1.421	0.41	8.9[h]	1.32	0.17[h]	1.32[h]	233
29	Anisole	154	1.514	0.9	4.3[h]	0.99			
30	*i*-Pentanol	130	1.405	3.5	14.7[h]	0.81	9.2	1.7	210
31	1,2-Dichloroethane	83	1.442	0.78	10.4[h]	1.25	0.16	0.87	230
32	*t*-Butanol	82	1.3851		12.5[h]	0.78	Miscible[i]	Miscible[i]	210
33	*n*-Butanol	118	1.397	2.6	17.5[h]	0.81	20.1	7.8	210
34	*n*-Propylamine	48	1.385	0.35	5.3	0.71	Miscible[i]	Miscible[i]	275
35	*n*-Propanol	97	1.384	1.9	20.3[h]	0.80	Miscible[i]	Miscible[i]	190
36	Tetrahydrofuran	66	1.405	0.46	7.6[h]	0.87	Miscible[i]	Miscible[i]	220
37	Ethyl acetate	77	1.370	0.43	6.0[h]	0.89	9.8[h]	9.7[h]	260
38	*i*-Propanol	82	1.384	1.9	20.3[h]	0.78	Miscible[i]	Miscible[i]	190
39	Chloroform	61	1.443	0.53	4.8	1.48		0.82	245
40	Acetophenone	202	1.532	1.64	17.4[h]	1.02			
41	Methyl ethyl ketone	80	1.376	0.38	18.5	0.80	23.4		330
42	Cyclohexanone	156	1.450	2.0	18.3	0.94		8.7	330
43	Nitrobenzene	211	1.550	1.8	34.8[h]	1.20			
44	Benzonitrile	191	1.526	1.2	25.2[h]	1.00			
45	Dioxane	101	1.420	1.2	2.2[h]	1.03	Miscible[i]	Miscible[i]	220

Table 2.2 (*Continued*)

Solvent		Boiling Point					Solubility[e] (weight percent)		UV Cutoff
Number	Name	(°C)	n^a	η^b	ε^c	d^d	Water/Solvent	Solvent/Water	(nm)
46	Tetramethylurea	175	1.449		23.0	0.96			
47	Ethanol	78	1.359	1.08	24.6[h]	0.79	Miscible[i]	Miscible[i]	190
48	Quinoline	237	1.625	3.37	9.0[h]				
49	Pyridine	115	1.507	0.88	12.4	0.98	Miscible[i]	Miscible[i]	305
50	Nitroethane	114	1.390	0.64			0.9	4.5	380
51	Acetone	56	1.356	0.30	20.7[h]	0.78	Miscible[i]	Miscible[i]	330
52	Ethylene glycol	197	1.431	16.5	37.7[h]				210
53	Benzyl alcohol	205	1.538	5.5	13.1	1.04		3.0	
54	Methoxyethanol	125	1.400	1.60	16.9[h]	0.96	Miscible[i]	Miscible[i]	
55	Aniline	184	1.02	3.77	6.9	1.02		3.8[h]	
56	*N*-Methyl formamide	182	1.447	1.65	182.0[h]	1.00	Miscible[i]	Miscible[i]	270
57	Acetic acid	118	1.370	1.1	6.2	1.04	Miscible[i]	Miscible[i]	
58	Acetonitrile	82	1.341	0.34	37.5	0.78	Miscible[i]	Miscible[i]	200
59	*N*,*N*-Dimethyl formamide	153	1.428	0.80	36.7[h]				270
60	Tetrahydrothiophene, 1,1-dioxide[j]	287	1.48	11.5	43.0[h]	1.27			

61	Dimethylsulfoxide	189	1.477	2.00	47.0[h]	1.10			
62	*N*-Methyl-2-pyrrolidone	202	1.468	1.67	32.0[h]	1.03			
63	Hexamethylphosphor triamide	233	1.457	3.0	30.0	1.0			
64	Methanol	65	1.326	0.54	32.7[h]	0.79	Miscible[i]	Miscible[i]	190
65	2-Aminoethanol	171	1.452	19.3	37.7[h]	1.01	Miscible[i]	Miscible[i]	
66	Nitromethane	101	1.380	0.61		1.13	2.1	11.0	380
67	*m*-Cresol	202	1.540	14.0	11.8[h]	1.03			
68	Formamide	210	1.447	3.3	111.0	1.13	Miscible[i]	Miscible[i]	270
69	Water	100	1.333	0.89	80.0	1.00	—	—	190

[a] Refractive index, 25°C.
[b] Viscosity (cP, 25°C).
[c] Dielectric constant, 20°C.
[d] Density, 25°C.
[e] g/g, Water in solvent or vice versa [6].
[f] Minnesota Mining and Manufacturing (3-M) fluorochemical.
[g] 2,2,4-Trimethylpentane.
[h] 25°C.
[i] In all proportions.
[j] Minimum wavelength for analysis of sample-solvent mixture by spectrophotometry.

There usually are many solvents available that have acceptable peripheral properties for a given application. Our approach is to select from this large group of solvents the best solvents from the standpoint of separation *per se*. Since binary or ternary solvent mixtures often can be employed in place of pure solvents, an enormous range of solvents (i.e., solvent mixtures) is available from which to choose.

Boiling Point

Normally we require a solvent whose boiling point is above the temperature of the separation process. In separations where temperature varies during the separation (e.g., Sohxlet extraction), a solvent is needed whose boiling point falls at some accessible higher temperature. Reference 1 also lists common solvents in order of boiling point.

We often want to remove the solvent from separated sample fractions on completion of the separation operation. For example, after fractional crystallization of a compound from some solvent, the recovered crystals of purified compound must be dried of solvent. Or following solvent extraction of a solid sample, the solvent must be removed from a recovered fraction. The easiest technique for removal of solvent from nonvolatile samples is simple solvent evaporation, which means that solvents boiling 10 to 50°C above the temperature of separation are preferable to higher-boiling solvents. In the case of volatile samples, fractional distillation can be used to separate solvent from final sample fractions. Again, the boiling points of solvent and sample can be used to select appropriate sample-solvent combinations.

Viscosity

A general rule is that low-viscosity solvents are preferable, other factors being equal. This is particularly true in liquid chromatography, where more viscous mobile phases mean poorer separations [4]. Fortunately low viscosities and low boiling points generally coincide for pure solvents, and these materials are generally used. Although solvent viscosity data are available for most pure solvents (e.g., see Refs. 1–3), one can usually estimate a rough viscosity value from the boiling point of a solvent. Thus consider the following values:

Solvent	Viscosity (cP at 20°C)	Boiling Point (°C)
n-Pentane	0.23	36
n-Octane	0.55	126
n-Dodecane	1.51	216
n-Hexadecane	3.34	287

These data are typical in showing a regular increase in solvent viscosity with boiling point. Other solvents usually fall close to this viscosity versus boiling point trend for the alkanes, as can be seen by comparing the data above with viscosity–boiling point values from Table 2.2 for other solvents. Exceptions are very polar solvents (e.g., alcohols) and compact molecules (e.g., cycloalkanes, aromatics, CCl_4), which have generally higher viscosities than predicted (e.g., the extreme example, cyclohexanol, with viscosity of 68 cP and boiling point of 161°C).

The viscosity of a solvent mixture is normally intermediate between the viscosities of the pure solvents composing the mixture. For a binary mixture of pure solvents A and B, the viscosity η of the mixture is given approximately by the relationship [6d]

$$\eta = (\eta_a)^{x_a} (\eta_b)^{x_b}$$

where η_a and η_b refer to the viscosities of pure A and B, respectively, and x_a and x_b refer to the mole fractions of A and B in the mixture. The primary practical significance of this relationship is that dilute solutions of a viscous solvent B in a nonviscous solvent A will have viscosities close to that of solvent A. Thus in applications where solvent viscosity must be as low as possible, it is nevertheless possible to use solutions of a relatively viscous solvent. This is illustrated by the following viscosity values calculated from the foregoing equation:

Mole Fraction B	Calculated Viscosity (cP)
0.00	0.3*
0.10	0.38
0.25	0.53
0.50	0.95
1.00	3.00*

* Assumed values.

Solvent Properties Affecting Detection

In some cases it is of interest to assay for one or more separated sample compounds in a solvent phase resulting from the separation (i.e., without separation of solvent from sample). This is true, for example, in liquid chromatography, where separated compounds in the mobile phase (solvent) go directly to a photometric, refractive index, or other detector. In this case the relative detector response of the solvent is often of interest. For example, a solvent that absorbs strongly at a given wavelength cannot be used for analysis at that wavelength (e.g., benzene as solvent at 254 nm). Table 2.2 lists the lower wavelength cutoff points for significant light transmission by various solvents. Alternatively (e.g., with refractive index detection), one may wish to maximize the difference in sample versus solvent refractive index values, for maximum detection sensitivity; solvent refractive index values are listed in Table 2.2. Reference 1 cites various other spectral characteristics of individual solvents but does not provide actual data.

Other Properties

Solvent density is important in phase separations based on gravity. Table 2.2 lists data for pure solvents. Solvent mixtures have densities close to the arithmetic average of the pure solvent components; for example, for binary solvent mixtures,

$$d \approx d_a \phi_a + d_b \phi_b$$

Here, d, d_a, and d_b refer, respectively, to densities of the mixture, of solvent A, and of solvent B; ϕ_a and ϕ_b refer to volume fractions in the mixture of solvents A and B, respectively.

Solvent toxicity is an important consideration, and this aspect of individual solvents is reviewed in Ref. 1. It should be noted that several solvents formerly regarded as relatively innocuous are now considered to be dangerous for long-term exposure. No attempt is made in this chapter to discuss this subject further.

Solvent flammability is of general interest in selecting solvents for some practical applications. Unfortunately low-boiling solvents—which are preferred for other reasons—tend to be the most flammable. Hydrocarbons boiling below 100°C generally have flash points less than 30°C, whereas oxygenated solvents such as alcohols, acids, and esters have somewhat higher flash points relative to hydrocarbons of similar boiling point. Halogenated solvents such as methylene chloride, chloroform, and carbon tetrachloride do not even have flash points and are therefore less flammable (but more toxic).

Solvent chemical reactivity is often an important consideration, since solvents that may react with the sample are generally undesirable. For this reason aldehydes are seldom used as solvents, and ketones are unsuitable in some applications. The known tendency of ethers to form peroxides that can then react with a sample suggests the use of these solvents with care (e.g., purified or freshly opened ether samples are preferred for many uses). Mixtures of pure solvents that can coreact are likewise undesirable.

Solvent miscibility with other solvents is of obvious interest in some applications (e.g., liquid–liquid partition and liquid chromatography, and Table

2.3 summarizes a limited number of related data. For a general scheme to predict the miscibility of different solvent pairs, see the treatment of (6e). The relative solubility of water in various solvents, and vice versa, is of special interest and is referred to further in following sections (data in Table 2.2).

Other solvent properties such as surface tension and freezing point can also play a role in special situations. Data are provided in Ref. 1 and are not discussed further. Reference 1 also treats the purification of solvents, which is often of interest in separation-related applications. Solvent costs are usually known from manufacturers' catalogs.

Table 2.3 Miscibility of Different Solvent Pairs [5]

	Paired Solvent[a]											
Solvent	*Phenol*	*Methanol*	*Ethanol*	*Ethylene glycol*	*Diethylene glycol*	*Glycerine*	*Acetic acid*	*Ethylene diamine*	*Formamide*	*Acetonitrile*	*Nitromethane*	*Water*
n-Hexane	i	i	m	i	—	i	m	—	i	i	i	i
n-Heptane	i	i	m	i	—	i	m	—	i	i	i	i
Isooctane	i	i	m	i	—	i	m	i	i	i	i	i
Cyclohexane	m	i	m	i	i	i	m	i	i	i	i	i
Benzene	m	m	m	i	i	i	m	m	i	m	i	i
Toluene	m	m	m	i	i	i	m	m	i	m	i	i
m-Xylene	m	m	m	i	i	i	m	m	i	m	—	i
CCl_4	m	m	m	i	i	i	m	m	i	m	m	i
$CHCl_3$	m	m	m	m	m	i	m	m	i	m	m	i
CH_2Cl_2	i	m	m	i	m	i	m	m	i	m	m	i
1,2-Dichloroethane	—	m	m	i	m	i	m	m	i	m	m	i
CS_2	—	i	m	i	i	i	m	—	i	i	i	i
Ethyl ether	—	m	m	i	i	i	m	—	i	—	m	i
Isopropyl ether	—	m	m	i	i	i	m	i	i	m	m	i
Methyl isobutyl ketone	—	m	m	i	m	i	m	i	—	—	m	i

[a] Key: m, miscible; i, immiscible (at room temperature); —, no data reported in Ref. 5.

2 FACTORS AFFECTING SOLUBILITY AND SEPARATION

In many separation processes the ability of the solvent to selectively dissolve certain sample components directly affects the resulting separation. In solvent extraction, for example, one sample component x may have a high solubility in the extraction solvent, whereas a second component y may have a very low solubility. As a result, compound x will be extracted to a much greater extent than y, thus effecting the separation of these two compounds. Similarly, in liquid-liquid column chromatography, compound x may be more soluble in the mobile phase, while compound y is more soluble in the stationary phase. Compound x will then move through the chromatographic column more rapidly than y, again resulting in the separation of x from y. The connection between solubility and separability in these and related examples should be obvious to the reader. The following discussion gives further clarification; for a general discussion and thermodynamic treatment of solubility and separation, see Ref. 7. If we recognize the importance of solubility in separation processes of the type listed in Table 2.1, we will want to know the factors that determine the relative solubility of a given compound in different solvents. This area is the subject of an enormous literature, one I review here primarily in intuitive and qualitative terms. The reader interested in more details or in mathematical rigor should consult Refs. 7 to 11a.

Let us begin with a simple description of the factors that make one compound (A) a better solvent than a second compound (B) for a given solute (x). For convenience, let us assume that A and B are immiscible, so that the two solvents can be shaken together with solute x. We can readily appreciate that if x is more soluble in A than in B, its concentration in phase A will be greater than in phase B, once the mixture of solvents and solute has equilibrated. In fact, the concentrations (better, mole fractions) of x in the two phases will be given as

$$\frac{C_{x,a}}{C_{x,b}} = e^{-\Delta G/RT} \tag{2.1}$$

where $C_{x,a}$ and $C_{x,b}$ are the concentrations of x in phases A and B, respectively, R is the gas constant, T is the temperature (°K), and ΔG is the free energy for transfer of one mole of compound x from phase B to phase A. Solution theories commonly ignore entropy effects, since these are usually subordinated to heat effects; thus we can rewrite (2.1) as

$$\frac{C_{x,a}}{C_{x,b}} \approx e^{-\Delta H/RT} \tag{2.1a}$$

Here ΔH is the enthalpy change for transfer of one mole of compound x from phase B to A. If ΔH is positive (interactions of x with solvent B stronger), the quantity on the right will be less than 1, and x will prefer phase B ($C_{x,b} > C_{x,a}$).

Let us next visualize the transfer of a molecule x from solvent B to solvent A, which corresponds to the quantity ΔH and therefore determines the relative solvency* of B versus A for x. Figure 2.1a portrays the interactions of a *part* of a molecule x (x_i) with surrounding molecules of solvent B; x_i might correspond, for example, to a specific functional group i in x. The interactions between x_i and surrounding solvent molecules B are shown in Fig. 2.1a by the indicated arrows. In a moment we examine these interactions more closely, but for now they can be thought of as loose chemical bonds.

In Fig. 2.1b we remove x_i from phase B, leaving a cavity that subsequently collapses (Fig. 2.1c) when the original interactions between molecules of B and x_i are replaced by interactions between adjacent molecules of B. In Figs. 2.1d to 2.1f the group x_i is added to solvent A, the reverse of the process shown in Figs. 2.1a to 2.1c: bond breaking between adjacent molecules of A with cavity formation (d, e), and insertion of x_i into the cavity (f). The overall process (a–f) corresponds to the transfer of the group x_i from solvent B to solvent A and gives us some insight into the factors that determine ΔH and the relative solubility of x_i in B versus A.

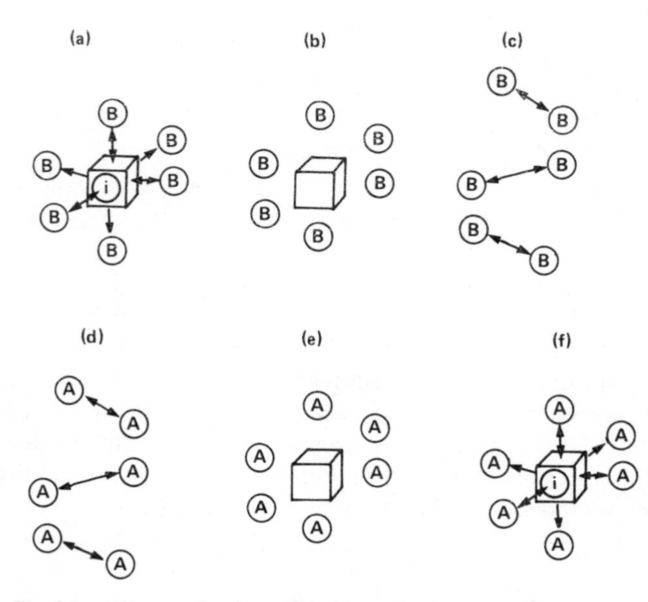

Fig. 2.1. The transfer of a solute group i from solvent B to solvent A.

* By "solvency" I mean the relative ability of a solvent to dissolve a particular sample.

In steps *a* and *b* of Fig. 2.1, bonds (or interactions) between B and x_i must be broken, requiring addition of heat to the system. The stronger are these interactions, the greater is the preference of x_i for solvent B, and the greater is the solubility of x_i in B. In step *c* interactions between like molecules B are formed, which releases heat from the system. The stronger are these interactions, the less is the preference of x_i for solvent B. In steps *d* and *e* interactions between like molecules A are broken, requiring addition of heat. The stronger are these interactions, the less is the preference of x_i for solvent A. In step *f* interactions between A and x_i are formed, releasing heat from the system. The stronger are these bonds, the greater is the preference of x_i for solvent A.

Thus it is clear that the value of ΔH for transfer of x_i from solvent B to A depends on the interactions between molecules of A (A–A), molecules of B (B–B), and between molecules of A or B and the group x_i (*A*–*x*, *B*–*x*). Let us next consider the nature and magnitude of these interactions between neighboring molecules.

Intermolecular Interactions

Intermolecular interactions exist in several different varieties, and these are important in affecting relative solvency and separation:

1. Dispersion interactions.
2. Dipole interactions: induction and orientation.
3. Hydrogen bonding.
4. Covalent bonding.

Dispersion Interactions

Dispersion or London forces exist between every pair of adjacent molecules, and these interactions normally account for the major part of the interaction energy that holds molecules together in the liquid phase. Dispersion forces arise as follows. Consider two unlike molecules (e.g., X and S), which are immediately adjacent. The electrons associated with each molecule are in constant, random motion, and at any instant in time the electrons of molecule X will have a certain configuration. In general this specific configuration is not symmetrical about the atomic nuclei, and an instantaneous dipole moment results for molecule X (Fig. 2.2*a*). This instantaneous dipole in X then *induces* an interactive dipole in molecule S, as in Fig. 2.2*b*. Because the resulting dipoles are aligned for electrostatic interaction (attraction of opposite charges), a net attractive interaction between molecules X and S results. These dispersion interactions are independent of the interactions of *permanent* molecular dipoles discussed below, and they occur in the case of both polar and nonpolar molecules.

The relative strength of this dispersion interaction between two molecules

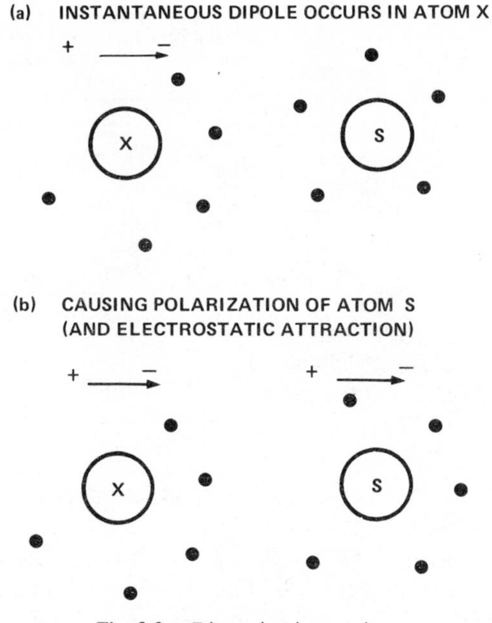

Fig. 2.2. Dispersion interactions.

—as in Fig. 2.2—depends on the number of electrons per unit volume of pure liquids X and S and on their polarizability. This should be apparent from Fig. 2.2, which reveals that the induced dipole formed in molecule S will be larger, the more electrons there are in S and the easier it is to displace (or *polarize*) each of these electrons. Since the overall tendency of compounds to interact by dispersion forces is closely related to the refractive index values of the compounds in question (see discussion of Ref. 7), dispersion interactions become stronger, the greater the refractive index values of compounds X and S.

Values of the refractive index of pure solvents and solutes are obtainable for most compounds from handbooks (e.g., Ref. 3) and provide a direct measure of the strength of dispersion interactions. In general, refractive index is greater for compounds with unsaturated bonds, and compounds with elements from the second and third rows of the periodic table. Table 2.4 presents some of these trends for representative compounds. Additional examples can be seen in Table 2.2.

Dipole Interactions

When a molecule possesses a permanent dipole moment (as opposed to a transient dipole as in Fig. 2.2), two additional interactions with adjacent molecules are possible.

Table 2.4 Refractive Index Values of Some Representative Compounds

Compound	*n*
n-Hexane	1.38
Cyclohexane	1.43
Benzene	1.50
1-Methylnaphthalene	1.62
1-Chlorohexane	1.42
1-Bromohexane	1.45
1-Iodohexane	1.49
Hexane-1-thiol	1.45

Dipole induction is the same type of interaction illustrated in Fig. 2.2*b* except that the transient dipole of molecule X is replaced with a permanent molecular dipole. This permanent dipole in X then induces a dipole in S, just as occurs in dispersion interactions. The net effect is an increase in the total interaction between X and S, due to the permanent dipole originally present in molecule X.

Dipole orientation involves the alignment of two adjacent molecules, each one possessing a permanent dipole moment, for maximum electrostatic attraction. For example, if X and S each refer to a molecule of acetonitrile, the molecules will line up as follows,

$$CH_3{-}\overset{+}{C}{\equiv}\overset{-}{N} \longleftrightarrow CH_3{-}\overset{+}{C}{\equiv}\overset{-}{N}$$

for maximum attraction between unlike charges.

Table 2.5 summarizes the dipole moments of a number of compound *functional groups* and simple solvent molecules. Because dipole interactions are quite short range, dipole interactions are determined by the sum of group dipoles within the molecule—*not* by the overall molecular dipole moment of the molecule. Thus in such compounds as ortho- and paradinitrobenzene,

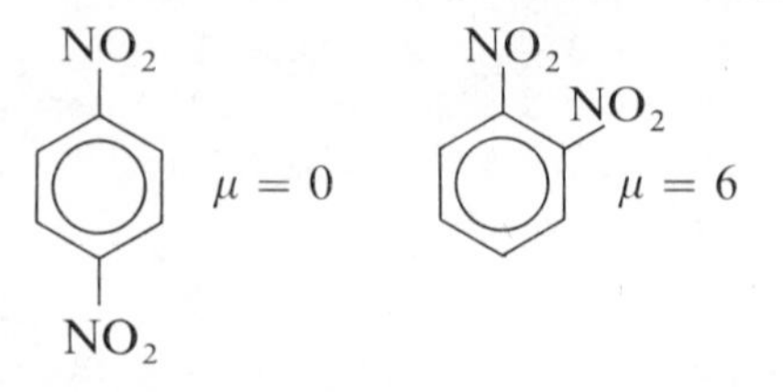

the molecular dipole moment μ (debyes) of the para compound is zero because of overall cancellation of group dipoles, whereas the group dipoles

Table 2.5 Dipole Moments[a] (debyes) of Functional Groups and Solvents

Functional groups (R–X)	
Tertiary amine, –N–	0.8
Secondary amine, –NH–	1.1
Ether, –O–	1.2
Primary amine, $-NH_2$	1.4
Sulfide, –S–	1.4
Thiol, –SH	1.4
Iodide, –I	1.6
Carboxylic acid, –COOH	1.7
Hydroxy, –OH	1.7
Fluoride, –F	1.7
Chloride, –Cl	1.8
Ester, –COO–	1.8
Bromide, –Br	1.8
Aldehyde, –CHO	2.5
Ketone, –CO–	2.7
Nitro, $-NO_2$	3.2
Nitrile, –CN	3.5
Sulfoxide, –SO–	3.5
Solvents	
Methylene iodide	1.1
Chloroform	1.2
Methylene bromide	1.4
Methylene chloride	1.5
Tetrahydrofuran	1.7

[a] Liquid phase values, data of Ref. 12; see also Ref. 1.

add in the ortho isomer to give a large molecular dipole. However the dipole interactions of these two compounds with surrounding molecules are in fact quite similar, because two nitro groups exist in each molecule, and the *individual* polar nitro groups undergo dipole interactions.

Dipole interactions often play an important role in affecting solubility and separation.

Hydrogen Bonding

The hydrogen bonding interactions between a proton-donor molecule A and a proton-acceptor molecule B are well known to most chemists and often play a dominant role in affecting solubility and separation. An example is provided by the donor chloroform and the acceptor trimethylamine:

$$Cl_3C–H \cdots :N–(CH_3)_3$$

The donor and acceptor molecules will align themselves to permit a hydrogen atom of the donor to interact with an electron pair of the acceptor, as shown. These interactions by hydrogen bonding can be quite strong, with interaction increasing for more acidic donors and more basic acceptors. However the acidity and basicity of different compounds in aqueous solution (i.e., their pK_a values) are generally poor guides to relative strength as hydrogen bonding donors or acceptors. A better index of hydrogen bonding acceptor strength is provided by actual experimental data for hydrogen bonding compounds. One such study [13] yields the relative ranking of different acceptors given in Table 2.6. Here compounds with larger values of the parameter H' perform better as proton acceptors (i.e., are more basic). Since H' is a logarithmic function like pH, negative values of H' are possible, and values near zero still imply some ability to function as a proton acceptor. As Table 2.6 indicates, values of H' are largely determined by the functional group serving

Table 2.6 Proton-Acceptor Strengths of Different Compounds in Hydrogen Bonding [13]

Base	*H′*
A. Aldehyde	
1. *p*-Nitrobenzaldehyde	0.36
2. Propionaldehyde	0.71
3. *n*-Butyraldehyde	0.78
4. *p*-Chlorobenzaldehyde	0.79
5. Benzaldehyde	0.83
6. *p*-Methoxybenzaldehyde	1.10
7. *p*-*N*,*N*-Dimethylaminobenzaldehyde	1.55
B. Alkyl halide	
8. Cyclohexyl fluoride	−0.05
C. Amine, Primary	
9. *β*,*β*,*β*-Trifluoroethylamine	0.59
10. Propargylamine	1.48
11. Cyclopropylamine	1.64
12. Benzylamine	1.75
13. *n*-Butylamine	2.11

Table 2.6 *(Continued)*

Base	*H′*
Tertiary	
14. Triallylamine	1.23
15. Tri-*n*-propylamine	1.45
16. Tri-*n*-butylamine	1.57
17. *N*,*N*-Dimethylbenzylamine	1.58
18. Triethylamine	1.91
19. *N*,*N*-Dimethyl-*n*-propylamine	1.98
20. *N*,*N*-Dimethylcyclohexylamine	2.07
D. Carboxylic acid fluoride	
21. Benzoyl fluoride	−0.30
22. *n*-Butyl fluoride	−0.16
23. *n*-Propionyl fluoride	−0.13
E. Carboxylic amide	
24. *N*,*N*-Dimethyltrifluoroacetamide	1.18
25. *N*,*N*-Dimethyl-*p*-nitrobenzamide	1.62
26. *N*,*N*-Dimethylchloroacetamide	1.68
27. *N*,*N*-Dimethylformamide	2.06
28. *N*,*N*-Dimethylbenzamide	2.22
29. *N*-Methyl-2-pyrrolidone	2.37
30. *N*,*N*-Dimethylacetamide	2.38
31. Tetramethylurea	2.42
F. Carboxylic ester	
32. Ethyl chloroacetate	0.69
33. β-Propiolactone	0.78
34. Ethyl benzoate	0.88
35. Methyl acetate	1.00
36. Ethyl propionate	1.02
37. Ethyl acetate	1.08
38. γ-Valerolactone	1.30
39. γ-Butyrolactone	1.33
G. Ether	
40. Anisole	0.02
41. Dioxane	0.71[a]
42. Di-*t*-butyl ether	0.71
43. Di-benzyl ether	0.72
44. Diethyl ether	0.98
45. Di-*n*-butyl ether	1.02

Table 2.6 (*Continued*)

Base	*H'*
46. Diisopropyl ether	1.05
47. Tetrahydrofuran	1.26
48. Tetrahydropyran	1.23
H. Ketone	
49. 1,1,1-Trifluoroacetone	−0.19
50. Biacetyl	0.33
51. Acetylacetone	0.48
52. *sym*-Dichloroacetone	0.48
53. Benzophenone	0.97
54. Acetophenone	1.13
55. Methyl *t*-butyl ketone	1.04
56. Di-*t*-butyl ketone	1.04
57. Methyl isopropyl ketone	1.04
58. Methyl ethyl ketone	1.18
59. Acetone	1.18
60. Cyclohexanone	1.30
60a. Xanthone	1.37
61. *p*-Methoxyacetophenone	1.40
62. Acetylferrocene	1.64
63. Flavone	1.99
64. 2,6-Dimethyl-γ-pyrone	2.50
I. Nitrile	
65. Benzonitrile	0.79
66. *p*-Methoxybenzonitrile	0.99
67. β-Ethoxypropionitrile	1.05
68. Acetonitrile	1.05
69. *p*-*N*,*N*-Dimethylaminobenzonitrile	1.38
J. Nitro compound	
70. Nitrobenzene	0.73
K. N-Oxides	
71. Pyridine *N*-oxide	2.76
L. Phosphine oxide	
72. Triphenyl phosphate	1.73
73. Diethoxytrichloromethylphosphine oxide	2.00
74. Dimethoxyphosphine oxide	2.25
75. Diethylfluorophosphine oxide	2.27

Table 2.6 (*Continued*)

Base	*H′*
76. Diethoxydichloromethylphosphine oxide	2.27
77. Diethoxyphosphine oxide	2.37
78. Trimethyl phosphate	2.40
79. Dimethoxyethylphosphine oxide	2.44
80. Diethoxychloromethylphosphine oxide	2.55
81. Triethyl phosphate	2.73
82. Diethoxymethylphosphine oxide	2.77
83. Diethoxy-*N*,*N*-diethylaminophosphine oxide	2.90
84. Triphenylphosphine oxide	3.16
85. Trimethylphosphine oxide	3.49
86. Hexamethylphosphoramide	3.56
87. Tri-*n*-propylphosphine oxide	3.60
88. Triethylphosphine oxide	3.64
M. Pyridine	
89. 3,5-Dichloropyridine	0.75
90. 2-Bromopyridine	0.94
91. Pyrimidine	1.05[a]
92. 3-Bromopyridine	1.26
93. Quinoline	1.85
94. Pyridine	1.88
95. 2-*n*-Butylpyridine	1.88
96. Isoquinoline	1.93
97. Acridine	1.97
98. 3-Methylpyridine	1.97
99. 2-Methylpyridine	2.01
100. 4-Methylpyridine	2.03
101. 2,6-Dimethylpyridine	2.13
102. 2,4-Dimethylpyridine	2.17
103. 4-Methoxypyridine	2.14
104. 2,4,6-Trimethylpyridine	2.30
105. 4-*N*,*N*-Dimethylaminopyridine	2.81
N. Sulfoxide	
106. Diethyl sulfite	0.98
107. Methyl *p*-nitrophenyl sulfoxide	1.58
108. Diphenyl sulfoxide	2.03
109. Methyl phenyl sulfoxide	2.15
110. Dibenzyl sulfoxide	2.28
111. Dimethyl sulfoxide	2.53
112. Di-*n*-butyl sulfoxide	2.60

Table 2.6 *(Continued)*

Base	*H′*
O. Sulfide	
113. Diethyl sulfide	0.11
114. Di-*n*-butyl sulfide	0.26
115. Di-*t*-butyl sulfide	0.21
P. Unsaturated hydrocarbons	
116. Azulene	0.02
117. 4,6,8-Trimethylazulene	0.33

[a] Statistical factor of 2 has been applied.

as proton acceptor (e.g., aliphatic ethers have $0.7 \leq H' \leq 1.3$, aliphatic sulfoxides have $2.5 \leq H' \leq 2.6$).

Relative donor strength has not been characterized in as much detail as for acceptors. However Table 2.7 provides a rough classification of donor strength for different classes of compounds. Strong proton donors are limited to chloroform and compounds with –OH groups. Strongly electron-withdrawing groups such as $-CF_3$, when substituted into alcohols or acids, further increase the donor strength of the –OH group. For many chemists the classification of Table 2.7 is perhaps more interesting in defining compounds that are *not* effective donors insofar as contributing to interaction between molecules and to solvency is concerned. Thus donor properties have often been ascribed to weakly acidic compounds such as sulfoxides, nitro compounds, ketones, and esters; even hydrocarbons have been postulated as

Table 2.7 Relative Proton-Donor Strengths of Different Compound Classes in Hydrogen Bonding

Strong	*Weak*	*Negligible*
Alcohols	Mono- or dialkyl amines	Ketones
Phenols	Amides[a]	Esters
Carboxylic acids		Nitro compounds
$CHCl_3$		Sulfoxides
		Other compounds

[a] Possessing N–H groups; amides are actually intermediate in donor strength between strong donors and amines.

having donor properties that can yield significant hydrogen bonding interactions. However it now appears that these latter, weakly acidic substances are very rarely significant as proton donors.

Covalent Bonding

Certain covalent interactions are often used in separation processes, mainly those that are readily reversible, allowing recovery of original sample components after separation. Included in this group of reactions are acid-base reactions and various complexation reactions. Compounds that are either acids or bases can be made ionic or nonionic, depending on solvent pH. Usually there is a large change in the relative solubility of the compound as a result of ionization, and this can greatly simplify its separation. For example, carboxylic acids can be extracted from an organic reaction mixture by aqueous or alcoholic NaOH solution. Complexation reactions that have been used in separation processes include a number of examples:

Complexed Solute	*Complexing Agent*
Olefinic derivatives	Silver ion
vic-Diols	Borate ion
Polycyclic aromatics	Picric acid

Such covalent reactions are usually simply superimposed on the normal separations that would result in the absence of acid-base reaction or complexation and are well understood by the practicing chemist. Further discussion is beyond the scope of this chapter.

Other Interactions

Ionic groups preferentially dissolve in solvents of higher dielectric constant, and Table 2.2 lists dielectric constants for different solvents as a rough guide to solvency for ionized compounds and salts.

Hydrophobic interactions are often mentioned when discussing aqueous solutions. This is not an additional type of interaction; rather, it is the consequence of interactions already discussed (see below).

The "Polarity" of Solvents and Solutes

Returning to our discussion of Fig. 2.1 and the relative solvency of compounds B and A for the group x_i, we see that the better solvent will be determined by the relative magnitude of the above-mentioned different interactions between molecules of A, B, and x_i. However this involves a large number of individual contributions to solvency, particularly if (as in the usual case) more than one type of interaction exists, and if several groups x_i are present in the solute molecule. Consequently, at this stage it would be difficult to make actual predictions on the relative solubility of a given solute x in actual solvent A versus B.

A practical expedient, which circumvents some of the difficulty just described, is to make use of the common chemical concept of *polarity*.

By "polarity" we mean here the relative ability of a molecule to engage in strong interactions with other "polar" molecules (not specifically the presence in a molecule of a large dipole moment). Thus water is commonly regarded as one of the most polar compounds—if not the most polar. Yet water has a relatively small dipole moment compared to less "polar" compounds such as ketones and nitriles. "Polarity," then, represents the ability of a molecule to enter into interactions of all kinds—dispersion, dipole, hydrogen bonding, and ionic—and "relative polarity" is the sum of all possible interactions. The actual polarity of different compounds has been defined in a number of ways, some of which are described in the following section.

Let us now return to the process of Fig. 2.1 and lump all the individual interactions between A, B, and x_i into a single "polar" interaction. I have previously identified four bond-breaking or bond-making steps in the transfer of x_i from solvent B to solvent A, and a contribution H to the total enthalpy change ΔH can be defined for each step:

Step	*Bonds Affected*	*Contribution H*
a, b	(B)–(x_i) bonds broken	$2H_{x,b}$
c	(B)–(B) bonds formed	$-H_{b,b}$
d, e	(A)–(A) bonds broken	$H_{a,a}$
f	(A)–(x_i) bonds formed	$-2H_{x,a}$

Note that half as many bonds or interactions are involved during cavity collapse or formation (steps *c* and *d, e*), as in removal or addition of x_i to a solvent; this accounts for the factor 2 shown in the H values for steps involving x_i. H is positive (energy required to be added to system) when bonds are broken and negative when bonds are formed.

Now we can add the H terms above to give an overall value of ΔH for the transfer of group x_i from solvent B to A:

$$\Delta H = (H_{a,a} - H_{b,b}) + 2(H_{x,b} - H_{x,a}) \tag{2.2}$$

We have defined solvent and solute "polarity" in terms of the strength of total interactions between adjacent molecules (i.e., the H values above). That is, a polar molecule i interacting with a polar molecule j should give a large value of H. Theoretical expressions for a specific interaction (e.g., dispersion) confirm this (see Refs. 7, 14, 15) and suggest that H can be related to the "polarity" P_i and P_j of molecules i and j as

$$H_{ij} = P_i P_j \tag{2.3}$$

Equation 2.3 is referred to as the "geometric mean approximation."

If we define the polarities of A, B, and x_i as P_a, P_b, and P_x, substitution of these values by way of (2.3) into (2.2) gives

$$\Delta H = (P_a^2 - P_b^2) + 2P_x(P_b - P_a) \tag{2.3a}$$

Equation 2.3a says that ΔH and the relative solvency of A and B for x_i depend on the polarities of A, B, and x_i. If x_i is exactly intermediate between A and B in terms of polarity (i.e., P_x = the average of P_a and P_b), (2.3a) becomes zero, and x_i is distributed equally between the two solvents A and B at equilibrium. This means that the solvencies of A and B for x_i are exactly equal. If the polarity P_x of the solute is closer to that of solvent A, ΔH becomes negative, and at equilibrium x_i will concentrate into solvent A (i.e., x_i is now more soluble in A than in B). Thus (2.3a) provides a quantitative statement of the old solubility rule "like dissolves like." Polar solutes preferentially dissolve in polar solvents, and nonpolar solutes preferentially dissolve in nonpolar solvents. A corollary to this rule and (2.3a) is that solute solubility in a given solvent is greatest when the polarities of solvent and solute are equal.

The next question is then, How do we define compound polarity? That is, How do we measure values of P_i for different compounds i? The answer to this question (following section) allows us to systematically vary solvent polarity, to achieve the required solvency for a given solute x. What about the polarity of solvent mixtures? The picture presented in Fig. 2.1 and in the foregoing text suggests that the interaction heats H_i will be averaged for the two solvents i and j of the mixture, according to their volume fractions ϕ_i and ϕ_j in the solvent mixture. This is in fact theoretically correct; thus for a mixture of solvents i and j, the polarity $P_{i,j}$ of the mixture is given as

$$P_{i,j} = \phi_i P_i + \phi_i P_j \tag{2.4}$$

Hydrophobic Interactions

Hydrophobic interactions are said to be associated with nonpolar solutes in polar solvents. What is meant can be better understood by the distribution of a nonpolar solute x between a polar solvent and a nonpolar solvent, B and A, respectively. The value of P_x will be small, so let it be negligible (this is never actually the case, but it makes the point more clearly). Now the ΔH value for transfer of x from solvent B to solvent A is given by (2.3a) as $(P_a^2 - P_b^2)$. If B is the polar solvent, $P_b \gg P_a$, and ΔH is seen to be negative; that is, x will concentrate into the nonpolar solvent A. The driving force is seen from (2.3a) to consist *not* of nonpolar interactions between x and the nonpolar solvent A, but rather comes from the term $(H_{a,a} - H_{b,b})$ of (2.2), which describes the heat required to form a cavity (into which the molecule x is then placed). In effect, the polar solvent "squeezes" out the nonpolar solute x because the interactions of x with B are much weaker than the interactions of molecules B with themselves. For a theoretical discussion of hydrophobic bonding, see Refs. 16 and 17.

Where the polar solvent is water, and nonpolar solutes are being squeezed out by hydrophobic interactions, the polarity of the water phase can be increased by addition of various salts ("salting out"). The effectiveness of different salts in increasing these hydrophobic interactions varies widely, leading to the use of the so-called lyotropic series of salts for the salting out of proteins from aqueous solutions (see discussion of Ref. 17b). This phenomenon is now well understood, and it is possible to predict the efficacy of different salts in the lyotropic series [17b].

Solvent Selectivity and the Failure of (2.3a)

If there were only one type of intermolecular interaction (e.g., dispersion forces), (2.3a) would be a reasonably reliable relationship. It would be possible to arrange all solvents in order of their polarity values P_i, and the solubility of a given solute would change regularly as P_i is changed (being a maximum when the polarities of solvent and of solute are the same). However this is far from the case in actual practice, since usually several types of interaction are important. Consequently it is possible to create only approximate listings of solvent polarity, the order of polarity changing somewhat from solute to solute, depending on the relative tendency of the solute to enter into specific intermolecular interactions.

It is still possible to estimate solubility or relative solubility by a process of successive approximation. Thus relative solvent polarity can be used to estimate a rough solubility (or relative solubility value). Then specific intermolecular interactions between solvent and solute can be considered, to correct this initial estimate of relative solubility. For example, the solute might be a strong proton acceptor and the solvent a strong proton donor. In this case, the solubility of the solute would be greater than predicted from the overall polarity values of solute and solvent. Later sections will develop this approach in some detail.

The failure of (2.3a) because of different intermolecular interactions is in fact a practical advantage. If only a single parameter P_i determined solute polarity, (2.3a) suggests that solutes of similar polarity could not be separated by distribution between two solvents A and B.* However differences in specific interactions between solute and solvent lead to corresponding differences

* Actually, as the following section discusses, solutes of different molecular size—and similar polarities—are theoretically separable according to (2.3a).

in solubility, and these can be exploited to achieve separation. Such differences in solubility for solvents of similar polarity are collectively referred to as *solvent selectivity*, meaning the ability of the solvent to discriminate or preferentially dissolve different solutes of similar polarity. The following section examines in detail the classification of different solvents according to selectivity as well as polarity.

3 SOLVENT CLASSIFICATION SCHEMES

Most chemists have a rough idea of solvent polarity differences and recognize hydrogen bonding as a possible contribution to solvent selectivity. In the past such solvent properties as dielectric constant and the solubility of water in the solvent were used as quantitative indices of solvent polarity, but these have been found to be at best rough measures, of limited general utility. Now we consider two more general and effective schemes for solvent classification, one based on the properties of pure compounds, the other derived from actual solubility measurements for model solutes. We first discuss the classification of solvents according to polarity by each approach, then broaden our examination to include solvent selectivity.

Solvent (and Solute) Polarity Scales

The Hildebrand Solubility Parameter δ

The solubility parameter δ is currently the most widely applied index of solvent or solute polarity, and in principle it can be used to make quantitative calculations of solubility and separation [7–11]. It is defined as

$$\delta = \left(\frac{-\Delta E_v}{V}\right)^{1/2} \tag{2.5}$$

where ΔE_v is the vaporization energy per mole of the compound in question and V is its molar volume; ΔE_v is in turn equal to $\Delta H_v - RT$, where ΔH is the heat of vaporization, and ΔH can be estimated from the compound boiling point T_b (°K at 760 mm) from the Hildebrand rule:

$$\Delta H_v\ (298°\text{K}) = 2950 + 23.7T_b \times 0.02T_b^2 \tag{2.5a}$$

Thus values of δ are easily calculated for any compound whose boiling point is known.

Let us first consider the significance of δ in terms of our previous discussion of Fig. 2.1. In Fig. 2.3, we consider the simple vaporization of a molecule B from pure B. The species B_i within the cavity of Fig. 2.3*a* now represents some fraction of a molecule of B, such that 1 mole of these B_i groups equals 1 ml.

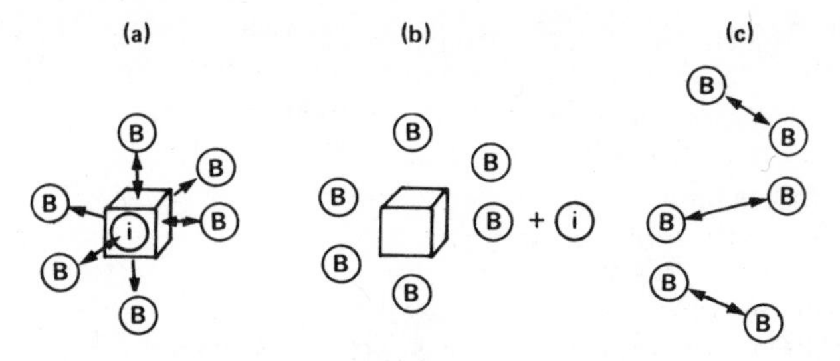

Fig. 2.3. Visualization of the vaporization of pure compound B.

In Fig. 2.3*b* the group B_i is removed from the cavity, to give vaporized B, and in Fig. 2.3*c* the cavity collapses. As before, the enthalpy changes can be listed:

Step	*Bonds Affected*	*Contribution of H*
b	(B)–(B_i) bonds broken	$2H_{b,b}$
c	(B)–(B) bonds formed	$-H_{b,b}$

The net enthalpy change ΔH is then equal to the sum of these two H values: $H_{b,b}$. But the value of ΔH is also the heat of vaporization of 1 ml of B, equal to $\Delta H_v/V \approx \delta^2$ (recalling that the term RT is small and tends to cancel in many solution processes). We previously defined $H_{b,b} = P_b^2$, and since in Fig. 2.3 $H_{b,b} \approx \delta^2$, we see that δ is essentially the polarity parameter P defined in Fig. 2.1 and in (2.3).

Now if we calculate ΔH for the transfer of the molecule (not group) x from solvent B to A as in Fig. 2.1 and in (2.3a), we must multiply the expression on the right-hand side of (2.3a) by V_x, the molar volume of x, and replace all P_i values by the corresponding δ values:

$$\Delta H = V_x(\delta_a^2 - \delta_b^2) + 2\delta_x(\delta_b - \delta_a) \tag{2.6}$$

Although our derivation of (2.6) has been essentially qualitative, this is the same result provided by a more rigorous application of solubility theory. At this point, the main significance of (2.6) versus (2.3a) is that the molecular size V_x of the solute x affects its relative solubility (as do the δ values for solvents and solute); the larger is V_x, the more affected will be the solubility of x by a change in solvent polarity.

Values of δ for Pure Compounds

Table 2.8 lists the solubility parameters δ of a number of common solvents at 25°C. Values of δ for a number of other compounds can be estimated from these data by noting that homologs of polar compounds tend to have similar

Table 2.8 Hildebrand Solubility Parameter δ and Related Selectivity Parameters [15]

Solvent						
Number[a]	*Name*	δ	δ_d	δ_o	δ_a	δ_b
—	Perfluoroalkanes	~6.0	~6.0	—	—	—
—	*n*-Alkanes	~8.0	~8.0	—	—	—
16	Diisopropyl ether	7.1	6.9	1.0	—	3.0
23	Diethyl ether	7.5	6.7	2.4	—	3.0
15	Triethylamine	7.5	7.5	—	—	4.5
7	Cyclohexane	8.2	8.2	—	—	—
11	Propyl chloride	8.4	7.3	2.9	—	0.7
10	Carbon tetrachloride	8.6	8.6	—	—	0.5
12	Diethyl sulfide	8.6	8.2	1.7	—	2.6
37	Ethyl acetate	8.9	7.0	4.0	—	2.7
34	Propylamine	8.9	7.3	1.7	1.8	5.5
13	Ethyl bromide	8.9	7.8	3.1	—	0.8
17	Toluene	8.9	8.9	—	—	0.6
36	Tetrahydrofuran	9.1	7.6	3.5	—	3.7
24	Benzene	9.2	9.2	—	—	0.6
39	Chloroform	9.3	8.1	3.0	6.5	0.5
41	Methyl ethyl ketone	9.5	7.1	4.7	—	3.2
51	Acetone	9.6	6.8	5.1	—	3.0
31	1,2-Dichloroethane	9.7	8.2	4.2	—	0.7
29	Anisole	9.7	9.1	2.1	—	1.7
19	Chlorobenzene	9.7	9.2	1.9	—	1.0
20	Bromobenzene	9.9	9.6	1.5	—	1.0
—	Methyl iodide	9.9	9.3	2.5	—	0.7
45	Dioxane	10.1	7.8	5.2	—	4.6
63	Hexamethylphosphortriamide	10.5	8.4	3.4	—	4.0
49	Pyridine	10.6	9.0	3.8	—	4.9
40	Acetophenone	10.6	9.6	2.7	—	3.3
44	Benzonitrile	10.7	9.2	3.4	—	2.3
—	Propionitrile	10.8	6.9	6.6	—	2.1
47	Quinoline	10.8	10.3	1.8	—	4.2
—	*N,N*-Dimethylacetamide	10.8	8.2	4.7	—	4.5
—	Nitroethane	11.0	7.3	6.0	—	1.0
43	Nitrobenzene	11.1	9.5	3.6	—	1.0
—	Tricresylphosphate	11.3	9.6	2.5	—	(?)
59	Dimethylformamide	11.8	7.9	6.2	—	4.6
38	Propanol	12.0	7.2	2.6	6.3	6.3
61	Dimethylsulfoxide	12.0	8.4	6.1	—	5.2
58	Acetonitrile	12.1	6.5	8.2	—	3.8
—	Phenol	12.1	9.5	2.3	9.3	2.3
47	Ethanol	12.7	6.8	3.4	6.9	6.9

Table 2.8 *(Continued)*

Solvent						
Number[a]	*Name*	δ	δ_d	δ_o	δ_a	δ_b
66	Nitromethane	12.9	7.3	8.3	—	1.2
—	γ-Butyrolactone	12.9	8.0	7.2	—	(?)
—	Propylene carbonate	13.3	9.8	5.9	—	(?)
—	Diethylene glycol	14.3	8.2	4.0	5.3	5.3
64	Methanol	14.5	6.2	4.9	8.3	8.3
52	Ethylene glycol	17.0	8.0	6.8	6.1	6.1
68	Formamide	19.2	8.3	(?)	Large	Large
69	Water	23.4	6.3	(?)	Large	Large

[a] Numbers are keyed to numbering of solvents in Table 2.2 (for convenience in cross-comparing solvent properties).

but slightly lower values of δ as compound molecular weight increases. Solubility parameters for ionic and nonvolatile compounds can be determined by methods described in Ref. 11.

The solubility parameter decreases slightly with temperature, but normally this effect can be ignored. It should be noted that solute solubility is a maximum when solute and solvent have the same polarity or value of δ. Now examination of (2.4) shows that two solvents whose values of δ are respectively higher and lower than that of a given solute can be blended to give a mixture whose δ value is equal to that of the solute, hence providing maximum solubility. Thus a *mixture* of two pure solvents can provide better solvency for a given solute than would either pure solvent. This unexpected result has been repeatedly confirmed by experiment.

The Rohrschneider Polarity Scale

Rohrschneider in a series of papers [18] has introduced the concept of measuring solvent properties (polarity and selectivity) through the use of model solutes (e.g., ethanol, nitromethane, pyridine), each of which exemplifies a particular type of solute-solvent interaction—for example, dipole interactions for nitromethane, acceptor interactions for pyridine, donor interactions for ethanol). In each case the relative solubility of the test solutes in a given solvent is determined (usually indirectly, by way of vapor pressure measurements) and is corrected for dispersion forces by comparing the result with an equivalent nonpolar solute and/or solvent. So far this work has been aimed primarily at characterizing stationary liquid phases for gas chromatography, but the principle is much more general.

More recently Rohrschneider has reported similar data for a wide variety of common solvents [19], and it is possible to treat these data similarly to arrive at solvent polarity and selectivity parameters [20]. The advantage of such a treatment over the solubility parameter approach is that the raw data are based on actual solutions rather than on pure compounds, and at this stage it appears that the Rohrschneider characterization parameters for different solutes *may* therefore be more reliable [20a].

As described in Ref. 20, the solubility data of Ref. 19 for the test solutes ethanol, dioxane, and nitromethane are corrected for dispersion interactions to arrive at a net solution energy for each solute in a given solvent: $\log(K''_g)_{\text{ethanol}}$, $\log(K''_g)_{\text{dioxane}}$, and $\log(K''_g)_{\text{nitromethane}}$. These three terms for each solvent are then summed to give an overall solvent polarity value P', which can be compared with the solubility parameter δ. An additional feature of P' versus δ is that dispersion interactions are essentially ignored in P', whereas they contribute substantially to δ. In practice it appears that dispersion interactions are less important in determining relative solubility than represented by δ, so this seems an additional advantage of P' versus δ as a measure of solvent polarity. Nevertheless, dispersion can be important in certain circumstances. The following sections discuss how dispersion interactions can be considered as a secondary correction when using P' for solvent polarity.

Values of P' for the solvents studied by Rohrschneider appear in Table 2.9. It is of interest to compare this ranking of solvent polarity with that given in

Table 2.9 Solvent Classification Parameters Derived from Rohrschneider Data [20]

Solvent					
Number[a]	*Name*[b]	P'[c]	x_e	x_d	x_n
	Squalane	−0.8			
2	Isooctane	−0.4			
	n-Decane	−0.3			
7	Cyclohexane	0.0			
4	*n*-Hexane	0.0			
9	Carbon disulfide (VIb)	1.0			
10	Carbon tetrachloride (VIb)	1.7	0.30	0.38	0.32
—	Dibutyl ether (I)	1.7	0.53	0.08	0.39
15	Triethylamine (I)	1.8	0.61	0.07	0.32
16	Diisoprophyl ether (I)	2.2	0.54	0.11	0.35
17	Toluene (VIb)	2.3	0.32	0.24	0.44
18	*p*-Xylene (VIb)	2.4	0.32	0.24	0.44

Table 2.9 (*Continued*)

Solvent					
Number[a]	*Name*[b]	P'^{c}	x_e	x_d	x_n
19	Chlorobenzene (VII)	2.7	0.24	0.34	0.42
20	Bromobenzene (VII)	2.7	0.24	0.34	0.42
21	Iodobenzene (VII)	2.7	0.24	0.36	0.40
22	Diphenyl ether (VII)	2.8	0.25	0.33	0.42
23	Ethoxybenzene (VIb)	2.9	0.27	0.29	0.44
	Diethyl ether (I)	2.9	0.55	0.11	0.34
24	Benzene (VIb)	3.0	0.29	0.28	0.43
—	Tricresyl phosphate (V)	3.1	0.35	0.18	0.47
13	Ethyl bromide (VIa)	3.1	0.32	0.28	0.40
25	*n*-Octanol (II)	3.2	0.61	0.14	0.25
26	Fluorobenzene (VII)	3.3	0.24	0.33	0.43
27	Dibenzyl ether (VIb)	3.3	0.27	0.27	0.46
28	Methylene chloride (V)	3.4	0.34	0.17	0.49
29	Methoxybenzene (VIb)	3.5	0.28	0.31	0.41
30	Isopentanol (II)	3.6	0.58	0.17	0.25
31	Ethylene chloride (V)	3.7	0.36	0.19	0.45
—	Bis(2-ethoxyethyl) ether (VIa)	3.9	0.35	0.19	0.46
32	*t*-Butanol (II)	3.9	0.55	0.23	0.22
33	Isopropanol (II)	4.3	0.54	0.20	0.26
34	Chloroform (VIII)	4.4	0.28	0.39	0.33
36	Acetophenone (VIa)	4.4	0.33	0.27	0.40
—	Methyl ethyl ketone (VIa)	4.5	0.36	0.17	0.47
37	Cyclohexanone (VIa)	4.5	0.35	0.23	0.42
38	*n*-Butanol (II)	3.9	0.53	0.21	0.26
39	*n*-Propanol (II)	3.9	0.53	0.21	0.26
40	Tetrahydrofuran (III)	4.2	0.41	0.19	0.40
41	2,6-Lutidine (III)	4.3	0.47	0.18	0.35
42	Ethyl acetate (VIa)	4.3	0.34	0.25	0.42
43	Nitrobenzene (VIb)	4.5	0.30	0.27	0.43
44	Benzonitrile (VIa)	4.6	0.35	0.26	0.39
45	Dioxane (VIa)	4.8	0.38	0.21	0.41
—	2-Picoline (III)	4.8	0.51	0.19	0.30
46	Tetramethylurea (III)	5.0	0.46	0.14	0.40
—	Diethylene glycol (IV)	5.0	0.43	0.24	0.33
—	Triethylene glycol (IV)	5.1[d]	0.43	0.24	0.33
47	Ethanol (II)	5.2 (6.2)	0.51	0.21	0.28

Table 2.9 (Continued)

Solvent					
Number[a]	Name	P' [c]	x_e	x_d	x_n
48	Quinoline (III)	5.2	0.40	0.27	0.33
49	Pyridine (III)	5.3	0.43	0.21	0.36
50	Nitroethane (VIb)	5.3	0.31	0.27	0.42
51	Acetone (VIa)	5.4[d]	0.36	0.24	0.40
52	Ethylene glycol (IV)	5.4[d] (8.5)	0.47	0.23	0.30
53	Benzyl alcohol (IV)	5.5	0.42	0.28	0.30
—	Tetramethylguanidine (I)	5.5	0.52	0.11	0.37
54	Methoxyethanol (IV)	5.7	0.39	0.25	0.36
—	Tris(cyanoethoxy)propane (VIa)	5.8	0.34	0.25	0.41
—	Propylene carbonate (VIb)	6.0	0.31	0.28	0.41
—	Oxydipropionitrile (VIa)	6.2	0.33	0.28	0.39
55	Aniline (VIa)	6.2[d]	0.34	0.30	0.36
56	*N*-Methyl formamide (III)	6.2 (7.5)	0.43	0.21	0.36
57	Acetic acid (IV)	6.2	0.41	0.29	0.30
58	Acetonitrile (VIa)	6.2	0.33	0.26	0.41
—	*N,N*-Dimethylacetamide (III)	6.3	0.43	0.20	0.37
59	Dimethylformamide (III)	6.4	0.41	0.21	0.38
60	Tetrahydrothiophene-1,1-dioxide (VIa)	6.5	0.35	0.27	0.38
61	Dimethylsulfoxide (VIa)	6.5	0.35	0.27	0.38
62	*N*-Methyl-2-pyrrolidone (III)	6.5	0.41	0.21	0.28
63	Hexamethylphosphortriamide (III)	6.6	0.49	0.15	0.36
64	Methanol (II)	6.6	0.51	0.19	0.30
66	Nitromethane (VIb)	6.8	0.28	0.30	0.42
67	*m*-Cresol (VIII)	7.0	0.39	0.36	0.25
68	Formamide (IV)	7.3[d] (8.5)	0.40	0.28	0.32
—	Dodecafluoroheptanol (VIII)	7.9	0.35	0.40	0.25
69	Water (VIII)	9.0[d]	0.40	0.34	0.26
—	Tetrafluoropropanol (VIII)	9.3	0.36	0.34	0.30
	Range	—	0.24–0.61	0.07–0.40	0.22–0.49

[a] Numbers are keyed to numbering of solvents in Table 2.2.

[b] Roman numerals in parentheses refer to a solvent group, defined as in Table 2.2 and Fig. 2.1.

[c] Corrected (more accurate) values.

[d] Incomplete data, estimated values.

Table 2.8 for the solubility parameter, and P' is plotted versus δ in Fig. 2.4. Generally there is a good correlation between these two polarity scales, indicating that each gives a reasonable representation of solvent polarity. One obvious exception is in the case of solvents that are strong proton acceptors but have no proton-donor ability (e.g., diethyl ether and triethylamine). These are moderately polar solvents on the Rohrschneider scale ($1.8 \leq P' \leq 2.9$), but they appear to be no more polar than alkanes on the Hildebrand scale. The reason is that values of δ are based on the properties of the pure solvent, where the strong acceptor properties of these solvents are not expressed. Most chemists would agree, however, that the P' values of diethyl ether and triethylamine more accurately reflect their "polarities" than do

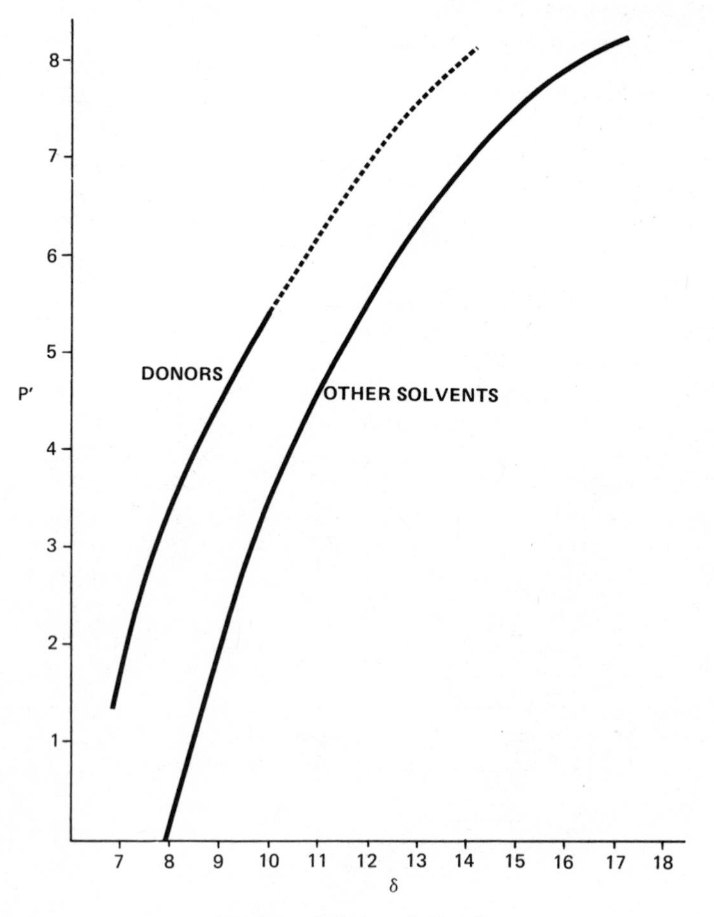

Fig. 2.4. Relation of P' to δ.

values of δ. Figure 2.4 shows a separate curve relating P' to δ for such donor solvents.

In a few cases the derived values of P' in Table 2.9 appear in error (by comparison with the plots of Fig. 2.4 plus elementary logic). When this is the case, a corrected value for P' is given in parentheses in Table 2.9 (obtained from Fig. 2.4 and the solvent δ value).

Solvent (and Solute) Selectivity Parameters

Subdividing the Solubility Parameter δ

A number of workers, most notably Hansen [21], have proposed that the overall value of δ for different solvents and solutes be subdivided into terms for the individual interactions (i.e., δ_d for dispersion, δ_p for dipole, and δ_h for hydrogen bonding interactions). These individual or specific solubility parameters then provide a better measure of how a given solvent will interact with a particular solute. Just as the overall δ value should be similar for solvent and solute for maximum solubility, so should the individual parameters δ_d, δ_p, and δ_h for solvent and solute be similar if maximum solubility is desired. Also, subdivision of δ in this way allows differences in solubility and solvent selectivity to be anticipated for solvents that have similar polarity but can exhibit differences in individual interactions. Thus if a change in solvent selectivity is desired, while maintaining solvent polarity roughly constant, a solvent with different values of δ_d, δ_p, and so on (but similar values of δ) should be chosen. This is illustrated below.

Although the approach of Hansen represented a major step forward in improving the ability of the solubility parameter to predict solubility, it suffered from certain fundamental limitations, as discussed in Ref. 14. Consideration of these limitations has led to the development of an improved solvent characterization scheme [14, 15], where now the division of δ is into the selectivity parameters δ_d (dispersion, as before), δ_o (orientation or dipole, similar to before), δ_a (acid or proton donor) and δ_b (base or proton acceptor). Table 2.8 lists these characterizing parameters for the solvents shown.

An important difference in the selectivity parameters of Table 2.8 versus the original parameters of Hansen is that maximum solubility now occurs when the product of δ_a for the solvent and δ_b for the solute (or vice versa) is a maximum, rather than when these values are equal for both solvent and solute. That is, maximum solubility is promoted by strong hydrogen bonding between solvent and solute molecules. For example, chloroform as solvent has a large value of δ_a and small value of δ_b. Hydrogen bonding with a solute will be strongest when the solute has a large value of δ_b (e.g., alkyl amines), rather than when the solute is also a strong donor and weak acceptor (like chloroform itself).

A similar solvent classification scheme (as in Table 2.8) has been proposed recently [22], and it is claimed to be more precise and reliable than previous solubility parameter schemes. This remains to be seen, however, as discussed in Ref. 23.

Subdividing the Polarity Parameter P'

The basis for a similar subdivision of P' into individual polarities representing different interactions is implicit in the derivation of P' as discussed earlier. Thus P' is the sum of terms $\log(K''_g)$ for ethanol (donor), dioxane (acceptor), and nitromethane (large dipole), and these individual terms can be directly compared to δ_o, δ_a, and δ_b. However it is more convenient, as we will see, to calculate the ratios of these specific terms to the total as follows:

$$x_e = \frac{\log(K''_g)_{\text{ethanol}}}{P'}$$

$$x_d = \frac{\log(K''_g)_{\text{dioxane}}}{P'}$$

$$x_n = \frac{\log(K''_g)_{\text{nitromethane}}}{P'}$$

Thus x_e, x_d, and x_n now represent the fractional polarity (or partial value of P') to be ascribed to proton acceptor, proton donor, or dipole interactions in the solvent. Note that a large value of x_e (donor solute) means a strong interaction with acceptor solvents; thus values of x_e, x_d, and x_n can be loosely compared with corresponding quantities δ_b, δ_a, and δ_o.

Values of the solvent selectivity parameters x_e, x_d, and x_n are given in Table 2.9 for several solvents. Solvents of similar selectivity can be recognized, if a triangular plot is made as in Fig. 2.5, which plots the individual solvents of Table 2.9 and groups them within the indicated, numbered circles —each of which defines a solvent selectivity class as noted in Table 2.9.

Because the relative importance of dipole, proton-donor, and proton-acceptor interactions for a solvent is determined by the functional groups in the solvent molecule, one would expect homologous solvents (e.g., the alcohols, the ethers) to have the same selectivity characteristics and to fall within the same solvent selectivity group of Fig. 2.5. This is indeed the case, as summarized in Table 2.10, which gives the kinds of compound falling within each selectivity group. Thus other solvents not included in Table 2.9 can be assigned to one of eight selectivity groups on the basis of similarity of molecular structure.

Note added in proof: In a more recent publication (23a) the solvent classification scheme of Table 2.9 and Figure 2.5 has been further refined. This has

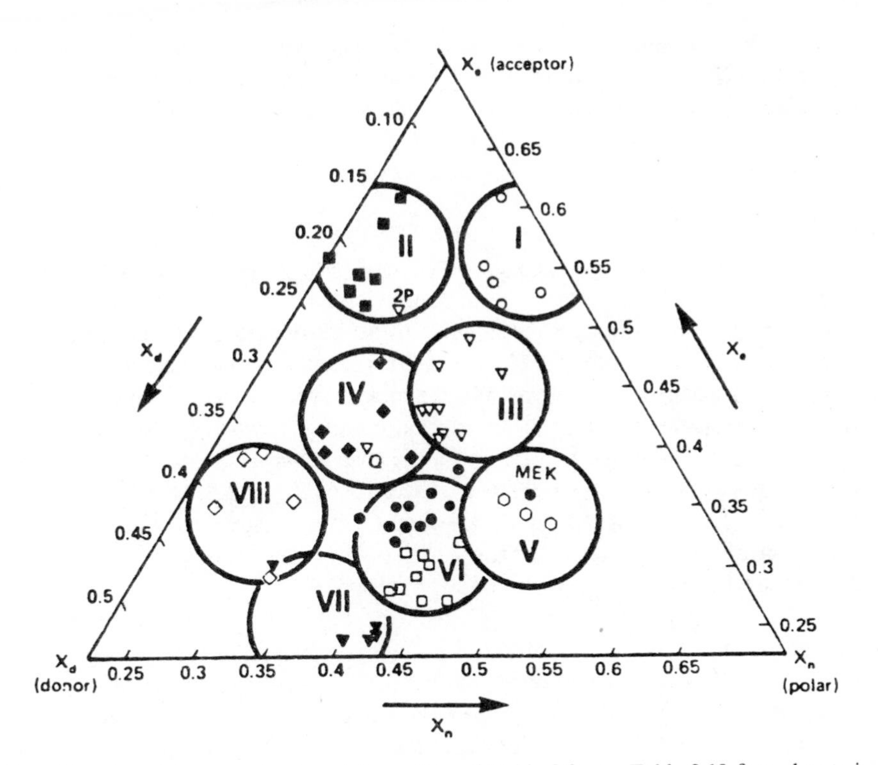

Fig. 2.5. Selectivity triangle plotted from data of Table 2.9; see Table 2.10 for solvents in various selectivity groups.

Table 2.10 Solvents Falling in Various Selectivity Groups of Fig. 2.5

Group	*Solvents*
I	Aliphatic ethers, trialkyl amines, tetramethylguanidine
II	Aliphatic alcohols
III	Pyridines, tetrahydrofuran, amides (except the more acidic formamide)
IV	Glycols, glycol ethers, benzyl alcohol, formamide, acetic acid
V	Methylene chloride, ethylene chloride, tricresyl phosphate
VIa	Alkyl halides, ketones, esters, nitriles, sulfoxides, sulfones, aniline, and dioxane
VIb	Nitro compounds, propylene carbonate, phenyl alkyl ethers, aromatic hydrocarbons
VII	Halobenzenes, diphenyl ether
VIII	Fluoroalkanols, *m*-cresol, chloroform, water

led to minor changes in certain of the P', x_e, etc. values, and to some shifting of solvents from one group to an adjacent selectivity group. However the overall utility of the data of Table 2.9 has not been significantly affected.

As a byproduct of this latter study (23a), it has been found that dispersion interactions in solutions of polar solvents appear to play an insignificant role in affecting solvent selectivity.

4 GENERAL APPROACH TO SOLVENT SELECTION

Either the solubility parameter or Rohrschneider polarity schemes can be used as a basis for solvent selection, in much the same way. I illustrate a general approach to solvent selection with the Rohrschneider scheme, because it is somewhat simpler to understand and because the Rohrschneider parameters appear to be more reliable. However data for the Rohrschneider parameters are limited to the 82 pure solvents of Table 2.9, and in some cases we may wish to consider additional solvent options. Since data for the Hildebrand parameters are available for other solvents, the Hildebrand parameters can be used to supplement the data of Table 2.9.

Another reason for considering the Hildebrand scheme, or the use of data on specific intermolecular interactions (e.g., Tables 2.5 to 2.7) becomes evident if one wants to apply chemical logic to solvents and solutes of known structure (e.g., maximize hydrogen bonding between solvent and a given solute, etc.). The Rohrschneider approach is linked only indirectly to specific interactions, and it is better used in a more or less empirical mode (see below).

Maximizing Solubility for a Given Solute

The most important step in solvent selection is often that of maximizing the solubility of a particular solute in a given solvent. Alternatively, one may wish simply to vary solute solubility in a controlled manner, as is commonly the case in various liquid chromatography methods. Solvent polarity P' is the most important factor affecting solubility, and P' can be systematically varied for maximum (or controlled) solubilities as follows. Select two solvents A (less polar) and B (more polar) of significantly different P' values but miscible in all proportions (e.g., see Table 2.3), and determine solubility as a function of P', using mixtures of A and B, with P' for a given mixture calculated as

$$P'_{\text{mixture}} = \phi_a P'_a + \phi_b P'_b \tag{2.7}$$

Here ϕ_a and ϕ_b are the volume fractions of A and B in the mixture, and P'_a and P'_b refer to P' values for the pure solvents. Normally an intermediate mixture of A and B will provide maximum solubility, indicating that P' values of the solvent mixture and of solute are the same. Occasionally, for very polar or nonpolar solutes, solubility is a maximum for one pure solvent or the other.

In this case the P' value of the solute lies outside the range bracketed by solvents A and B, meaning that maximum solubility requires the use of a third solvent. For example, assume that maximum solubility with mixtures of A and B occurs for pure B; that is, the solute is quite polar. In this case select a third pure solvent C, which is more polar than B, and repeat the measurement of solubility for different mixtures of B and C. Presumably maximum solubility will now occur for a mixture of B and C. In any case, the P' value of the solvent that gives maximum (or the desired) solubility should be noted.

Usually we can further increase solubility for a given solute by trying solvents of similar polarity (value of P'), but differing selectivity. Now the selectivity of a solvent mixture is determined largely by the *selectivity group* of the more polar component B, and selectivity can be changed by substituting B in the solvent mixture by another solvent D, chosen from a different selectivity group. This can be illustrated with a hypothetical example. Assume that we begin our tests of solute solubility with the two solvents, hexane (A, $P'_a = 0.0$) and chloroform (B, $P'_b = 4.4$). Also assume that maximum solubility is provided by a 30 volume percent chloroform-hexane mixture, so that from (2.7) and Table 2.9, P' for the solute is

$$(0.7 \times 0.0) + (0.3 \times 4.4) = 1.3$$

Now chloroform (solvent B) is from selectivity class VIII of Fig. 2.5, and a large difference in selectivity would be expected by moving to a selectivity group far removed from VIII in Fig. 2.5 (e.g., II, I, or V). If we pick group I, from Table 2.9 we can select several possible substitutes for chloroform in the original mixture (e.g., ethyl ether, $P' = 2.9$). From (2.7) we calculate that $P' = 1.3$ for ethyl ether–hexane as solvent would require 45 % ethyl ether in the mixture. We should therefore try 45 % ethyl ether–hexane as solvent and determine solute solubility. Since (2.7) is at best approximate, we would want to further vary the volume percentage of ether in our solvent to confirm the solvent composition for maximum solubility (e.g., 35 and 55 % ether–hexane). This change from chloroform as polar solvent to ethyl ether might or might not provide an increase in solute solubility. If further increase in solubility is desired, other substitutions for solvent B in the mixture can be tried, in the same way that ethyl ether was investigated (i.e., 25 volume percent ethanol–hexane (II), 38 volume percent methylene chloride–hexane (V), etc.).

Maximizing the Separation of Two Solutes

Suppose that two solutes x and y are present in some sample matrix (e.g., some solvent, an insoluble solid, etc). Assume that we desire the selective extraction or leaching of solute x from the mixture, with minimum coextraction of y, using some solvent that is immiscible with the sample matrix. We begin by studying the extraction of x and y as a function of solvent polarity, just as we determined maximum solute solubility earlier—for example, using mixtures of chloroform (B) and hexane (A). Generally it is found that the extraction of x and y versus P' for the extraction solvent yields plots such as that of Fig. 2.6. Because of the differences in P' values for x and y, some intermediate solvent polarity (P'_1 in Fig. 2.6) may yield significant extraction of x (60 % in the example) and negligible extraction of y. Repeated extraction of the sample with solvent of $P' = P'_1$ will then achieve complete separation of x from the sample matrix, without contamination by y.

If the two curves for x and y in Fig. 2.6 were reversed, several strategies would be possible. First, we might simply extract all y from the sample, using solvent of $P' = P'_1$. Then we could reextract the sample with solvent of $P' = P'_2$ for complete extraction of pure x. Second, at large values of P' for the extraction solvent, the curves for x and y are reversed; thus a solvent of greater polarity ($P \gg P'_x$) may provide selective extraction of x free of y. Third, we may try some other strong solvent in place of chloroform, as in the previous example of maximizing solute solubility. Repetition of the plots of

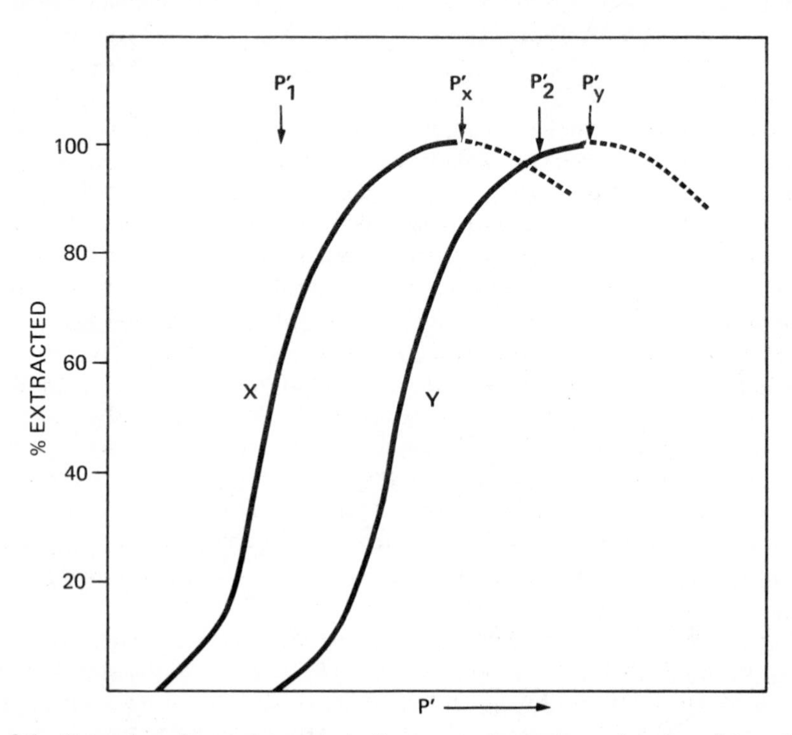

Fig. 2.6. Extraction of two solutes x and y from a sample matrix as a function of the polarity P' of the extraction solvent: P'_x and P'_y refer to polarities of solutes x and y; for P'_1 and P'_2 see text.

Fig. 2.6 with different selectivity group solvents in place of chloroform might in some cases reverse the two curves, allowing the simple extraction of x free of y with a less polar solvent. This approach can also be used to maximize the spacing of the two extraction curves, for maximum separation of x and y.

Other examples of the application of the foregoing solvent selection strategy to enhance separation will probably be apparent to the reader, once the fundamental concept of varying solvent polarity and selectivity is grasped. Specific applications for individual separation methods are discussed in the following section.

One final element in this overall strategy should be considered. So far we have ignored the possibility of deliberately varying dispersion interactions so as to enhance selectivity. Normally differences in dispersion interactions among different solvents will play at best a minor role in affecting solvent selectivity. However dispersion interactions occasionally are worth considering in the separation of solutes that differ in refractive index. Then a change in solvent refractive index can produce a useful further change in solvent selectivity. Dispersion interactions can be changed in such cases as follows. Select another *nonpolar* pure solvent E, where the refractive index of E is quite different from that of A, but where $P'_a \approx P'_e$. Substitute E for A in the mixture A–B previously optimized for P' and selectivity group of B (I, II, etc.). If necessary, investigate further small changes in the concentration of E in the mixture, to achieve final reoptimization of P'. Since the nonpolar component (A or E) plays little role in affecting dipole or hydrogen bonding interactions, and the final P' value of the mixture B/E will be similar to that of B/A, all the advantages achieved during optimization of P' and selectivity group initially will be retained during this final alteration of dispersion selectivity.

For the case of water-soluble solutes, we may add to the preceding approach the use of covalent interactions as discussed in Section 2. Thus the pH of the solvent can be varied, complexation reactions can be explored as a means of selectively removing desired solutes into the solvent phase, and so on. The utility of such techniques is evident.

To summarize, we have the following steps in solvent selection:

1. Choose two pure solvents A (less polar) and B (more polar) and determine what mixture of A/B provides maximum solute solubility or the best separation of two solutes x and y.
2. If solubility or separation is inadequate, substitute other polar solvents for B in the mixture, maintaining P' constant from step 1 [by way of (2.7)]; select these polar solvents from different selectivity groups (e.g., I, II, V, VIII).
3. If necessary, readjust P' for the new mixture by changing the proportions of the two solvents in the mixture.

4. If further adjustment in selectivity is necessary, change the nonpolar solvent A for another nonpolar solvent of different refractive index [keep P' constant by way of (2.7)].

5. Consider covalent interactions (via changes in pH, addition of complexing agents, etc.), if appropriate, as a means of creating pronounced solvent selectivity for some solutes.

In many separations step 1 suffices; that is, a simple optimization of solvent polarity may be adequate.

5 APPLICATION IN DIFFERENT SEPARATION METHODS

The previous discussion and guidelines for solvent selection are now applied to several individual separation methods.

Solvent Extraction (Leaching)

The general approach to separation or purification by solvent extraction was discussed specifically in the preceding section. In the simplest application, it is desired to remove some solute or group of solutes from an insoluble matrix (e.g., a synthetic polymer, dried paint or ink, animal or vegetable tissue). In this case the solvent that provides the highest solvency for the solutes in question is desired, and this solvent should be matched in terms of polarity and selectivity to the extractable solutes.

Where selective separation of one solute from another is desired during solvent extraction, the solvent must be selected to maximize the solubility of the desired solute, and to minimize the solubility of other solute(s). Section 4 provides a detailed description of how we proceed in this case.

In some cases, complete extraction of compounds of interest from an insoluble sample matrix requires effective wetting of the matrix by the solvent. Thus, the extraction of hydrophilic samples such as animal tissue is best done with water-miscible solvents such as methanol or acetone, modified if necessary by the admixture of other solvents into the solvent mixture. Similarly, the extraction of oily solids such as chopped nuts is best done with water-immiscible solvents such as benzene or chloroform.

Liquid-Liquid Partition

In liquid-liquid partition the sample is normally shaken with two immiscible liquids (e.g., in a separatory funnel). The liquids are chosen to provide maximum distribution of one solute or solutes into one solvent phase, and maximum distribution of undesired solutes into the second solvent phase. The approach is essentially the same as described in Section 4 for selective extraction of one or more solutes from an insoluble sample matrix. Now, however, a second solvent replaces the sample matrix as the phase into which it is desired to concentrate sample constituents that are not of interest (of course in some cases we are interested in the solutes recovered from both phases). This second solvent phase is selected to maximize the solubility of undesired sample components, in the same way that the first solvent is chosen to maximize the solubility of desired sample components. By changing the polarity and/or selectivity group of each solvent in the manner outlined previously, the optimum separation of sample components can be effected. For a detailed listing of immiscible solvent pairs for liquid-liquid partition, see Ref. 5, 6e.

Liquid-Liquid Chromatography

Selection of solvents for liquid-liquid chromatography (LLC) is conceptually the same as selecting solvents for liquid-liquid partition, since the same separation mechanism prevails for each of these techniques. We seldom wish to maximize solute solubility in either liquid phase in LLC, however. Rather, we wish to *control* the solute distribution to give two- to tenfold concentration of solutes of interest into the stationary liquid phase. This is normally achieved by first adjusting the polarity of the mobile phase solvent to give appropriate retention times for solutes of interest (capacity factor or k' values of 2 to 10, see Ref. 4), then varying solvent selectivity group as previously. The basic process is the same as already outlined for various extraction and partition processes.

Fractional Crystallization

In fractional crystallization work we desire a solvent that will provide good solubility for the solute of interest at one temperature T_1, poor solubility at some lower temperature T_2, and good solubility for impurities at T_2. The ratio of solubilities of our desired compound at two temperatures in any solvent is primarily a function of the heat of fusion of the compound and of T_1/T_2, and more or less independent of the solvent (this is assumed in most elementary texts on physical chemistry). Therefore our primary concern is the selection of a solvent that will be able to dissolve adequate amounts of our sample in a convenient solvent volume and at a convenient temperature T_1. We can choose such a solvent in the same way we maximized solute solubility in the preceding section: by varying first polarity, then selectivity. At the same time we must keep in mind temperature as a variable, and the limitations imposed by solvent boiling and freezing points (see Table 2.2 and Ref. 1 for actual data).

Normally a solvent that allows adequate solution of the sample and yields crystallization at a lower temperature will provide purification of sample from minor impurities—regardless of the selectivity of the solvent for impurities versus the compound of interest. However enhanced purity of crystallized solute can be achieved by selecting a solvent that dissolves impurities to a greater extent than the compound of interest (i.e., is a selective solvent for the impurities).

Extractive Distillation

The normal separation of two or more solutes by distillation can be enhanced by adding a nonvolatile solvent to the sample before distillation. To the extent that one sample constituent is more strongly held by this added solvent than are other sample constituents, its vapor pressure at any temperature will be less and its apparent boiling point will be increased. Thus we can vary the apparent boiling points—and ease of separation—of the various sample compounds by choosing the right solvent. Here both solvent polarity and selectivity group affect the apparent boiling points of different compounds, and selection of the right solvent is achieved in the same way that solvents are chosen for solvent extraction (preceding section).

Solvent-Adsorbent Partition and Liquid-Solid Chromatography

The two above-normal processes—batch contacting of a sample solution by an adsorbent, or adsorption chromatography—are basically similar, in the same way that liquid-liquid partition and liquid-liquid chromatography are similar. The selection of an optimum solvent proceeds in the same manner for each, and liquid-solid chromatography is discussed here as example.

Because the interactions at the adsorbent surface are normally all-important in controlling the relative adsorption of different compounds from some solvent, relative adsorption is determined mainly by the energy of adsorption of the solvent. Strongly adsorbed solvents decrease sample adsorption and give lower retention times in liquid-solid chromatography, whereas weakly adsorbed solvents behave in the opposite manner. Usually an intermediate solvent strength (*strong* solvents are strongly adsorbed and give less retention of sample, and vice versa for *weak* solvents) is required to provide some adsorption of desired sample compounds and yield acceptable separations (see discussion of Refs. 4 and 24).

Solvent strength in adsorption chromatography increases with solvent polarity for polar adsorbents such as silica, alumina, and other inorganic oxides, and decreases with solvent polarity for charcoal as adsorbent. Solvent strength can be characterized by an empirical parameter ε^0, values of which appear in Table 2.11 for alumina as adsorbent. Values of ε^0 for silica and other polar adsorbents are similar to the values given in Table 2.11 (see discussion of Ref. 23). Table 2.12 presents a solvent strength series—a so-called eluotropic series—for charcoal as adsorbent.

Table 2.11 Solvent Strength in Adsorption Chromatography [24]: Data for Alumina as Adsorbent

Solvent	$\varepsilon^0(Al_2O_3)$
Fluoroalkanes	−0.25
n-Pentane	0.00
Isooctane	0.01
Petroleum ether, Skellysolve B, etc.	0.01
n-Decane	0.04
Cyclohexane	0.04
Cyclopentane	0.05
Diisobutylene	0.06
1-Pentene	0.08
Carbon disulfide	0.15
Carbon tetrachloride	0.18
Amyl chloride	0.26
Xylene	0.26
i-Propyl ether	0.28
i-Propyl chloride	0.29
Toluene	0.29
n-Propyl chloride	0.30
Chlorobenzene	0.30
Benzene	0.32
Ethyl bromide	0.37
Ethyl ether	0.38
Ethyl sulfide	0.38
Chloroform	0.40
Methylene chloride	0.42
Methyl isobutyl ketone	0.43
Ethylene dichloride	0.44
Methylethylketone	0.51
1-Nitropropane	0.53
Triethylamine	0.54
Acetone	0.56
Dioxane	0.56
Tetrahydrofuran	0.57
Ethyl acetate	0.58
Methyl acetate	0.60
Amyl alcohol	0.61
Aniline	0.62
Diethyl amine	0.63
Nitromethane	0.64
Acetonitrile	0.65

Table 2.11 *(Continued)*

Solvent	$\varepsilon^0(Al_2O_3)$
Pyridine	0.71
Butyl cellusolve	0.74
Dimethyl sulfoxide	0.75
i-Propanol, *n*-propanol	0.82
Ethanol	0.88
Methanol	0.95
Ethylene glycol	1.11
Acetic acid	Large

Solvent selectivity in adsorption chromatography on polar adsorbents is determined mainly by so-called solute localization and by hydrogen bonding between solvent and sample molecules. A detailed discussion is beyond the scope of this chapter, but for details see Refs. 4, 24, and 24a, and prior references. Data presented in preceding sections can be used to estimate hydrogen bonding between solute and solvent, leading to an understanding of solvent selectivity in adsorption chromatography.

An interesting problem in adsorption chromatography on polar adsorbents sometimes arises with poorly soluble polar samples. Solvents with the appropriate strength (ε^0 value) are often too nonpolar to allow adequate solubility of the sample in the mobile phase, thus limiting the maximum amount of sample that can be separated under conditions that optimize the separation. In this case we want to maximize sample solubility, or the solvent P' value, while holding ε^0 constant. It is of interest to consider the general relationship

Table 2.12 Solvent Strength in Adsorption Chromatography [24]: Data for Charcoal as Adsorbent

Water
Methanol
Ethanol
Acetone
Propanol
Ethyl ether
Butanol
Ethyl acetate
n-Hexane
Benzene

between P' and ε^0 for different solvent mixtures, as in Fig. 2.7. Here values of P' for various mixtures of hexane ($P' = 0$) with more polar solvents are plotted versus the ε^0 value (on silica) of the solvent mixture. We see for a given value of ε^0 that P' for the solvent mixture can vary widely (i.e., yielding differences in sample solubility). In general, mixtures of less polar solvents (e.g., CCl_4, benzene) are advantageous, relative to mixtures of more polar solvents (e.g., ethyl ether). Of course, to maintain ε^0 constant, this requires higher concentrations of B in the solvent mixture in the case of solvents B (CCl_4, benzene, etc.) with smaller P' values.

Gas Chromatography

In GC the relative solubility or retention of different compounds on the column (i.e., in the stationary phase liquid) is controlled primarily by column temperature. An increase in temperature results in decreasing retention for all sample components. The selection of a particular stationary phase solvent is determined largely by the selectivity requirements of a given sample. A major distinction is made between polar and nonpolar phases, and the "polarity" of the stationary phase liquid is of major importance in determining separation selectivity. Selectivity differences in stationary phases also exist for phases of similar polarity and are the result of different intermolecular interactions as described above. The Rohrschneider approach to characterizing solvent selectivity has been widely applied to the classification of

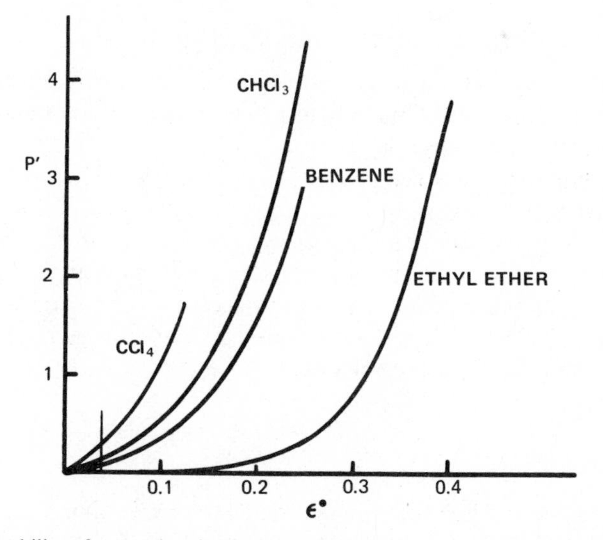

Fig. 2.7. The ability of a sample solvent to dissolve the sample P' versus the strength ε° of that solvent in liquid-solid chromatography. Silica as absorbent.

solvents for GC stationary phases (e.g., Refs. 18, 25, 26, and recent books on GC). In view of the similarity of presently used schemes for classifying GC phases to that described here for solvents in all separation processes, the reader is simply referred to the latter references.

Another facet of GC stationary phases that is of interest is the use of binary solutions, rather than pure solvents. Although mixtures have not been widely used in the past, Purnell and co-workers (e.g., Ref. 27) have suggested that selectivity control for complex mixtures is best achieved through the use of mixed solvents. One significant feature of this work is the empirical observation that GC partition coefficients K_R vary linearly with the composition of the mixed solvent:

$$K_R = \phi_a K_a + \phi_b K_b \tag{2.8}$$

Here ϕ_a and ϕ_b refer to volume fractions of pure solvents A and B, and K_a and K_b refer to K_R values for the two solvents as pure GC phases. Equation 2.8 does not agree with previous theories of solution behavior, including the theories discussed in preceding sections of this chapter. Its apparent validity in a limited number of GC systems may represent the fortuitous combination of several factors [adsorption effects, the approximate equivalence of (2.8) and theoretical relationships when K_a is not greatly different from K_b, etc.]. In any case, to the extent that (2.8) is empirically valid in a given GC system, it represents a useful basis for estimating and controlling separation selectivity.

Dialysis, Ultrafiltration, and Thermal Diffusion

The solvent does not normally play an important role in affecting separation when the techniques of dialysis, ultrafiltration, and thermal diffusion are used. Rather, the solvent is selected to provide good solubility for the sample and to be compatible with the physical requirements of the system (does not dissolve membranes, has right boiling point for thermal diffusion, etc.).

The foregoing examples and discussion are not intended to be complete, nor to anticipate every separation problem that may arise. However the general approach outlined in the preceding section appears applicable, in one way or another, to many separaton problems. The same general logic can be brought to each individual problem as it is perceived.

ACKNOWLEDGMENT

I acknowledge the helpful comments of Drs. D. L. Saunders (Union Oil Company) and J. J. Kirkland (E. I. Du Pont de Nemours) in the preparation of this chapter.

SYMBOLS

A, B	refers usually to pure solvents A (nonpolar) and B (polar); also, donor compound (A) and acceptor compound (B)
B_i	some part of a solvent molecule B, such that 1 mole of the groups B_i has a volume of 1 ml
$C_{x,a}$, $C_{x,b}$	concentrations (mole fractions) of solute x in pure solvents A and B, respectively
d	density of a solvent (ml/g)
d_a, d_b	values of d for solvents A and B, respectively
H'	parameter characterizing basicity or proton-acceptor ability of a solvent or solute
H_i	a contribution to ΔH from some process i
$H_{a,a}$, $H_{b,b}$	values of H_i due to interactions between like compounds (solvents) A and B, respectively
$H_{x,a}$, $H_{x,a}$	values of H_i due to interactions of solute group x_i with solvent molecules A and B, respectively
i	in Fig. 2.1, the group x_i; in Fig. 2.3, the group B_i
$\log(K_g'')$	net solution energy for indicated solute in a given solvent [e.g., $\log(K_g'')_{\text{ethanol}}$ is for solute ethanol]
P_a, P_b, etc.	value of polarity P for indicated species (a, b, etc.)
P'	Rohrschneider solvent polarity parameter
P_1', P_2',	particular values of P', indicated in Fig. 2.6
P_a', P_b', etc.	values of P' for compounds A, B, etc.
R	the gas constant (cal/°K)
T	absolute temperature (°K)
V	molar volume (ml) of a given compound (2.5)
V_x	molar volume of solute x
x, y	solutes or compounds
x_a, x_b	mole fractions of pure solvents A and B in mixture of A and B
x_d, x_e, x_n	solvent selectivity parameters of Rohrschneider; Table 2.9 and defined in text
x_i	some part of a solute molecule x (e.g., a functional group)
δ	Hildebrand solubility parameter in (2.5)
δ_a, δ_b	values of δ for solvents A and B; (2.6) only
δ_d, δ_o, δ_a, δ_b	contributions to δ due to dispersion, dipole, proton-donor, and proton-acceptor interactions, respectively

ΔE_v	energy of vaporization (cal/mole) of a compound
ΔG	free energy for transfer of one mole of solute x from solvent B to solvent A
ΔH	enthalpy for transfer of one mole of solute x from solvent B to solvent A
ΔH_v	enthalpy of vaporization (cal/mole) for a compound; equal to $\Delta E_v + RT$ [see (2.5a)]
η	viscosity (cP)
η_a, η_b	values of η for pure solvents A and B
ϕ_a, ϕ_b	volume fractions of compounds A and B in mixture of A and B

References

1. J. A. Riddick and W. B. Bunger, *Organic Solvents*, 3rd ed., Wiley-Interscience, New York, 1970.
2. F. D. Rossini et al., *Selected Values of Physical and Thermodynamic Properties of Hydrocarbons and Related Compounds*, Carnegie Press, 1953.
3. R. C. Weast, Ed., *Handbook of Chemistry and Physics*, 52nd ed., Chemical Rubber Co., Cleveland, 1971.
4. L. R. Snyder and J. J. Kirkland, *Introduction to Modern Liquid Chromatography*, Wiley-Interscience, New York, 1974.
5. F. A. V. Metsch, *Angew. Chem.*, **65**, 586 (1953).

6a. I. Mellan, *Industrial Solvents Handbook*, Noyes Data Corp., Park Ridge, N.J., 1970.

6b. B. A. Englin et al., *Chem. Abstr.*, **63**, 146 (1965).

6c. C. McAuliffe, *Nature*, **200**, 1092 (1963).

6d. W. R. Gambill, *Chem. Eng.*, 151 (1959).

6e. N. B. Godfrey, *Chem. Tech.*, **360**, (1972).

7. B. L. Karger, L. R. Snyder, and C. Horvath, *An Introduction to Separation Science*, Wiley-Interscience, New York, 1973, Chapters 2, 10.
8. J. H. Hildebrand and R. L. Scott, *The Solubility of Nonelectrolytes*, 3rd ed., Dover, New York, 1964.
9. J. H. Hildebrand and R. L. Scott, *Regular Solutions*, Prentice-Hall, Englewood Cliffs, N.J., 1962.
10. M. J. R. Dack, Ed., *Solutions and Solubilities*, Part I, Wiley-Interscience, New York, 1975.
11. A. F. M. Barton, *Chem. Rev.*, **75**, 731 (2975).

11a. H. M. N. H. Irving, in *Ion Exchange and Solvent Extraction*, Vol. 6, J. A. Marinsky and Y. Marcus, eds., Marcel Dekker, New York, 1974, Chap. 3.

12. C. P. Smyth, *Dielectric Behavior and Structure*, McGraw-Hill, New York, 1955.
13. R. W. Taft, D. Gurka, L. Joris, P. von R. Schleyer, and J. W. Rakshys, *J. Am. Chem. Soc.*, **91**, 4801 (1969).

14. R. A. Keller, B. L. Karger, and L. R. Snyder, in *Gas Chromatography*, 1970, R. Stock, Ed., Institute of Petroleum, 1971, p. 125.
15. B. L. Karger, L. R. Snyder, and C. Eon, *J. Chromatogr.*, **125**, 71 (1976).
16. C. Tanford, *The Hydrophobic Effect*, Wiley-Interscience, New York, 1973.
17a. C. Horvath, W. Melander, and I. Molnar, *J. Chromatogr.*, **125**, 129 (1976).
17b. W. Melander and C. Horvath, *Biochemistry*, in press.
18. L. Rohrschneider, *J. Chromatogr.*, **22**, 6 (1966); *Adv. Chromatogr.*, **4**, 333 (1967).
19. L. Rohrschneider, *Anal. Chem.*, **45**, 1241 (1973).
20. L. R. Snyder, *J. Chromatogr.*, **92**, 223 (1974).
20a. J. J. Kirkland, unpublished data.
21. C. M. Hansen, *J. Paint Technol.*, **39**, 104, 505, 511 (1967).
22. R. Tijssen, H. A. H. Billiet, and P. J. Schoenmakers, *J. Chromatogr.*, **122**, 185 (1976).
23. B. L. Karger, L. R. Snyder, and C. Eon, *Anal. Chem.*, in press.
23a. L. R. Snyder, *J. Chromatogr.*, in press.
24. L. R. Snyder, *Principles of Adsorption Chromatography*, Dekker, New York, 1968, Chapter 8.
24a. L. R. Snyder, *Anal. Chem.*, **46**, 1384 (1974).
25. W. O. McReynolds, *J. Chromatogr. Sci.*, **8**, 685 (1970).
26. C. E. Figgins, T. N. Risby, and P. C. Jurs, *J. Chromatogr. Sci.*, **14**, 453 (1976).
27. R. J. Laub and J. H. Purnell, *Anal. Chem.*, **48**, 799 (1976).

Chapter **III**

LIQUID–LIQUID EXTRACTION

Edward G. Scheibel

1 **Introduction 78**

2 **Phase Equilibria Relationship 79**

Representation of Data for Binary Mixtures 79
Representation of Data for Ternary Mixtures 82
Phase Equilibria Data for Multicomponent Systems 89
Interpretation of Phase Equilibria Data for Nonionizing Solutes 90
Interpretation of Distribution Data for Ionizing Solutes 92

3 **Theory of Liquid Extraction Processes 103**

Extraction for Removal 104
Multiple Extractions with Fresh Solvent 105
Countercurrent Extraction 107
Batchwise Simulation 107
Continuous Column Operation 109
Fractionation Processes 115
Batchwise Techniques 115
Continuous Countercurrent Fractionation with Two Solvents 125
Fractionation with Single Solvent and Reflux 134
Dual-Solvent Fractionation with Reflux 139

4 **Equipment and Procedures 142**

Analytical Applications 143
Simulation of Continuous Processes 148
Preparation of Small-Scale Samples 155
Mixer-Settler Devices 155
Agitated Columns 161
Pulse Columns 167
Centrifugal Extractors 173
Operation of Continuous Countercurrent Extraction Equipment 175
Pilot Plant Extraction Studies 179

5 **Applications of Liquid Extraction 180**

Extraction for Removal 180
Fractionation Processes 182

Metallurgical Processes 185
Biologically Active Compounds 188

Nomenclature 192

---

1 INTRODUCTION

The resolution of mixtures into pure compounds based on differences in their distribution coefficients was considered as early as 1925 by Frenc [1], who concluded that the laboratory technique utilizing batchwise operation was not very practical. Jantzen subsequently developed a small glassware device for continuous countercurrent operation and demonstrated that it was possible to isolate many more fractions from coal tar distillates by liquid extraction than could be obtained by fractional distillation. He summarized his work in a publication in 1932 [2], but very little general interest developed in this process over the next decade. Tuttle [3] patented the Duo-sol process for solvent refining of petroleum in 1933, and the commercial installations utilized seven mixer-settler stages. Other developments of this decade can be found in the numerous patents of Van Dijck, many of which are mentioned subsequently.

The use of liquid extraction as a separation technique received its major impetus with the development of a practical device for multistage operation by Craig in 1944 [4]. The first model contained 25 stages, and the same design was subsequently employed in a 54-stage unit. As applications of this process expanded, difficult separations were attempted, requiring more stages than were obtainable in the original design. A new all-glass unit was designed [5] that could be assembled to provide any desired number of stages. At the peak of the interest in this process as an analytical tool, automated racks containing thousands of stages were used, and they were available from several suppliers.

With the advent of gas chromatography and liquid chromatography, which were more convenient to operate and required much smaller samples, the use of the Craig apparatus as an analytical tool diminished. However the factor that endowed the chromatograph with one of its advantages—namely, the small sample size—makes it impractical for the preparation of moderate quantities of a desired fraction for extensive studies. The Craig device provides some limited utility in this area.

For the preparation of appreciable quantities of a product purified by liquid extraction, a continuous countercurrent operation similar to that employed by Jantzen can be used. Various equipment designs are suggested in the patents of Van Dijck, but it is not clear exactly how many were actually fabricated and operated. In 1948 Scheibel [6] described the design and performance of a multistage continuous countercurrent fractional liquid extraction column. Interest in liquid extraction as a commercial process for both purification and fractionation accelerated, and several different designs have been developed such as the RDC (rotating disc *c*ontactor) column patented by Reman [7], the Mixco column described by Oldshue and Rushton [8], and the Karr column [9]. At present columns handling more than 2000 liters of solution per minute are in commercial use.

Laboratory units are available in different sizes, from the original 1-in. diameter Scheibel column, which can process as little as 1 liter/hr of solvents, up to columns 6-in. in diameter, which can handle more than 100 liters of liquids per hour and might be considered to be small pilot plant units.

The following sections discuss the basic principles of extraction processes for both purification and separation, and the design and operation of laboratory and small-scale production equipment. Present and future applications of the fractional liquid extraction process are also considered, to enable the research chemist to recognize potential uses of this process when he encounters purification and separation problems that cannot be conveniently solved by distillation.

2 PHASE EQUILIBRIA RELATIONSHIP

The design of liquid extraction processes depends on the distribution coefficients of different compounds between immiscible solvents. A thorough understanding of the phase behavior of the two solvents is necessary for defining the operating limits of such processes. The influence of one or more solutes on the mutual solubility must also be recognized, as well as the effect of diluents that may be added to one or both of the solvents to obtain more favorable properties. Graphical representation becomes increasingly complex with multicomponent mixtures, but the fundamental characteristics are similar to those of the binary system in that they all have a maximum and a minimum temperature at which two liquid phases exist. Liquid extraction processes must be carried out between these limits.

Representation of Data for Binary Mixtures

Phase relationships for binary systems can be fully represented on two-dimensional figures and are the simplest to visualize and understand. In general, the mutual solubility of all immiscible solvents increases with temperature until a critical solution temperature is reached at which the two immiscible liquid phases become identical, and above this temperature only one liquid phase can exist. In a few systems the solubility of one component in the other appears to decrease with increasing temperature over a limited temperature before finally increasing to the critical solution temperature. Thus the critical solution temperature normally represents the maximum operating temperature for liquid extraction with the given pair of solvents.

The lowest temperature is that at which the solubility limit of the solid is reached in one of the phases; Fig. 3.1 is a typical phase diagram for the phenol-water system [10]. In this diagram point *A* represents the melting point of pure phenol, and the curve *AB* shows the solubility of solid phenol in water. The critical solution temperature (CST) and the critical solution composition (CSC) are shown at point *C*. At any temperature below the CST, compositions of the two equilibrium phases are represented by curves *BC* and *DC* on each side of the CSC. At point *B* the heavy, phenol-rich phase is saturated with solid phenol and is in equilibrium with the water-rich light phase of composition *D*. When one recognizes that all compositions on this diagram have a particular vapor pressure and are in equilibrium with a vapor phase at all temperatures, it is apparent that four different phases are in equilibrium with this composition at this temperature. It is thus designated as a quadruple point, and since all four phases are in equilibrium with the second liquid phase at point *D*, this also becomes a quadruple point.

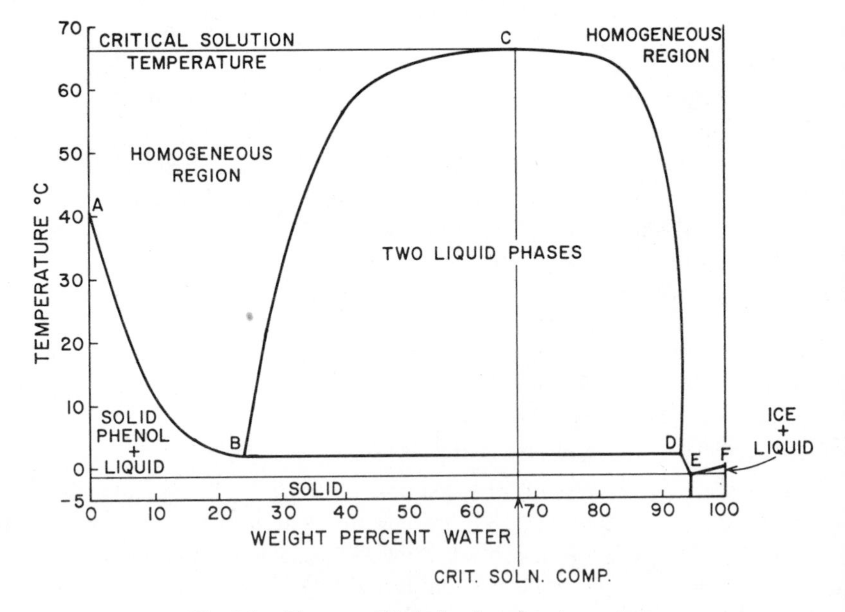

Fig. 3.1. Phase equilibria in phenol-water system.

Line *DE* shows the solubility of solid phenol in water, and point *E* represents the eutectic temperature and composition for the phenol-water system. This is the lowest temperature at which a liquid phase can exist. The line *EF* represents the solubility of ice in phenol solutions.

Figure 3.1 illustrates the usual type of phase equilibria diagram for immiscible solvents. These mixtures have a quadruple point, a eutectic point, and a critical solution point. In many cases the critical solution point is at a temperature above the atmospheric boiling point of the mixture and must be determined under pressure, but a maximum CST exists for all mixtures.

One other type of immiscible system contains a minimum critical solution temperature, and all the examples of this behavior consist of mixtures of water, or polyhydroxy compounds forming strong hydrogen bonds, with amines and other nitrogen-containing compounds, such as pyridine derivatives, which are strongly basic. The 3,4-dimethyl pyridine–water system (Fig. 3.2) has a minimum critical solution point at −3.6°C and 75.5 % water by weight and a maximum critical solution point at 162.5°C and 64 % water by weight [11]. In this system the eutectic temperature is sufficiently below the minimum CST that the immiscible region does not overlap with the solubility or freezing point curves.

There are sufficient binary systems that exhibit increasing solubility with decreasing temperature in the low-temperature region similar to curve *DC* in Fig. 3.1 to suggest that all immiscible mixtures have a minimum critical

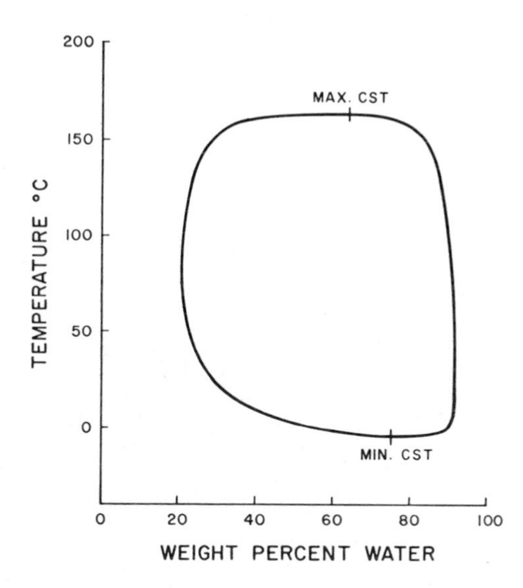

Fig. 3.2. Solubility curve for 3,4-dimethylpyridine–water system.

solution point as well as a maximum critical solution point, but only in the previously mentioned cases is it above the eutectic temperature of the freezing point curve so that it can be detected.

Representation of Data for Ternary Mixtures

When a third component is added to a mixture of two immiscible solvents, it distributes between the two liquid phases and usually increases the mutual solubility of the solvents. When the third component is liquid and completely miscible with both solvents in all proportions, the equilibrium phase diagram at a given temperature is normally, as in Fig. 3.3. The solubility of heavy solvent in the light solvent is shown at point A and the vice versa solubility at point B. The mutual solubility curve is indicated as $ACPDB$. A ternary mixture obtained by adding solute to a heterogeneous mixture of immiscible solvents of composition E to give a composition F, lying inside the mutual solubility curve, will separate into two liquid phases of compositions C and D. The line between these equilibrium phases is called a tie line, and the light phase compositions lie along ACP whereas the heavy phase compositions lie along BDP. Point P, at which the equilibrium phases become identical, is defined as the plait point.

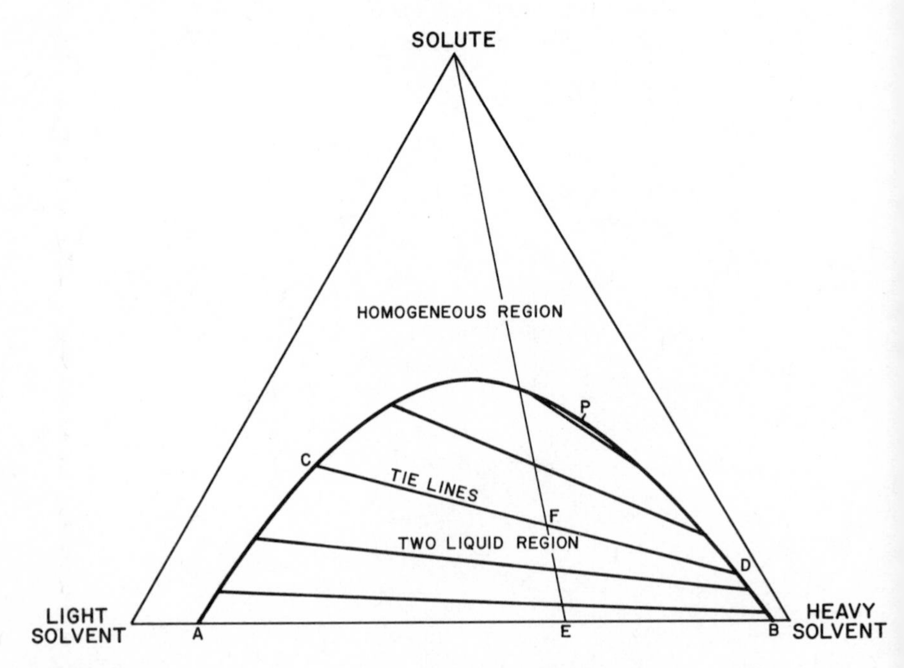

Fig. 3.3. Typical ternary diagram for a completely miscible solute in immiscible solvents.

Reference to Fig. 3.1 reveals that at higher temperatures points A and B will approach each other, and the area enclosed by the mutual solubility curve usually decreases until at the critical temperature for the solvents, the plait point P approaches the same CSC and no immiscible region exists. A different behavior can be expected when the freezing point curve for one of the binary pairs in the ternary system shows two or more eutectic points with maxima between them, indicating the formation of addition compounds. Thus the freezing point curve for phenol-acetone mixtures exhibits a maximum as well as a sufficient distortion of the freezing point curve at high acetone concentrations to indicate a series of addition compounds consisting of one phenol with one or more acetone molecules. Figure 3.4 is the ternary diagram for the acetone-water-phenol system [12]. At temperatures above the CST for the phenol-water system, the immiscibility region falls inside the ternary diagram and the mutual solubility curve has two plait points, one at higher and one at lower acetone concentrations as indicated by points P_1 and P_2 on the solubility curve. The immiscible region decreases with increasing temperature to a single point at the critical solution temperature of the ternary mixture, which in this case is about 90°C.

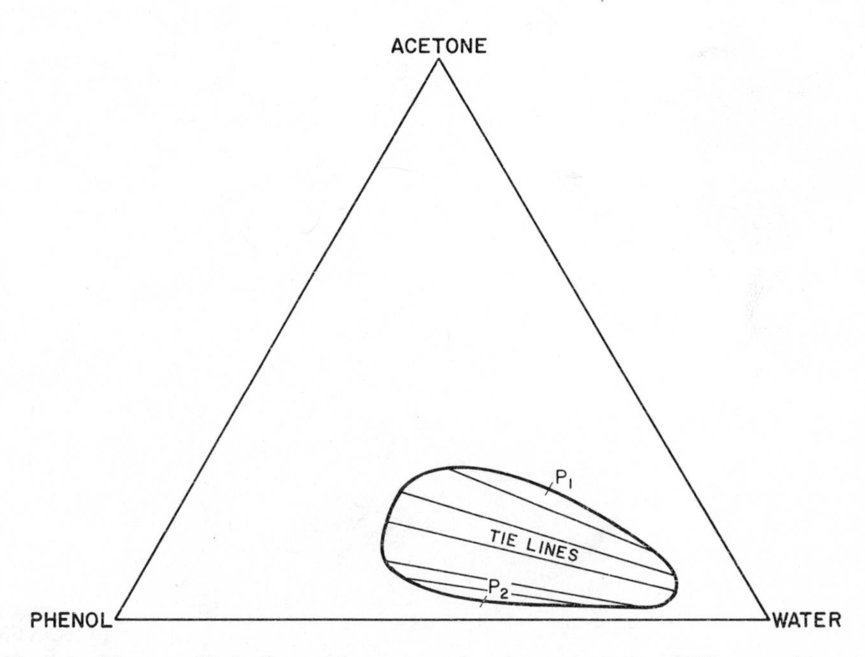

Fig. 3.4. Phase equilibria diagram for acetone-phenol-water system at 80°C compositions in weight fractions.

Ternary systems showing phase diagrams similar to Fig. 3.4 are not very common because the tendency to form complex addition compounds must be relatively strong to produce the effect displayed. In many cases, where this tendency can be recognized from the freezing point curves of the binary, it merely distorts the mutual solubility curve as the CST for the solvents is approached, but does not produce immiscibility above this temperature.

When the third component is only partially miscible with one of the solvents, the phase diagram will resemble Fig. 3.5. This important type of system is considered in detail later because if component B is less soluble than component C in solvent A, it will be possible to separate them by fractional liquid extraction with a single solvent. Such a system requires that both compounds be only partially soluble in the solvent. Practical applications are generally limited to the separation of isomers, members of the same homologous series, or similar, closely related compounds. This figure shows a discontinuity in the solubility curves at *X* and *Y* that is more likely to occur when components B and C are dissimilar. When these two components are isomers or closely related compounds, such as members of the same homologous series, the discontinuity becomes so slight that it is generally unrecognizable until temperatures approach the CST of the immiscible binary mixtures, AB or AC, whichever is lower.

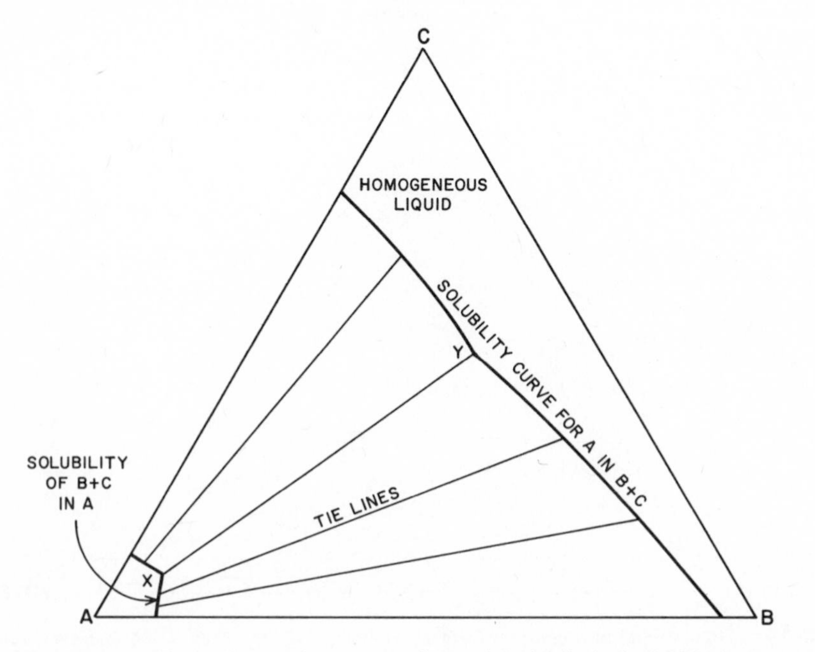

Fig. 3.5. Phase equilibria diagram for compounds B and C with limited solubility in solvent A.

Figure 3.3 can also assume different forms as the temperature is decreased below the freezing point of one of the compounds. Thus if the heavy solvent has the highest freezing point of the three components, the phase diagram will look like Fig. 3.6, where the two liquid phases exist in the area enclosed by the curves between points *B*, *P*, and *C*, and the solubility of the heavy phase solid in the liquid is given by curves *AB* and *CD*. In the area bounded by the lines between the heavy solvent corner of the ternary diagram and points *B* and *C*, the mixture will separate into a solid phase and two liquid phases of compositions corresponding to the compositions at points *B* and *C*.

When the solute has the highest freezing point of the three components, the immiscible region of Fig. 3.3. will be truncated below the plait point at temperatures below the saturation temperature in the solvent mixture, as indicated by the solid lines in Fig. 3.7. Two liquid phases exist within the area *ACDB* only. In the area included between the two lines from *C* and *D* to the solute corner of the ternary diagram, the mixture consists of solid solute and two liquids, one of composition corresponding to point *C* and the other of composition given by point *D*. These liquid compositions are fixed, and only the area *ACDB* is available for liquid extraction operations.

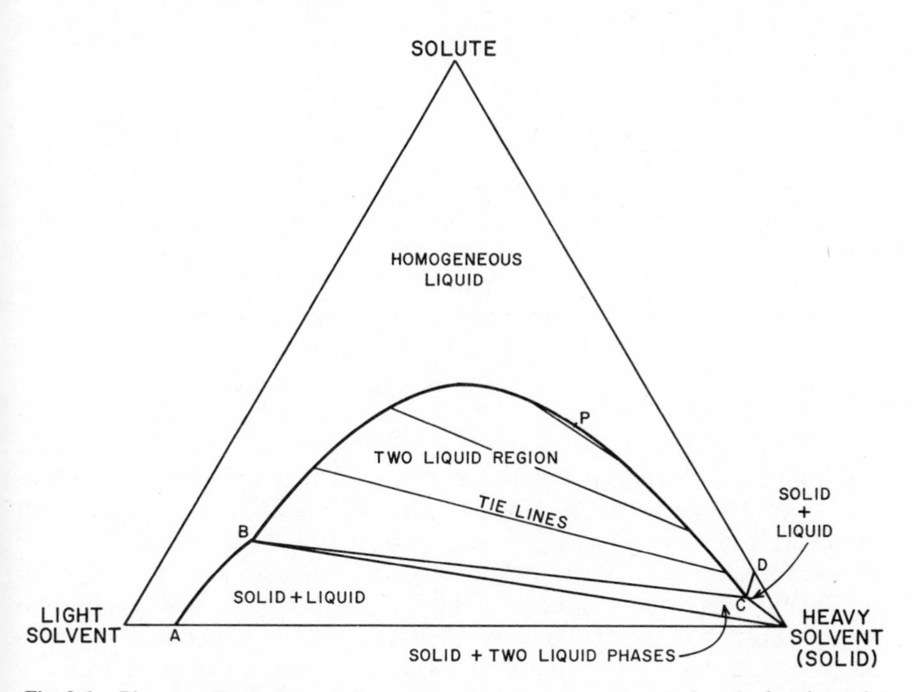

Fig. 3.6. Phase equilibria diagram for ternary system at temperatures below the freezing point of the heavy solvent.

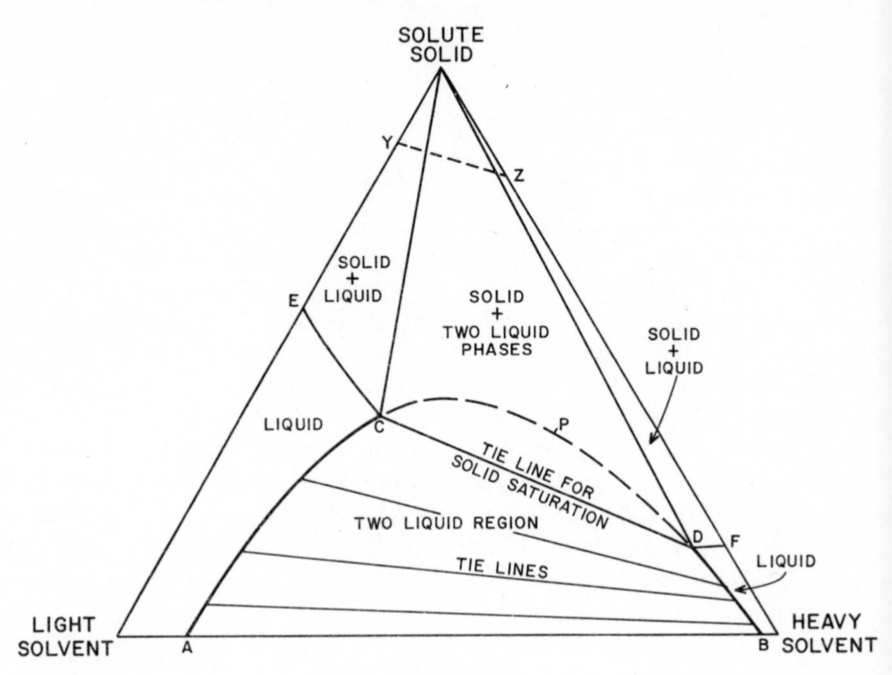

Fig. 3.7. Phase equilibria diagram for ternary system at temperatures below the freezing point of the solute.

At temperatures below the freezing point of the solute and above the saturation temperature at the plait point, the dashed curve *YZ* indicates the solubility of the solid in the solvent mixture. As the temperature is decreased, this solubility curve becomes tangent to the mutual solubility curve at the plait point, and at lower temperatures the tie line for saturation with solid solute approaches the solvent side of the ternary diagram, as shown.

When the temperature is also below the freezing point of one of the solvents, the solubility limit of the solid solvent in the mixture will be reached. The phase equilibria diagram will then resemble that in Fig. 3.8, in which a two-liquid region occurs only in the area *BGFC*. Other heterogeneous regions are identical to those of Figs. 3.6 and 3.7 as indicated. The two-liquid region decreases with decreasing temperature until the tie lines *BC* and *GF* coincide at a ternary eutectic temperature, at which the curve for solute solubility in the light solvent intersects the solubility curve for the solid heavy solvent in the light solvent at the final tie line relationship *BC*, as in Fig. 3.9. The solubility lines in Fig. 3.8 (*FE* and *CD*) have been drawn such that the ternary two-liquid-phase eutectic temperature occurs below the eutectic temperature for the solute and heavy solvent. If this latter binary eutectic temperature

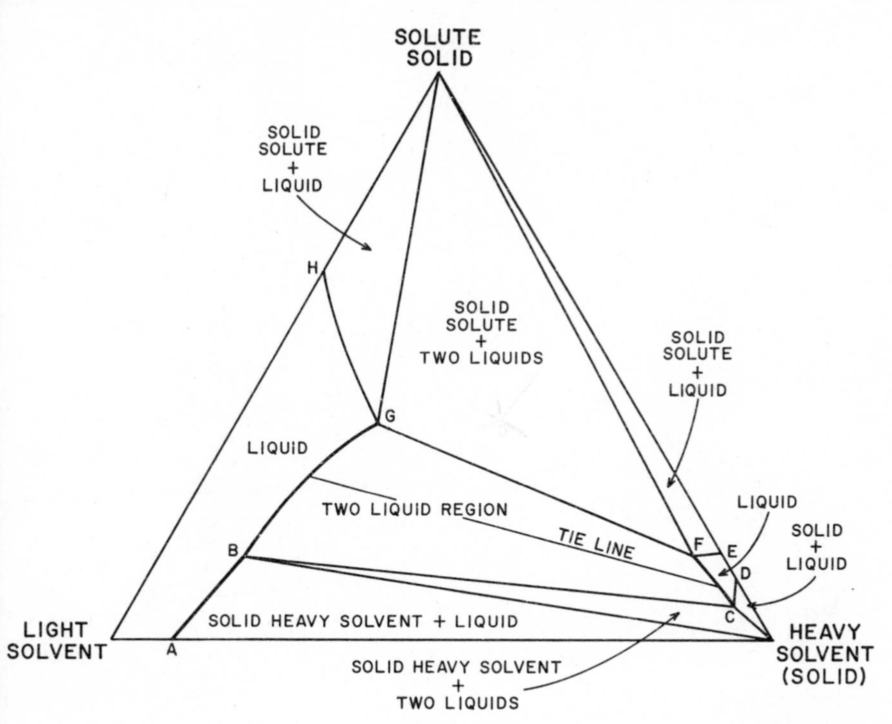

Fig. 3.8. Phase equilibria diagram for ternary system at temperatures below freezing points of solute and one solvent.

were lower than the ternary eutectic, the solubility lines *FE* and *CD* would be divergent at low light solvent concentrations, as shown by the dashed lines *CE* and *CF* in Fig. 3.9. At lower temperatures these lines become a smooth curve, with the enclosed area decreasing to a eutectic point along the solute–heavy solvent boundary of the ternary diagram.

In Fig. 3.9 the freezing point of the light solvent is assumed to be below the temperature of the two-liquid ternary eutectic. At lower temperatures the solubility curve *ABD* becomes a smooth curve displaced toward the light solvent corner of the ternary diagram. The dashed curve *XY* represents the solubility of solid light solvent at a temperature below its freezing point. At some lower temperature the solubility curve *ABD* will intersect the latter curve at *X* if the eutectic temperature of the two solvents is higher than that of the light solvent and solute. The lowest temperature at which liquid exists in such a system is the solute–light solvent eutectic. Conversely, if the ternary eutectic temperature is the lowest temperature at which liquid exists in such a system, point *Y* will approach *D* at the solute–light eutectic temperature,

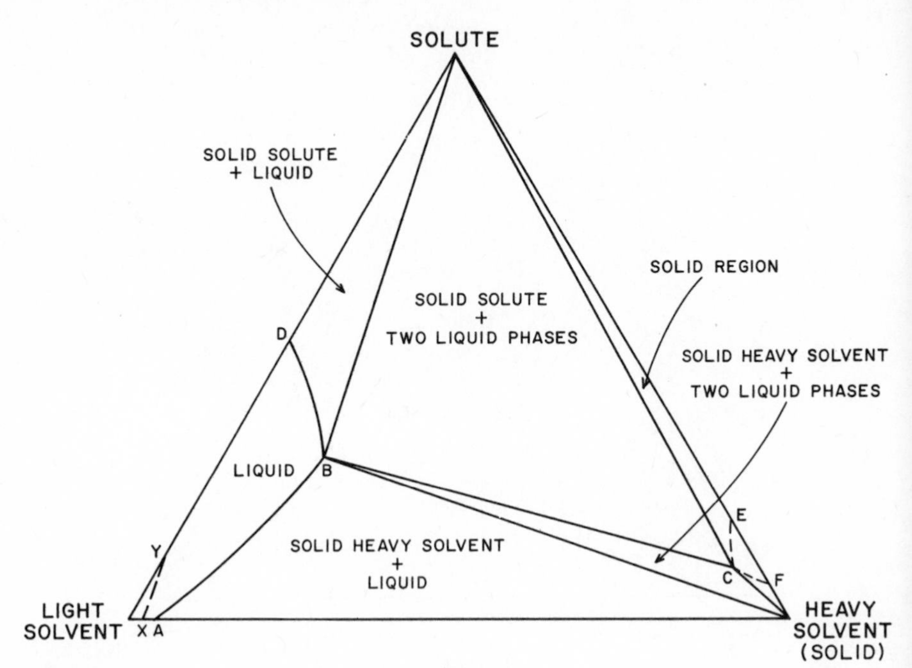

Fig. 3.9. Solubility curves at ternary eutectic temperature for two liquid phases and two solid phases.

and at lower temperatures this point will approach the ternary eutectic composition. On the solute–heavy solvent side of the diagram, the area between the corners and point *C* will represent solid phase if the ternary eutectic temperature is below the solute–heavy solvent eutectic temperature. The dashed curves *EC* and *CF* represent the alternate case of a binary eutectic temperature below that of the ternary eutectic so that the area *EDF* represents a liquid region, and similar to Fig. 3.8, the other areas represent liquid–solid regions.

Similar techniques can be applied to evaluate other systems for the existence of two-liquid regions. A thorough understanding of phase behavior of immiscible liquids is essential to the development of new liquid extraction processes involving conditions different from those encountered in standard techniques. A knowledge of temperature effects can also provide the basis for selection of optimum extraction conditions.

One other type of ternary system may be encountered, in which all three binary mixtures are only partially immiscible. The phase equilibria for such a system appears in Fig. 3.10. This type of system is of limited value in liquid extraction processes because only the three two-phase regions are applicable.

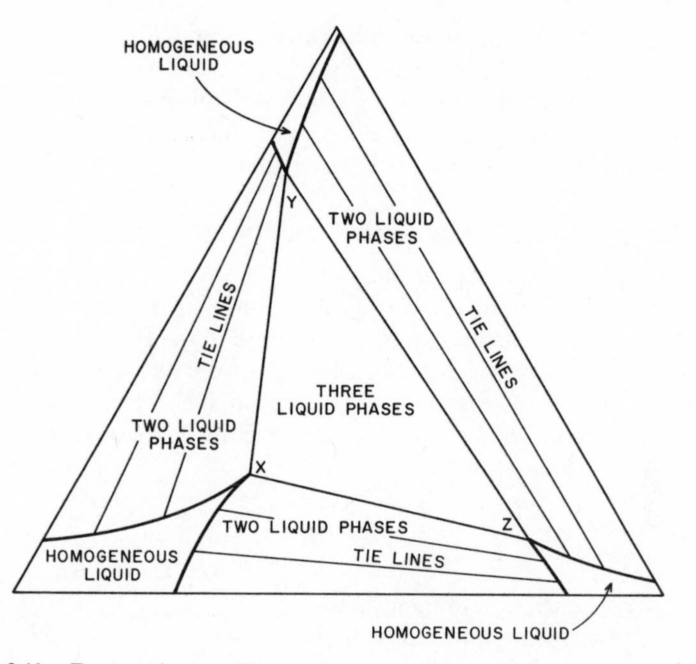

Fig. 3.10. Ternary phase equilibria diagram for three immiscible binary mixtures.

All compositions lying within the triangle XYZ separate into three phases of compositions corresponding to the corners of the triangle. As the mutual solubilities of all three binary pairs increase with increasing temperature, the two-phase regions become smaller, separating into individual areas, each with its own plait point.

Phase Equilibria Data for Multicomponent Systems

In the case of ternary mixtures it is possible to visualize the effect of temperature in the third dimension, to be able to interpolate and predict trends in the phase equilibria data. In all the foregoing figures this three-dimensional effect can be represented as a family of curves projected on the ternary diagram, and some cases have been discussed on this basis. A four-component system can be represented by an equilateral tetrahedron, and projection of the phase equilibria boundaries to a ternary diagram along any face of the tetrahedron can provide some measure of interpretation. Theoretically, temperature variations could be indicated by different families of projection curves, but interpreting such curves is difficult. There is no simple technique for representing compositions of such systems.

Thus all systems involving more than three components are generally represented by a series of curves selected to be consistent with reasonable assumptions used for interpretation of data and design calculations on the separation process. These curves may also be reduced to mathematical equations for use in computer programs for process design calculations.

Interpretation of Phase Equilibria Data for Nonionizing Solutes

The experimental tie line data illustrated in Figs. 3.3 to 3.10 may be represented mathematically by defining a distribution coefficient as the ratio of the concentration of the solute in the light phase to the concentration in the heavy phase. In some previous publications this ratio has been designated as K, but in this work it is represented by D, to distinguish it from the ionization equilibrium constant, which is used subsequently in correlating data on ionizing solutes. Thus

$$D = \frac{C_1}{C_2} \tag{3.1}$$

where C_1 = concentration in light phase
C_2 = concentration in heavy phase

and the concentrations may be expressed in any consistent units, such as moles or grams per liter of solution, mole fraction, weight fraction or volume fraction, or any concentration ratio, such as moles/1000 g of solvent, moles per mole of solvent, or grams per gram of solvent.

The value of D varies with concentration, temperature, and to a small extent with the external pressure on the system. The latter effect becomes most noticeable close to the critical solution temperature of the solvent mixture.

In dilute solutions the distribution coefficient can be considered to be constant, and in nonionizing solutions it may be essentially constant up to concentrations as high as 10 mole percent. The distribution coefficient defined in terms of mole fraction concentrations in each phase is related to the thermodynamic activity coefficients, such that

$$D = \frac{x_1}{x_2} = \frac{\gamma_2}{\gamma_1} \tag{3.2}$$

where x and γ are, respectively, the mole fraction and the activity coefficient of the solute in a given phase, and subscripts 1 and 2 refer to the light and heavy phases, respectively. Activity coefficients are unity in ideal systems, higher in systems showing positive deviations from ideality, and lower in systems showing negative deviations.

Activity coefficients vary from unity for the pure compound to a limiting value at infinite dilution. A quantitative review of the thermodynamics of nonideal solutions can be found in standard textbooks and is beyond the scope of this chapter; however certain qualitative relationships will prove useful in the search for suitable solvents for an extraction process. Chemical compounds having an affinity for each other, usually through the formation of strong hydrogen bonds, will exhibit negative deviations from ideality, and such compounds are always completely miscible with each other. Immiscibility occurs only in systems where one of the components forms appreciably stronger hydrogen bonds with itself than with the other compound. In all immiscible systems the limiting value of the activity coefficient of at least one of the components at infinite dilution must be greater than 10.

It is thus apparent that, when solvents are sufficiently dissimilar to be immiscible, they should have different affinities for different solutes distributed between them. And conversely, no two different compounds should have identical distribution coefficients between immiscible solvents. This was the principle for which Craig originally developed his countercurrent distribution apparatus.

The variation of the distribution coefficient D results from the variation of the activity coefficients with concentration in accordance with (3.2). The classical Gibbs-Duhem relationship of phase equilibria shows that this variation cannot be linear with concentration over any appreciable range, and all mathematical solutions of the Gibbs-Duhem equation that are consistent with experimental data show larger variations with concentration when the actual deviations from ideality are greater (i.e., at infinite dilution). Thus when a solute is distributed between two immiscible solvents, one in which it exhibits small deviations from ideality and the other showing large deviations, the distribution coefficient can be expected to vary significantly with concentration, even at concentrations below 1 mole percent. Conversely, if the deviations from ideality are similar in magnitude, the distribution coefficients may be reasonably constant up to concentrations in excess of 10 mole percent.

All ternary systems in which the solute is completely miscible with both solvents will have a plait point as in Fig. 3.3, and at this point the distribution coefficient is unity. Thus all such systems will have distribution coefficients that decrease with concentration if they are greater than unity and increase with concentration if they are less than unity. And if the coefficient should be close to unity at infinite dilution, they might vary in both directions over different concentration ranges before reaching the plait point concentration. In these cases in particular, the variation also depends on the concentration units used in defining the distribution coefficient, and the choice of proper units may make it appear very nearly constant.

Interpretation of Distribution Data for Ionizing Solutes

In the case of solutes that ionize in solution, such as acids or bases, the overall or apparent distribution coefficient varies significantly with the pH of the solution. Thus when an organic acid is distributed between an organic solvent and a basic solution, the major portion of the acid in the aqueous phase is present in the ionized form. Distribution coefficients for such systems can be calculated at different pH values from the ionization constant defined as

$$K_1 = \frac{[H^+][A^-]}{[HA]} \tag{3.3}$$

where $[H^+]$ = concentration of hydrogen ion (moles/liter)
$[A^-]$ = concentration of acid ion (moles/liter)
$[HA]$ = concentration of free acid (moles/liter)

If the distribution coefficient for the free acid between the solvent and the aqueous phase is expressed as

$$D' = \frac{[HA]_s}{[HA]} \tag{3.4}$$

where $[HA]_s$ is the concentration of the acid in the solvent (moles/liter), the apparent distribution coefficient, D, obtained by determining the total acid, both free and ionized, in the aqueous phase can be derived as

$$D = \frac{[HA]_s}{[HA] + [A^-]} = \frac{D'}{1 + K_1/[H^+]} \tag{3.5}$$

The apparent distribution coefficient of a diacid between a solvent and an aqueous phase can also be derived on the same basis as

$$D = \frac{[H_2A]_s}{[H_2A] + [HA^-] + [A^{2-}]} = \frac{D'}{1 + K_1/[H^+] + K_1K_2/[H^+]^2} \tag{3.6}$$

where the first and second ionization constants are defined as

$$K_1 = \frac{[H^+][HA^-]}{[H_2A]} \tag{3.7}$$

$$K_2 = \frac{[H^+][A^{2-}]}{[HA^-]} \tag{3.8}$$

and D' is the distribution coefficient for the free acid between the solvent and the aqueous phase as in the previous case.

Similarly the apparent distribution coefficient for a triacid can be shown to be

$$D = \frac{[H_3A]_s}{[H_3A] + [H_2A^-] + [HA^{2-}] + [A^{3-}]}$$

$$= \frac{D'}{1 + K_1/[H^+] + K_1K_2/[H^+]^2 + K_1K_2K_3/[H^+]^3} \quad (3.9)$$

where the first, second, and third ionization constants are defined as

$$K_1 = \frac{[H^+][H_2A^-]}{[H_3A]} \quad (3.10)$$

$$K_2 = \frac{[H^+][HA^{2-}]}{[H_2A^-]} \quad (3.11)$$

$$K_3 = \frac{[H^+][A^{3-}]}{[HA^{2-}]} \quad (3.12)$$

and other terms have the same corresponding significance as in the previous equations.

These relationships provide the basis for the separation of complex mixtures of organic acids into pure compounds because they supply a new dimension to the separation process. Nonionizing compounds depend on finding a pair of immiscible solvents in which the distribution coefficients for the given compounds show the greatest difference and yield the most efficient separation. With ionizing solvents, the separation can be effected by the differences in the ionization constants and these differences are frequently large enough to overcome an adverse ratio of distribution coefficients; however the maximum advantage results when both ratios are effective in the same direction.

Figure 3.11 illustrates this effect for a mixture of acids, one with pK = 8 and the other with pK = 9. The solid curves are based on a distribution coefficient of 10 for both acids, and their apparent coefficients differ by a factor approaching 10 at higher pH, making a separation that would be impossible at low pH (i.e., below 6) relatively easy at pH values above 8. Obviously if D' for the acid with the lower pK were less, separation would be possible in neutral or low pH solutions but would be enhanced at higher pH. The dashed curve, based on a value of D' = 40 for this acid, is interesting because this factor makes it more soluble in the organic solvent at low pH and less soluble above pH = 8.7, permitting fractionation to be achieved at both ends of the pH scale; the pure products, however, will be in the opposite phases. No separation will occur at pH 8.7. Practical considerations, such as total solubility in the solvents, will indicate that the optimum pH for this particular

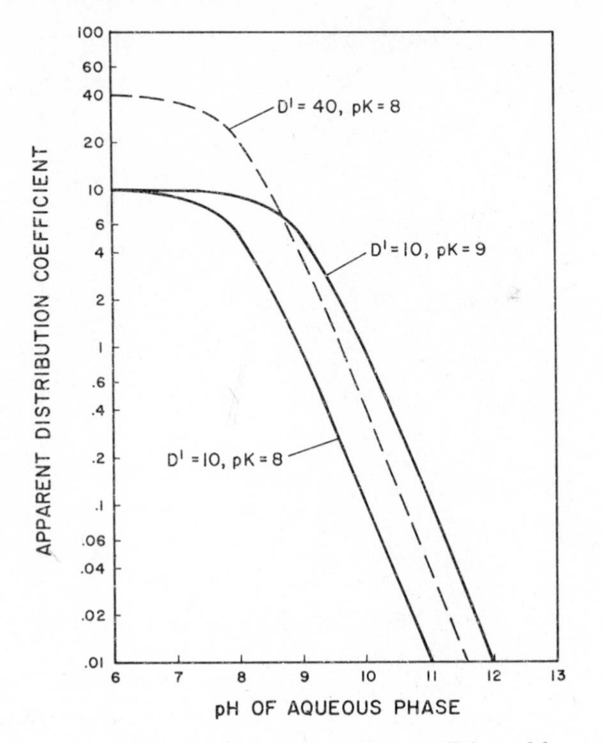

Fig. 3.11. Effect of ionization constant and distribution coefficient of free acid on apparent distribution coefficients at different pH values.

separation is between pH 9.5 and 10.5, where distribution coefficients are close to unity.

The effect of pH on the distribution coefficients for mono-, di-, and tri-acids is shown in Fig. 3.12 for acids of identical D' values and the same pK values for all the first, second, and third ionization constants, where they exist. Recognizing that one cycle on the logarithmic distribution coefficient scale corresponds to one unit on the pH scale, it is clear that at high pH values, the slope of the distribution coefficient curve for the monoacid approaches -1, while the slope of the diacid curve approaches -2 and the slope of the triacid curve approaches -3. Thus for these compounds, which cannot be separated at low pH, the monoacid can be separated from the di- and triacids above pH 9, and the di- and triacids can be separated easily at pH values above 10. From this figure it can be recognized that when the distribution coefficients and ionization constants for mono-, di-, and triacids are equal, the different acids can be separated at pH values equal to the pK values of their distinguishing functional group.

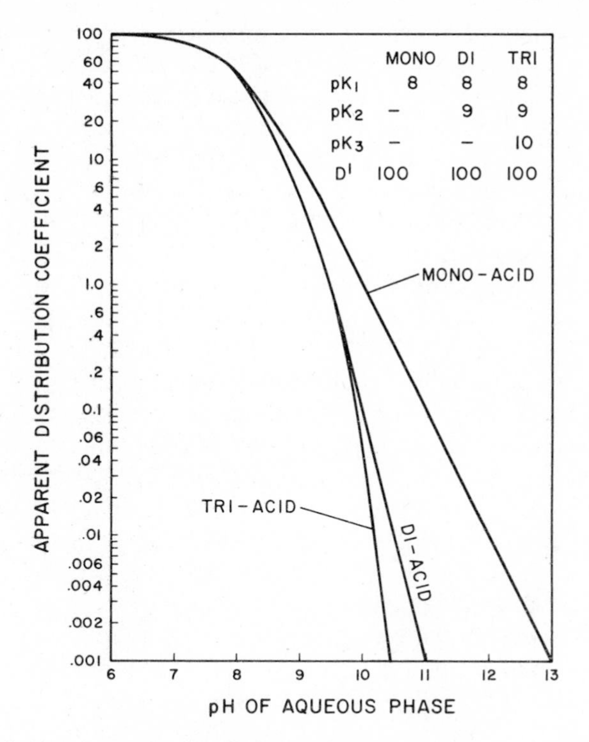

Fig. 3.12. Variation of apparent distribution coefficients of mono-, di-, and triacids with pH.

Many interesting combinations of the constants in (3.5), (3.6), and (3.9) exist. For example, if all pK values are as given in Fig. 3.12 but the distribution coefficient for the free diacid is greater than 100, pure diacid can be separated from the mixture in the light phase at low pH, pure triacid can be separated from the mixture in the aqueous phase at intermediate pH, and pure monoacid can be separated from the mixture in the light phase at high pH. It thus becomes possible to extract any desired component from the mixture without complete resolution, as would be required to isolate pure diacid from the mixture when all distribution coefficients are identical, as in Fig. 3.12.

Acids generally exhibit different D' and pK values for the respective ionizations, and when these values are known it is frequently possible to extract a desired acid from a complex mixture in a single step. Normally, no more than two steps are required, and the optimum pH can be determined for each of the steps by considering the distributions curve such as that in Fig. 3.12.

However the generation of curves of this type from fundamental concepts is not as simple in real systems as the previous equations indicate. These equations have been derived on the basis of the assumption that the aqueous solutions are "ideal" in the sense that the electrochemist defines this term. The positive and negative ions in solution have an ionic atmosphere that retards their mobility in solution, and all ionization equilibria data are extrapolated to infinite dilution because only under this limiting condition are the ion activity coefficients unity. The effect of ion density (i.e., concentration) on ion activity coefficients can be expressed by the semitheoretical Debye-Hückel equation based on the concept of an ionic atmosphere of fixed size about every ion. This so-called limiting law is accurate only up to 0.0002 mole fraction (ca. 0.01 mole/liter). Between these concentrations and saturation, which may be higher than 0.05 mole fraction in many cases, the ion activity coefficients depart from the fractional values predicted by the Debye-Hückel law and may even become greater than unity. No theoretical explanation has been advanced to explain this change from negative deviations from ideality predicted by the ionic interactions according to the Debye-Hückel theory, to positive deviations. Empirical estimates are essential for the thermodynamic interpretation of ionic equilibria at the higher concentrations. This region is the most important in the purification of appreciable quantities of product because the most economic operation is obtained at high concentrations.

The thermodynamic equilibrium constant for the ionization of an acid is actually defined as

$$K_1' = \frac{a_{H^+} a_{A^-}}{a_{HA}} \tag{3.13}$$

where a denotes activity and the subscripts identify the components of the equilibrium mixture. The activity coefficient is the correction factor applied to the concentration to obtain the activity of each component

$$a_i = \gamma_i x_i \tag{3.14}$$

and the thermodynamic ionization constant becomes

$$K_1' = \frac{(\gamma_{H^+} x_{H^+})(\gamma_{A^-} x_{A^-})}{\gamma_{HA} x_{HA}} \tag{3.15}$$

The relation between the thermodynamic equilibrium constant K_1' and the constant obtained by extrapolating electrochemical measurements to infinite dilution can be shown to be

$$K_1 = \frac{1000}{18} \frac{\gamma_{H^+}^0 \gamma_{A^-}^0}{\gamma_{HA}^0} K_1' \tag{3.16}$$

where the superscripts designate infinite dilution and the numerical value adjusts for the different concentration units.

Since $\gamma^0_{H^+}$ and $\gamma^0_{A^-}$ are defined as unity, the thermodynamic equilibrium constant can be expressed in terms of the electrochemical ionization constant as

$$K_1' = \frac{\gamma^0_{HA}}{55.55} K_1 \tag{3.17}$$

This is the thermodynamic constant from which the molar free energies of ionization can be calculated by the relation

$$\Delta G'_{ion} = -RT \ln K_1' \tag{3.18}$$

and free energies of ionization, normally calculated by electrochemists by the analogous equation

$$\Delta G_{ion} = -RT \ln K_1 \tag{3.19}$$

are not applicable to thermodynamic calculations. They must be corrected at infinite dilution by adding the term for the coefficient in (3.18) to give

$$\Delta G'_{ion} = -RT \ln K_1 - RT \ln \frac{\gamma^0_{HA}}{55.55} \tag{3.20}$$

Since the activity coefficient of the free acid as previously noted has its maximum variation with concentration in this region, and the total number of moles in a liter (i.e., the denominator in the second term) will also vary to some extent with concentration, the assumption of a constant value for K_1 at all concentrations in electrochemical calculations is not consistent with classical thermodynamics.

The electrochemical definition of K_1 takes into account only the effect of interionic attractions on ion activities as

$$K_1 = \frac{\gamma_{H^+}[H^+]\gamma_{A^-}[A^-]}{[HA]} \tag{3.21}$$

where γ_{H^+} = activity coefficient of the positive ions
γ_{A^-} = activity coefficient of the negative ions

The deviations of divalent and trivalent ions from ideality are greater than those of the monovalent ions because of their stronger interionic attractions. Thus their activity coefficients are smaller than those of the monovalent ions. Equations 3.6 and 3.9 can be derived from ion activities to give, respectively,

$$D = \frac{D'}{1 + \dfrac{K_1}{\gamma_{H^+}\gamma_{A^-}[H^+]} + \dfrac{K_1 K_2}{\gamma^2_{H^+}\gamma_{A^{2-}}[H^+]^2}} \tag{3.22}$$

$$D = \frac{D'}{1 + \dfrac{K_1}{\gamma_{H^+}\gamma_{A^-}[H^+]} + \dfrac{K_1 K_2}{\gamma^2_{H^+}\gamma_{A^{2-}}[H^+]^2} + \dfrac{K_1 K_2 K_3}{\gamma^3_{H^+}\gamma_{A^{3-}}[H^+]^3}} \tag{3.23}$$

Ion activity coefficients may be estimated from the Debye-Hückel relationship

$$\log \gamma = -\frac{A\,|Z_1 Z_2|\sqrt{I}}{1 + Ba\sqrt{I}} \tag{3.24}$$

where I = ionic strength of solution $= \frac{1}{2}\Sigma Z_i^2 C$
Z = valence of the individual ion
C = ion concentration (moles/liter)
a = ionic radius (Å)

and A and B are constants derived from the dielectric constant for water and the absolute temperature. Values of these constants at temperatures in the range of 0 to 100°C are tabulated in electrochemical textbooks. For data reduction in computer programs, the following equations are useful:

$$A = 0.4919 + 0.006508T + 0.000005135T^2 \tag{3.25}$$

$$B = 0.3248 + 0.0001516T + 0.000000880T^2 \tag{3.26}$$

Kielland has estimated the ionic radii for 130 inorganic and organic ions [13], and values range from 2.5 to 11. The values represent the ionic atmosphere rather than the molecular size. In the case of inorganic ions, the monovalent ones have the lowest radii and the tetravalent ions, the largest radii. However the ionic radius of a multivalent acid ion changes by less than 0.5 Å, and this is probably within the accuracy of data and the evaluation technique. The outstanding exception to this general relationship is the hydrogen ion, for which the value of a is 9, possibly indicating a high degree of solvation with water.

The radii of organic ions vary between 3.5 and 8, with the smaller molecules at the lower end of the range and large molecules of more than 10 carbon atoms at the upper end. In these cases, also, multivalent ions of the same acid have essentially identical radii. Figure 3.13 compares the calculated individual ion activity coefficients for sodium salts of organic acids with the measurements of average activity coefficients from the tabulation of Robinson and Stokes [14]. The calculated curves for the sodium cation and the acid anion were based on values of 4.5 and 6.0, respectively, and agreement is good up to concentrations of 0.1 mole of salt per liter. The activity coefficient for the hydrogen ion also appears in this figure for reference. The hydroxyl ion, with $a = 3.5$, will lie slightly below the calculated curve for the Na^+ ion. At the higher concentrations in the region of 1 mole/liter, the observed activity coefficients initially increase with the size of the organic molecule, then decrease with molecular size. Practical operation of a fractional liquid extraction process generally requires operation at the maximum concentration, and fundamental theory is lacking to predict activity coefficients at these concentra-

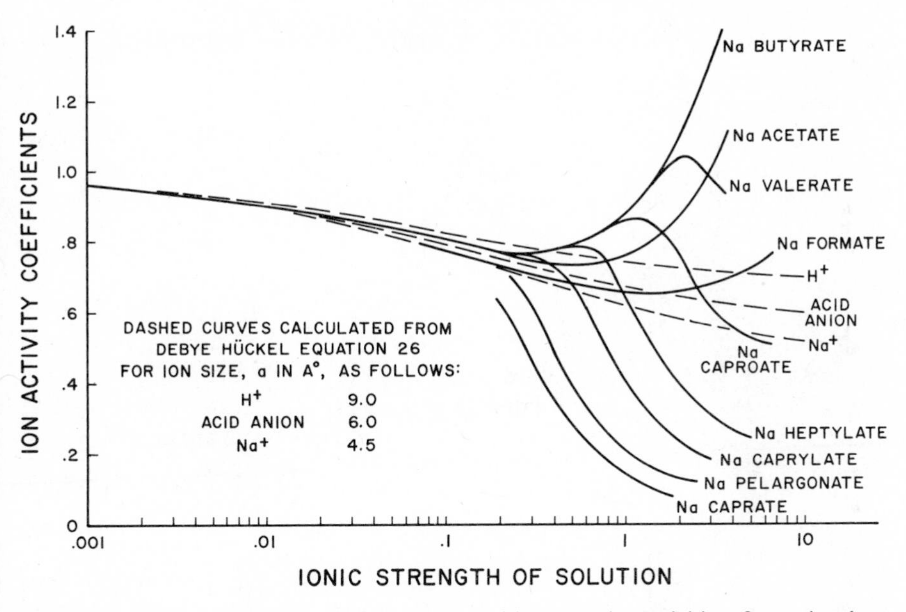

Fig. 3.13. Comparison of Debye-Hückel theory with average ion activities of organic salts.

tions. Rigorous interpretation of liquid distribution data thus requires empirical correlation of activity coefficient data for use in (3.21) to (3.23).

Figure 3.14 represents the effect of using the Debye-Hückel equation to correlate distribution data obtained for monovalent acids of ionic radius of 6.0 Å. The curves are based on a distribution coefficient of 100 for the free acid, and apparent distribution coefficients theoretically calculated for adding an acid with pK = 8 to a heterogeneous mixture of the solvent and caustic solutions of different concentrations. It is assumed that the total concentrations in both phases and the pH of the aqueous phase have been measured. It must also be noted that since a pH meter measures hydrogen ion activity rather than hydrogen ion concentration, the abscissa is actually $-\log(\gamma_{H^+}[H^+])$. The individual ion concentrations in the solution are related by the ionization constants for the acid and for water

$$\frac{K_1}{\gamma_{H^+}\gamma_{A^-}} = \frac{[H^+][A^-]}{[HA]} \tag{3.27}$$

$$\frac{K_2}{\gamma_{H^+}\gamma_{OH^-}} = [H^+][OH^-] \tag{3.28}$$

the ion charge balance

$$[H^+] + [Na^+] = [OH^-] + [A^-] \tag{3.29}$$

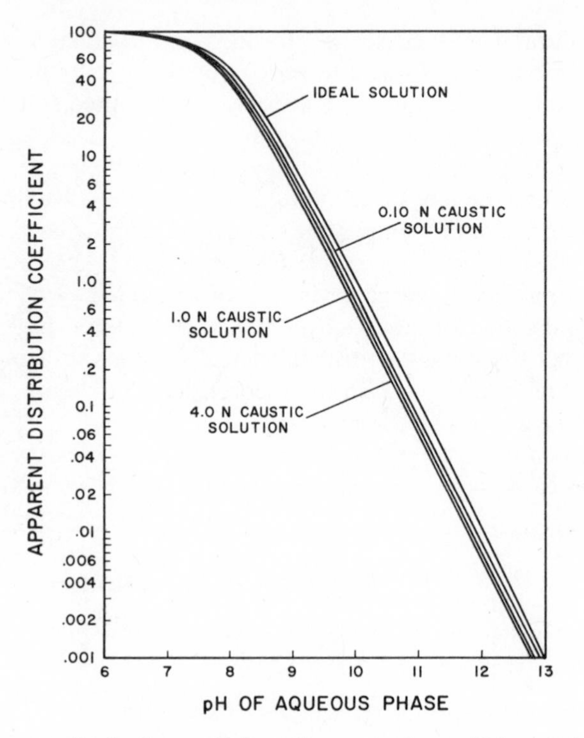

Fig. 3.14. Apparent distribution coefficients for monovalent acid in solutions of different caustic concentrations; curves based on ionization constant of 10^{-8}.

and the material balances for the total acid T observed in solution

$$T = [HA] + [A^-] \tag{3.30}$$

and the known amount of $[Na^+]$ added, assuming the caustic and the salt to be completely ionized so that no undissociated NaOH or NaA salt exists in solution.

The equations above have four unknowns; namely, $[H^+]$, $[OH^-]$, $[A^-]$, and $[HA]$, and they can be solved by trial and error. Thus assuming a value for I equal to the initial caustic concentration and calculating the left-hand terms of (3.27) and (3.28) from the activity coefficients, it becomes possible to solve the equations above for the four unknowns by trial and error. Recalculation of the ion activity coefficients based on the new calculated ionic strength will permit a repeat of the procedure until the newly calculated concentrations agree with the previous ones to a specified tolerance.

Figure 3.14 shows that the apparent distribution coefficients may differ by a factor greater than 1.5 between infinite dilution and a 4 N caustic solution.

If one could accurately measure concentrations over a sufficiently wide range of pH values, this would not lead to serious errors in the interpretation of data. At high and low distribution coefficients, the limited solubility in one of the phases will result in low concentrations in the other phase. Analytical accuracy will limit the range of pH values over which the apparent distribution coefficients may be measured. Thus if accurate data are limited to pH values of about 3, the uncertainty in the evaluation of the activity coefficients can reduce the credibility of the correlation for extrapolation to other conditions. Considering also the large deviations of actual activity coefficients of the different acid salts from the Debye-Hückel equation at high concentrations could affect the apparent distribution coefficients to a much greater extent; thus correlations based on the theoretical relationships of (3.27) to (3.30) may not be entirely fundamental.

The difficulty in developing fundamental data on di- and trivalent acids becomes even greater. Figure 3.15 presents the theoretical calculations for the

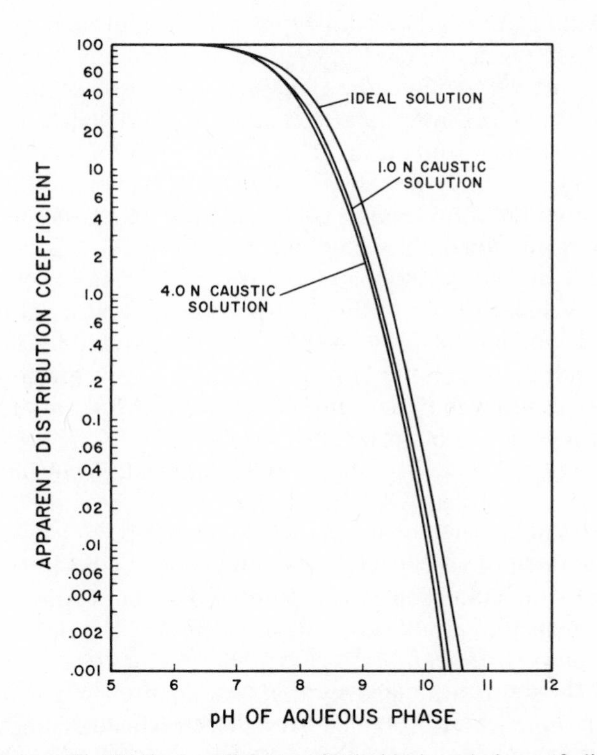

Fig. 3.15. Apparent distribution coefficients of triacids in solutions of different caustic concentrations; curves based on ionization constants of 10^{-8}, 10^{-9}, and 10^{-10} for the first, second, and third hydrogen ions, respectively.

apparent distribution coefficients for a trivalent acid for which the respective ionization constants K_1, K_2, and K_3 are 10^{-8}, 10^{-9}, and 10^{-10}, similar to the triacid considered in Fig. 3.12 on an ideal basis. Here the effect of concentration is greater than for the simple acid in Fig. 3.14, and the technique of applying the Debye-Hückel equation to a multicomponent mixture of ions of different valences is also questionable. In the calculation of activity coefficients by (3.26), all the acid anions of the different valences were assumed to have a radius of 6 Å, and since all cations were monovalent, the values of the $|Z_1Z_2|$ term in the numerator were 2.0 and 3.0 for the di- and trivalent anions, respectively. The analogous equations for the ionic equilibria, charge balance, and total acid balance can be solved by a trial-and-error technique. In this case the equations involve higher order terms in $[H^+]$, and instead of different approaches possible in the previous case, the best computer program for the trial-and-error solution in this case is obtained by assuming values of the pH in the equilibrium relations to match the material balance and the charge balance.

The calculations for Fig. 3.12 show that the distribution data below a pH equal to the first pK can be used to evaluate this first equilibrium constant. Thus only the small variation in the apparent distribution coefficient between pH 6 and 8 may be used to evaluate pK_1. When this is known, the value of pK_2 can be determined from the difference between the observed distribution coefficients and the extrapolation based on the pK_1 value alone. Since the trivalent ion becomes significant at pH 9, only the data between pH 8 and 9 are useful for evaluation of the second pK. The third ionization constant may be evaluated from the deviations of the observed distribution coefficients from the values calculated for the first two constants alone at pH values above pK_2. Theoretically this provides a greater range, but distribution coefficients become so small that the limitations of salt solubility in the aqueous phase and analytical accuracy in the solvent phase restrict the pH range over which meaningful data can be obtained.

Obviously if the pK values for the successive ionization constants differ by larger amounts, the problem will be minimized, but all the multivalent organic acids studied by Scheibel have shown pK values differing by less than 2. It is necessary to resort to a successive-trial approach to establish the best values for two or more ionization constants, using the distribution data at the lower pH values primarily to establish the value of D_1' and the initial trial values for pK_1.

Ionization constants evaluated by this technique from distribution data usually differ from those obtained from electrochemical measurements, which as previously noted are generally determined by extrapolation to infinite dilution. Some of the deviation could be attributed to the difference in the activity coefficient of the free acid at infinite dilution and at the concentrations used in the distribution data. This could not account for differences as large as unity in the pK values, as are sometimes found. But in spite of the uncertainty in the fundamental nature of the correlations, they are adequate for the determination of operating conditions for laboratory and pilot plant studies. When modified to be consistent with pilot plant performance data, they are suitable for optimization of the full-scale design. The most significant feature of a complete study of this kind is the indication that selectivities predicted from ionization constants evaluated by electrochemical measurements may sometimes be several times those determined from distribution data. Thus distribution coefficients predicted from literature values of ionization constants on two compounds that differ by a factor of less than 3 (equivalent to a pK difference of 0.5) will not necessarily indicate a practical separation process.

Organic amines may also be separated by the same extraction process based on their ionization constants, and the analogous relationships apply to the separation of mono-, di-, and triamines. Thus the apparent distribution coefficient for a triamine can be derived from the respective ionization constants as

$$D = \frac{D_1}{1 + \dfrac{K_1}{\gamma_{OH^-}\gamma_{M^+}[OH^-]} + \dfrac{K_1 K_2}{\gamma_{OH^-}^2 \gamma_{M^{2+}}[OH^-]} + \dfrac{K_1 K_2 K_3}{\gamma_{OH^-}^3 \gamma_{M^{3+}}[OH^-]}} \tag{3.31}$$

and analogous to the previous equations, the apparent distribution coefficient for a diamine will omit the fourth term in the denominator. The corresponding equation for the monoamine will omit the third term as well.

Distribution data for organic amines can be interpreted and correlated by the same techniques previously described for organic acids and the same limitations apply. Published ionization constants will provide the clues to a practical liquid extraction process for separation but the theoretical calculations must be verified with actual distribution data and the constants in the correlating equation modified accordingly.

When organic compounds ionize in aqueous solution to form either acids or bases, it is usually possible to utilize the differences in the ionization equilibria to develop a practical process for their separation and purification.

3 THEORY OF LIQUID EXTRACTION PROCESSES

Liquid extraction involves the transfer of solute from one liquid phase to another. When the solute is an individual component or a group of components, all of which are transferred in the same direction, the process may be designated "extraction for removal," to distinguish it from "fractional liquid extraction" in which the solute components are separated in the different solvent phases. In the latter process, the net transfer of the components must be in opposite directions to effect a separation.

All extractions involve contacting the two liquid phases to bring them to equilibrium, but the arrangement of the extraction steps depends on the objectives. Thus the processes are considered separately.

Extraction for Removal

Extraction for removal is the simplest type of extraction process. In commercial applications the incentive for the transfer is economic, and in laboratory operations it may be simple convenience. Extraction can purify a compound either by removing it preferentially from a solution containing impurities or by removing the impurities from the original solution. Typical applications involve the removal of a nonionizing or very weakly ionizing compound from an aqueous solution of acids or bases by neutralizing the solution and extracting with an organic solvent. The aqueous solution will thus be purified, either for recovery of the acid or base or to prepare it for the next processing step.

In addition to purification, liquid extraction may be used to transfer a solute to a different solvent from which it may be more readily recovered. Thus when a solute forms an azeotrope with the original solvent or boils at very nearly the same temperature, rendering fractional distillation difficult and costly, it can be extracted into a solvent that has a boiling point sufficiently different to facilitate the fractional distillation. Economics normally dictates that the solvent be higher boiling than the solute removed or that it effect a concentration of the solute so that the heat requirements of the subsequent distillation will be minimized. Exceptions arise when the solute has very little volatility or when it is unstable at its boiling point under reasonably attainable pressures. In such cases a volatile solvent is required, and a low latent heat will improve the economics of the recovery process. Another purpose may be to transfer the solute to a solvent suitable for the next step of the process. Then the solvent is selected for its chemical properties rather than for its thermodynamic properties.

Extraction may also be used to purify a solution by removing pollutants or any undesirable compounds that would adversely affect the subsequent use of the solution. Typical applications can be found in the removal of phenolic compounds from the water effluents from chemical plants and in the removal of catalyst poisons from a recycle solvent. On a laboratory scale extraction may not always be as convenient or simple as adsorption for these purposes, but on a commercial scale the economics generally favor the extraction process. Thus it must be considered for pilot plant and subsequent large installations whenever purifications of this type are involved.

Multiple Extractions with Fresh Solvent

The basic principles for multiple extractions with fresh solvent can be derived from a material balance and the definition of the distribution coefficient. Thus when a quantity of light solvent L is used to extract a quantity of heavy solvent H containing x_0 concentration of solute, the residual concentration x_1 in the heavy phase is given as

$$x_1 = \frac{x_0}{1 + LD/H}. \tag{3.32}$$

where the function LD/H in the denominator is defined as the extraction factor. The units of L and H must be consistent with the units in defining D, to ensure that the extraction factor will be dimensionless. Thus if the distribution coefficient is given as the ratio of the concentrations in moles per unit volume of solution, L and H must be expressed in the same volume units. If concentrations are dilute, the volume change accompanying the transfer of solute from one solvent to the other may be negligible. When concentrations are large, however, it is more convenient to define D in terms of concentrations per unit quantity of solvent rather than per unit quantity of solution. Then, when the solvents are essentially immiscible with each other, the ratio of L to H will be the same for successive extractions with the same volume of light solvent. Thus the amount of residual solute in the heavy solution after successive extractions with n portions of light solvent of quantity L becomes

$$x_n = \frac{x_0}{(1 + LD/H)^n} = \frac{x_0}{(1 + E)^n} \tag{3.33}$$

where E = extraction factor defined as LD/H.

If D varies over the concentration range from x_0 to x_n, the rigorous derivation will give

$$x_n = \frac{x_0}{(1 + E_1)(1 + E_2)(1 + E_3)\cdots(1 + E_n)} \tag{3.34}$$

where the subscripts identify the extraction factors in the successive steps evaluated from the distribution coefficients at the given concentrations. When the variation is small (i.e., of the order of 10 to 20 %), an average value, selected by assigning greater weight to the lower concentrations, can be readily estimated for use in the simpler equation (3.33) to evaluate the number of extractions required with fresh solvent to remove a desired fraction of the solute. Thus

$$n = \frac{\log x_0/x_n}{\log(1 + \mathrm{E_{av}})} \tag{3.35}$$

where E_{av} = average extraction factor = LD_{av}/H.

When the mutual solubility of the solvents varies appreciably with solute concentration, the estimate of an average extraction factor may be subject to appreciable uncertainty in the average solvent ratio L/H, as well as in the average value of the distribution coefficient. For such cases a graphical technique can provide a rigorous calculation subject only to the scale limitations of the diagram. Figure 3.16 gives the graphical calculations for a three-stage extraction of a solute using a light solvent quantity in each stage equal to that of the heavy solution. If the solubility curve and tie line data are represented on the basis of volume fraction, the quantities of both liquids may be on a volume basis; however the solubility curve and tie line data are normally plotted on a weight or mole basis. The figure in this case is assumed to be a weight fraction basis, and the weight of light solvent used in each stage is equal to the weight of the heavy solution.

Point F designates the composition of the initial solution. The total mixture will be located on the line between F and the light solvent corner at the point where the light solvent concentration is 50 %, as shown at point M_1. The tie line through this point is located either by direct interpolation of the tie lines on the figure, or through the use of an equilibrium curve of solute concentration in the light solvent against solute concentration in the heavy solvent.

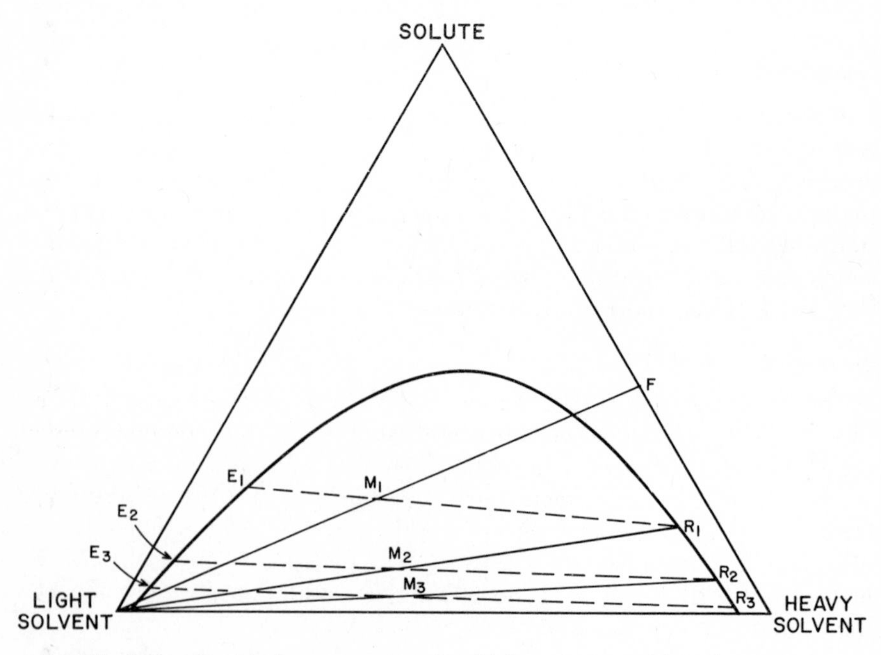

Fig. 3.16. Graphical representation of multiple extractions with fresh solvent.

The latter curve can be interpolated by trial to find the tie line through M_1 more precisely than is possible by estimating it from the variation of the slopes of the different experimental tie lines on the ternary diagram. The extract composition from the first stage is thus located at the light solvent side of the tie line at E_1 and the raffinate composition is located at R_1. The quantities of the respective phases are in the inverse ratio to their distances from M_1, and the quantity of raffinate of composition R_1 may be found by multiplying the total weight of the mixture of solvent and solution by the ratio of the distance E_1–M_1 to the distance E_1–R_1. The mix point for the second extraction is similarly located on the line between R_1 and the light solvent corner of the diagram such that the distance M_2–R_1 divided by the total length of the line is equal to the fraction of fresh solvent in the total mixture. The tie line through point M_2 defines the extract composition E_2 and the raffinate composition R_2 in this stage. The procedure can be repeated until the desired degree of removal is obtained. When solute concentrations are less than a few percent, the mutual solubility and the distribution coefficient will vary so little that an average value for the extraction factor may be estimated with sufficient accuracy to permit the use of (3.35) to determine the stages required for high degrees of removal. In the low concentration range, construction inaccuracies on the diagram usually are greater than the uncertainty in estimating the average extraction factor.

Countercurrent Extraction

When a high degree of solute removal is desired, a large amount of solvent is required by the procedure described in the previous section. Such a procedure is used primarily in laboratory work to prepare samples of purified material for further study. It is rarely used in pilot plant or commercial operation where efficient and economic extraction is required. Even in relatively simple laboratory investigations, a repetitive procedure can be developed that will achieve a more efficient removal with less solvent.

BATCHWISE SIMULATION

The procedure of Bush and Densen [15] is schematized in Fig. 3.17, where the extraction steps along the upper right-hand side of the diagram represent the extraction of a single portion of the feed solution with six portions of fresh solvent. The successive extracts are then used to extract a second portion of fresh feed in the same sequence employed with the first portion of feed. In the diagram the same procedure is repeated four more times, to ensure that the first extract portion will contain a solute concentration very nearly equal to the equilibrium concentration with the feed. Successive portions of extract will contain less solute but always more than that of the first extraction for

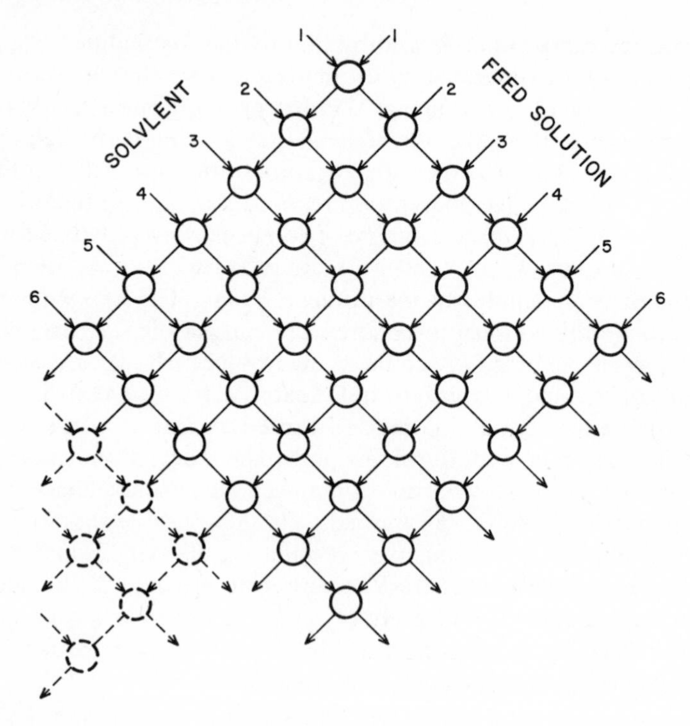

Fig. 3.17. Extraction pattern for laboratory technique for removal of solute from solution by simulated countercurrent flow.

each portion. If an extract pattern of this type is carried out by dividing the feed solution into six portions and operating at an extraction factor of unity, the sixth extract portion will contain essentially the same solute concentration as the first extract obtained in a single extraction stage. All other extracts will contain more solute, approaching the concentration in equilibrium with the feed, which is the theoretical maximum concentration attainable. Combining all six extracts for solute recovery will result in a total solvent solution containing almost five-sixths of the total amount of solute in the six feed portions, and combining all the raffinate portions will give a raffinate containing slightly more than one-sixth of the solute in the feed. This compares with extracting the total feed with the total solvent in a single stage in which only half the solute will be recovered in the solvent phase.

The greater the number of portions of the feed and solvent used along the sides of the extraction pattern in Fig. 3.17, the greater will be the degree of removal of the solute with a given amount of solvent, and a 10-portion pattern will give close to 90 % removal after combining all the raffinates. This pattern will require 100 extraction operations and provides only about 7 % greater removal than the 36 extractions shown in Fig. 3.17. One might even consider a four-portion pattern, giving about 73 % removal in 16 extractions, or a five-portion pattern, giving about 79 % removal in 25 extractions, which are more practical than the six-portion pattern shown.

When an extraction factor greater than unity is used in a pattern similar to that of Fig. 3.17, the degree of removal is significantly increased; under these conditions, dividing the feed into about four portions for a similar extraction pattern is usually the most practical. Obtaining maximum yield of product will require combining all solvent portions from the pattern in Fig. 3.17. If the raffinate solution is to be studied with maximum removal of the solute and the one-sixth portion from the extreme left-hand extraction stage is not adequate, the extraction pattern can be extended by extracting the second feed portion with an additional portion of fresh solvent and the third portion with another additional portion of fresh solvent, as illustrated by the dashed lines in the lower left-hand corner. This will provide about two-thirds of the feed with almost the same degree of solute removal as in the extraction with six successive portions of fresh solvent. Obviously the pattern could be completed to provide a raffinate sample of the same volume as the original feed, which has also been subjected to six extractions.

This particular extraction technique is a segment of a more general procedure developed for fractional liquid extraction. The calculation techniques are derived in Section 4 for the general case, and the application to a pattern of the type shown in Fig. 3.17 then becomes apparent.

CONTINUOUS COLUMN OPERATION

The batchwise extraction techniques described in the previous section can be carried out in separatory funnels, and the interpretation of the data is relatively simple. However manual operations are tedious and time-consuming, and procedures based on more than 10 stages require special equipment. Countercurrent flow provides the most efficient utilization of the solvent. By operating at extraction factors of unity or higher, it is theoretically possible to remove all the solute from a solution if sufficient stages are provided. The number of stages required decreases with larger extraction factors, but the maximum solute concentration in the solvent is obtained at an extraction factor of unity. Thus all extraction processes, whether on a laboratory or commercial scale, are compromises between minimum stages and maximum solvent concentrations.

Figure 3.18 is a schematic diagram of a continuous countercurrent extraction process for solute removal. The solute quantities in the different streams between the stages are indicated, based on a unit quantity of solute in the final raffinate and a constant extraction factor of E in all stages.

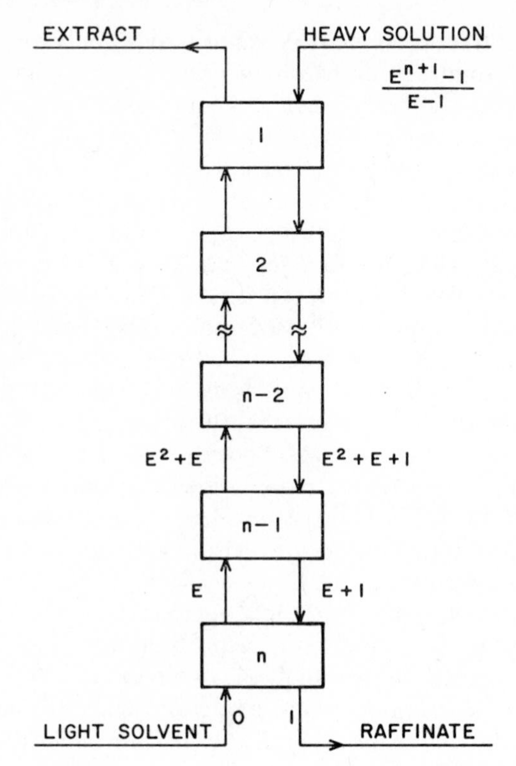

Fig. 3.18. Schematic diagram of multistage continuous extraction for removal.

For this extraction factor, the material balance for the solute in any stage shows that the fraction of the solute in the light phase is expressed as

$$p = \frac{E}{E + 1} \tag{3.36}$$

and the fraction of solute unextracted becomes

$$q = \frac{1}{E + 1} \tag{3.37}$$

while the ratio of p/q is equal to E. Figure 3.18 includes the solute quantities in the streams between the different stages, and the ratio between the residual and the initial solute concentrations after n stages can be derived from the material balances as

$$\frac{x_n}{x_0} = \frac{E - 1}{E^{n+1} - 1} \tag{3.38}$$

where the fresh solvent is solute free. When the recycle solvent contains y_{n+1} concentration of solute, it can be shown that

$$\frac{x_n - y_{n+1}/D}{x_0 - y_{n+1}/D} = \frac{E - 1}{E^{n+1} - 1} \tag{3.39}$$

Figure 3.19 depicts the relation between the left-hand term of (3.39) and the extraction factor for different numbers of stages. For most laboratory operations in which fresh solvent is used, $y_{n+1} = 0$, but in pilot plant and commercial operations where recycle solvent is used, the residual solute must be considered.

Equation 3.39 is based on a constant extraction factor in all stages and is generally applicable only to dilute solutions. At high solute concentrations the mutual solubility of the solvents is affected by the solute concentration, and stagewise calculations can be made by the graphical technique of Hunter and Nash [16] as illustrated in Fig. 3.20. The feed, raffinate, extract, and fresh solvent compositions are located at F, R_n, S_1, and S_{n+1}, respectively, on the

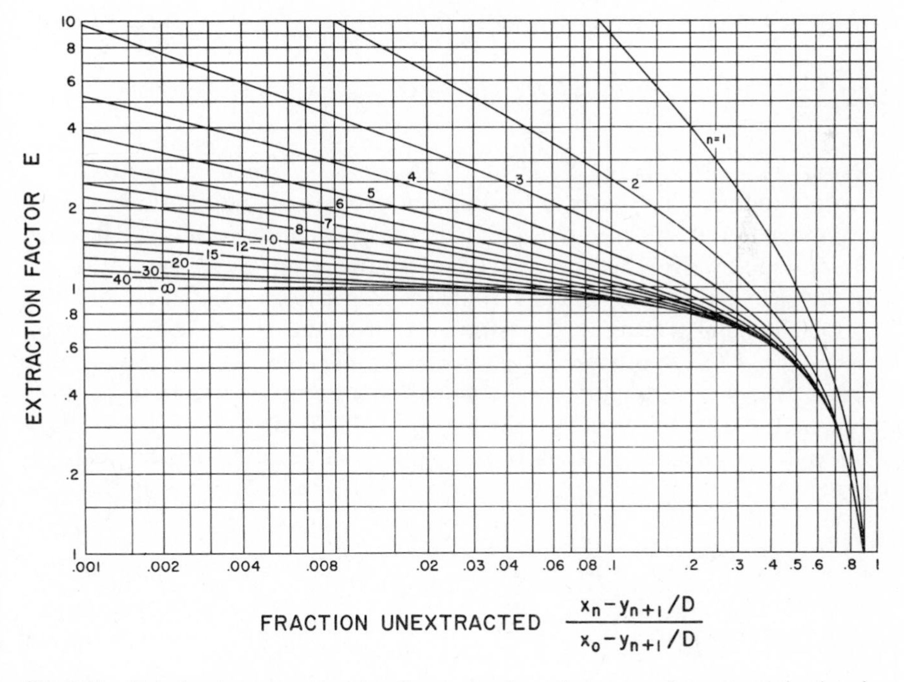

Fig. 3.19. Relation between extraction factor, number of stages, and unextracted solute in continuous countercurrent extraction

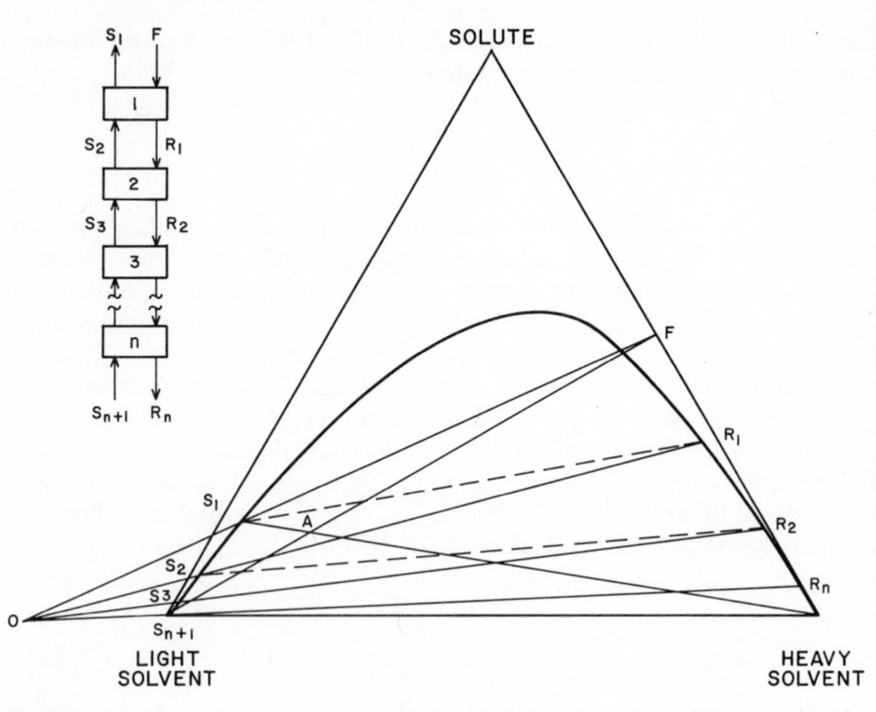

Fig. 3.20. Continuous countercurrent extraction calculations by method of Hunter and Nash.

ternary diagram in Fig. 3.20. The lines *FS* and *RS* are then extended to intersect at point O. This point can be shown by geometry to lie on the line satisfying the material balance at all points in the extraction operation. The point R_1, in equilibrium with the extract composition S_1, is located at the other end of the tie line, through S_1. A line from O to R_1 intersects the mutual solubility curve at S_2 and gives the composition satisfying the material balance on the first stage. The solvent concentration in equilibrium with S_2 is again located from the tie line through S_2 to R_2, and alternate equilibrium determinations and material balances are continued until the final material balance line through S_{n+1} passes through the raffinate concentration R_n, or a slightly lower solute concentration along the same side of the diagram. In the latter case the fractional amount of a theoretical stage necessary to reach the desired raffinate composition must be interpolated.

It should be noted that when the compositions of all streams have been specified, the ratio of solvent to feed quantities can be determined. The composition of the total amount of material introduced into the column must lie on the line between F and S_{n+1}, and the composition of the total amount leaving the column must lie on the line between R_n and S_1. The only point satisfying both conditions is the intersection of the lines; thus the relative amounts of solvent and feed can be determined from the ratio of the distance AF to $S_{n+1}A$, and the relative amounts of extract and raffinate are in the ratio of the distances R_nA to AS_1.

On the other hand, the usual manner for specifying an extraction operation is to give the solvent-to-feed ratio and either the degree of extraction or the raffinate concentration. In this case the total mix composition of feed and solvent is first located, and the extract composition is obtained by extending the line from R_n through this mix point to the mutual solubility curve. It is obvious that since each of the product streams leaves the column after contacting the other phase, their compositions must lie along the mutual solubility curve.

It is also important to recognize that the other end of the tie line through S_1 must lie below the line S_1F or the desired extraction with the given solvent quantity cannot be achieved. The solvent quantity giving a tie line through S_1 that coincides with S_1F is known as the minimum solvent rate and will require an infinite number of stages to effect the separation. Smaller solvent quantities could not give the desired result, even if the infinite number of stages could be provided.

Upon application of the Hunter-Nash method, it becomes immediately apparent that interpolation of the tie line data on the ternary diagram is not very precise. Varteressian and Fenske [17] suggested a more convenient method whereby the tie line data are plotted as an equilibrium curve as in Fig. 3.21. The curvature of the operating line is determined by constructing a series of lines arbitrarily spaced through the point O in Fig. 3.20 and plotting the solute concentrations at the two intersections with the mutual solubility curve as the appropriate coordinates of the figure. The number of theoretical stages required is then determined by stepping them off between the operating line and the equilibrium curve as shown.

This type of diagram is particularly useful if the distribution coefficients vary with solute concentration, but the mutual solubility of the solvent is not appreciably affected. Under these conditions the operating line is straight, with a slope equal to $(y_1 - y_{n+1})/(x_0 - x_n)$, and the ternary diagram construction of Fig. 3.20 is not required. Also, when both the equilibrium curve and the operating line are straight, the graphical calculation technique of this figure becomes identical to the relationship of (3.39). Consequently, when a high degree of removal is involved that would require redrawing the operating line and equilibrium curve in the dilute region to a larger scale for adequate accuracy, (3.39) usually can be solved for n with greater accuracy than would be attainable by the graphical technique.

Equation 3.39 is always applicable over the concentration range in which the equilibrium curve is linear through the origin, and it could be used to

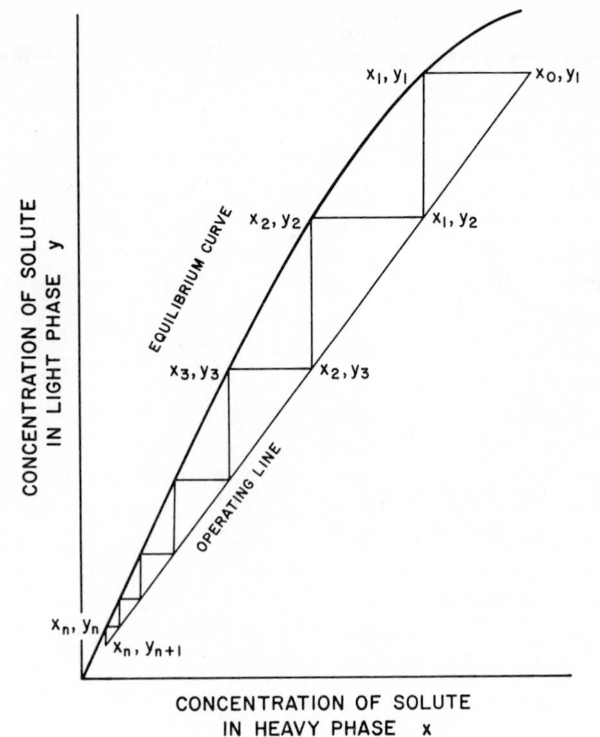

Fig. 3.21. Graphical equilibrium stage calculations.

evaluate additional stages required after the first three had been determined graphically. The value of D to be used in the equation is the limiting slope of the equilibrium curve at $x = 0$. This type of plot can also be used to evaluate the number of theoretical stages from the compositions of the input and product streams in a packed or spray column and the stage efficiency in a column with discrete mixing stages. The minimum solvent quantity for any given specifications, usually a critical factor in the design and operation of extraction processes, can also be evaluated from Fig. 3.21.

The minimum solvent quantity is indicated by the intersection of the operating line and equilibrium curve at y_0. Any smaller solvent quantity will give an intersection at a value of y less than y_0, and it will not be possible to step off the number of necessary stages. The intersection of the operating line and equilibrium curve is called a "pinch," and the pinch compositions are reached after infinite stages. The concept of pinch compositions proves useful in the calculations of fractional liquid extraction, which is discussed later.

Fractionation Processes

Extraction for removal involves the transfer of a component of a mixture from one phase to another. It may be considered to be analogous to evaporation, with the only distinction that this operation involves a vapor phase and a liquid phase and extraction involves two liquid phases. Fractionation by liquid extraction between two solvents that exhibit different distribution coefficients for different compounds is analogous to fractional distillation in which components are separated by their different vapor pressures, which also results in different distribution coefficients between the vapor and liquid phases. Fractional distillation can be modified by changing the character of the liquid phase as in azeotropic and extractive distillation, but little can be done to alter the character of the vapor phase beyond the minor effect of changing the pressure. In liquid extraction it is possible to change or to modify both solvents within the limits of retaining their immiscibility; thus fractionation by liquid extraction is, fundamentally, a more powerful separation technique than distillation. Development was originally inhibited by the lack of suitable equipment for simple operation. History does not record when the use of liquid extraction was first considered for fractionation; but as previously noted, Frenc [18] mentioned it in 1925 as an impractical procedure. Furthermore, the work of Jantzen [19] on his delicate glass multistage agitated extraction column, which was something of a glass blower's masterpiece, did not immediately stimulate interest in this process.

Batchwise Techniques

It was not until 1944, when Lyman C. Craig developed his relatively simple countercurrent fractionation device for the analysis of mixtures [20], that the practical utility of this separation process began to be appreciated. Craig introduced the sample into the top stage of the triangular extraction pattern appearing in Fig. 3.22 and subjected it to the countercurrent flow of the solvents. Components will distribute differently in the stages of any given line parallel to the base of the triangle, depending on their distribution coefficients.

The fractions of the total amount of any given solute in the light and heavy phases in any stage may be designated as p and q, respectively. These fractions are related to the extraction factor as

$$p = \frac{E}{E + 1} \tag{3.40}$$

$$q = \frac{1}{E + 1} \tag{3.41}$$

and the quantities in the different stages of the pattern can be related by these functions as indicated by the terms within the different stages.

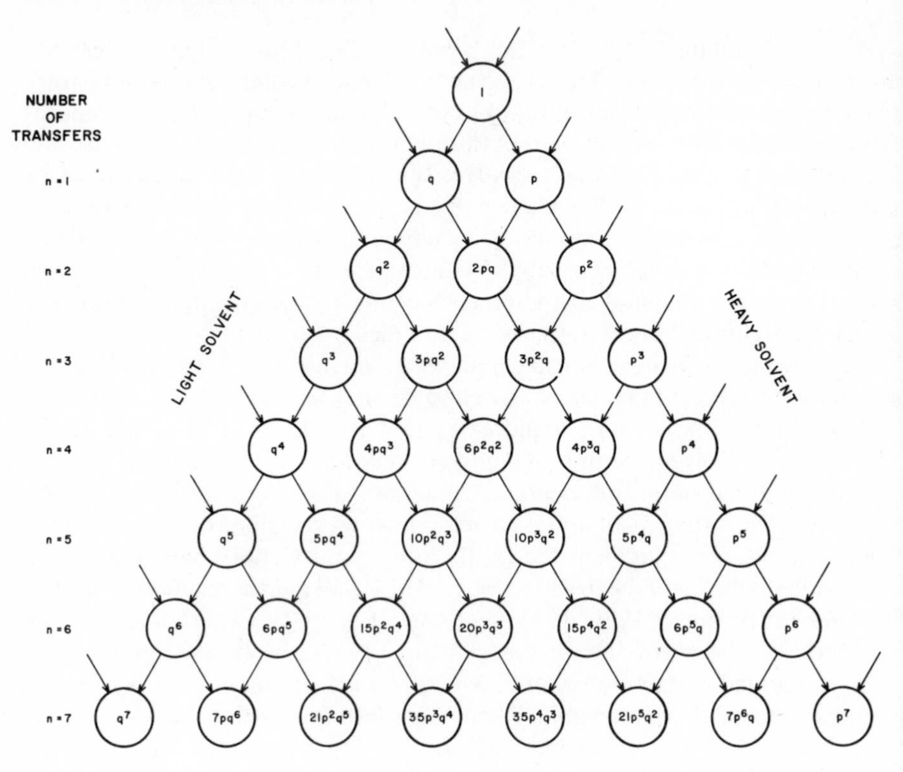

Fig. 3.22. Countercurrent distribution pattern developed in Craig apparatus.

The functions in any horizontal line in this pattern can be recognized as the binomial expansion of $(p + q)^n$. If T represents the total amount of a solute in any stage, the stages are designated by n as the horizontal line number below the top, and r is the stage location in the line designating the stage on the right as zero, the quantity in any stage is given by

$$T_{n,r} = \frac{n!}{r!(n-r)!} p^r q^{n-r} \tag{3.42}$$

The term n also represents 1 less than the total number of stages in any horizontal row. In the mechanics of operation of the Craig apparatus, this number is identical to the number transfers as defined in development of the theory of countercurrent fractionation for his device. Substitution of (3.40) and (3.41) in (3.42) gives

$$T_{n,r} = \frac{n!}{r!(n-r)!} \frac{E^r}{(E+1)^n} \tag{3.43}$$

Almost simultaneously with Craig's publication of his equipment design and calculation technique, Stene [21] also derived fundamental relationships for the quantities in the different stages. Stene's numbering system for identifying the stages is different from that of Craig. Since the concept of a transfer is inherent in the operation of his equipment, Craig's numbering system is used in this chapter. Stene's identification is consistent with continuous countercurrent operation, and differences are noted later.

When the values obtained from (3.43) are plotted against the stage location r for any number of transfers n, the total quantity of a given solute will show a maximum that is related to the extraction factor.

Furthermore, as the value of n increases, the values of the binomial expansion terms approach the normal probability curve such that

$$T_{r,n} = \frac{E + 1}{\sqrt{2\pi n E}} \exp\left[\frac{(r_m - r)^2}{2nE/(E+1)^2}\right] \tag{3.44}$$

where r_m designates the stage number showing the maximum solute concentration. This curve is symmetrical, and it now becomes possible to derive the equation for calculating the value of E, thus the distribution coefficient of a compound from the location of the maximum concentration.

Since this maximum will be halfway between any two stages having equal solute concentrations, consider two adjacent stages r and $r + 1$ such that $r_m = r + 0.5$. Equating the values of T given by (3.43) and solving for E in terms of r_m gives

$$E = \frac{r + 1}{n - r} = \frac{r_m + 0.5}{n - r_m + 0.5} \tag{3.45}$$

and since this symmetry is best at large values of n and r_m, the numerical value of 0.5 has a negligible effect on the value of the fraction. Thus

$$E = \frac{r_m}{n - r_m} \tag{3.46}$$

and one can either evaluate E from the location of the maximum or, given the value of E, the location of the maximum can be predicted as

$$r_m = \frac{E}{E + 1} n \tag{3.47}$$

It is interesting to note that the coefficient of n is the fraction of the solute in the light phase in a single-stage equilibrium, previously designated p. Thus the fraction of the total stages between the light solvent feed end of the countercurrent distribution pattern and the point of maximum concentration is identical to the fraction of the particular solute in the light phase in a single-stage equilibrium. This is the basis for the separation of mixtures by the countercurrent distribution technique developed by Craig.

The improved resolution of mixtures with increasing numbers of stages can be readily explained on the basis of (3.44). It is not readily obvious from the polynomial expansion of the Fig. 3.22 as given by (3.42) and (3.43), and comparative calculations would be required to verify this effect.

Figure 3.23 gives the probability curve calculated by (3.44). Integration of the area under this curve will show that about 90 % of the total area is centered in the region between −1.65 and +1.65. If this yield is acceptable, it is possible to determine the number of stages required for the separation of compounds of different distribution coefficients into products of given purity. In this particular case, since 5 % of the compound is on each side of the limiting values just given, product purities will be 95 % if the initial feed contains equal amounts of two components.

Table 3.1 illustrates the calculation of the bandwidth for 90 % yield of each component and the peak location for different extraction factors. The relative distribution coefficients for the components has been taken as 1.2 for this study, and the solvent ratio has been selected to give extraction factors of 1.0 and 1.2, respectively. The bandwidth on each side of the maximum has been calculated for a value of the abscissa of 1.65 in Fig. 3.23 such that

$$r_m - r = \pm 2.33 \frac{\sqrt{E}}{E + 1} \sqrt{n} \tag{3.48}$$

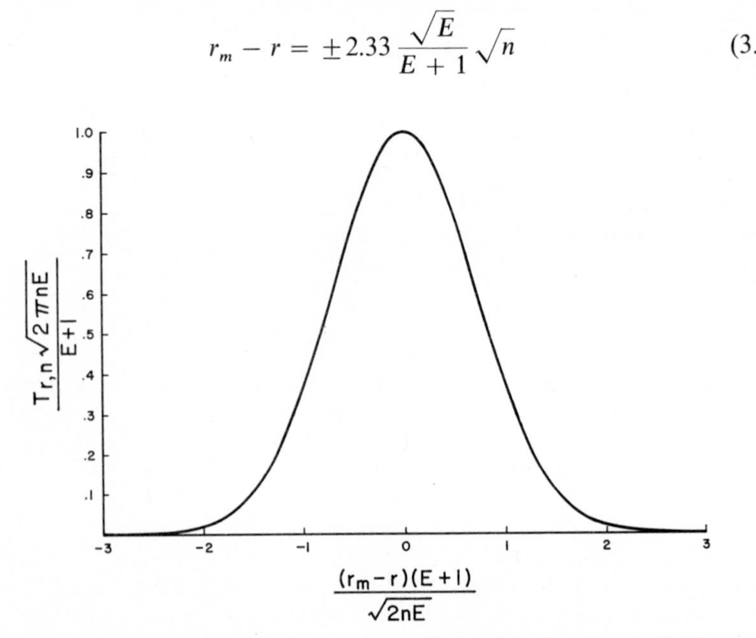

Fig. 3.23. Probability distribution curve calculated for functions in (3.44).

and the location of the peak has been calculated by (3.47). Since the first transfer gives solute in two stages, the number of stages is actually one greater than the number of transfers.

Inspection of the values in Table 3.1 reveals that the peak separation is less than one-tenth of the peak width when 10 stages are used; thus essentially no separation is obtained in this number of stages. Since the peak separation is about one-quarter of the bandwidth at 100 transfers, the separation might be apparent from the variation of total concentration of solutes with stage number at this point; the overlap, however, is too great to provide a practical separation. At 1000 transfers the peak separation is almost 90 % of the bandwidth, permitting the attainment of a resolution into products of either lower yield or lower purity.

At 10,000 transfers the peak separation is almost twice the bandwidth, which indicates that resolution into essentially pure products at complete yield has been achieved. Interpolation shows that about 2600 transfers would be required to meet the original specifications of 90 % yield at 95 % purity. In this case the yields could be increased to 95 % if the contents of the stages on opposite sides of the two peaks were collected. If the original mixture contained other components with extraction factors of 0.833 and 1.44, this would not be possible. If the original mixture also contained equal amounts of all four such components, the fractions of the two previous ones would contain an additional 5 % of the other adjacent compound; therefore net purity would be 90 % and more stages would be required to obtain greater yields and purities.

Figure 3.24 illustrates the character of the countercurrent distribution curves developed by the different numbers of transfers for two compounds

Table 3.1 Effect of Number of Stages on Peak Location and Peak Width for Compounds of Different Distribution Coefficients

Total Number of Transfers, n	*Component A* ($E = 1.0$)		*Component B* ($E = 1.2$)	
	Peak Location, r_m	*Bandwidth,*[a] $r_m - r$	*Peak Location,* r_m	*Bandwidth,*[a] $r_m - r$
10	5.000	3.68	5.454	3.67
100	50.000	11.65	54.545	11.60
1000	500.00	36.8	545.4	36.7
10,000	5000.0	116.5	5454.5	116.0

[a] The total bandwidth extends on both sides of the maximum and is twice the value of $r_m - r$.

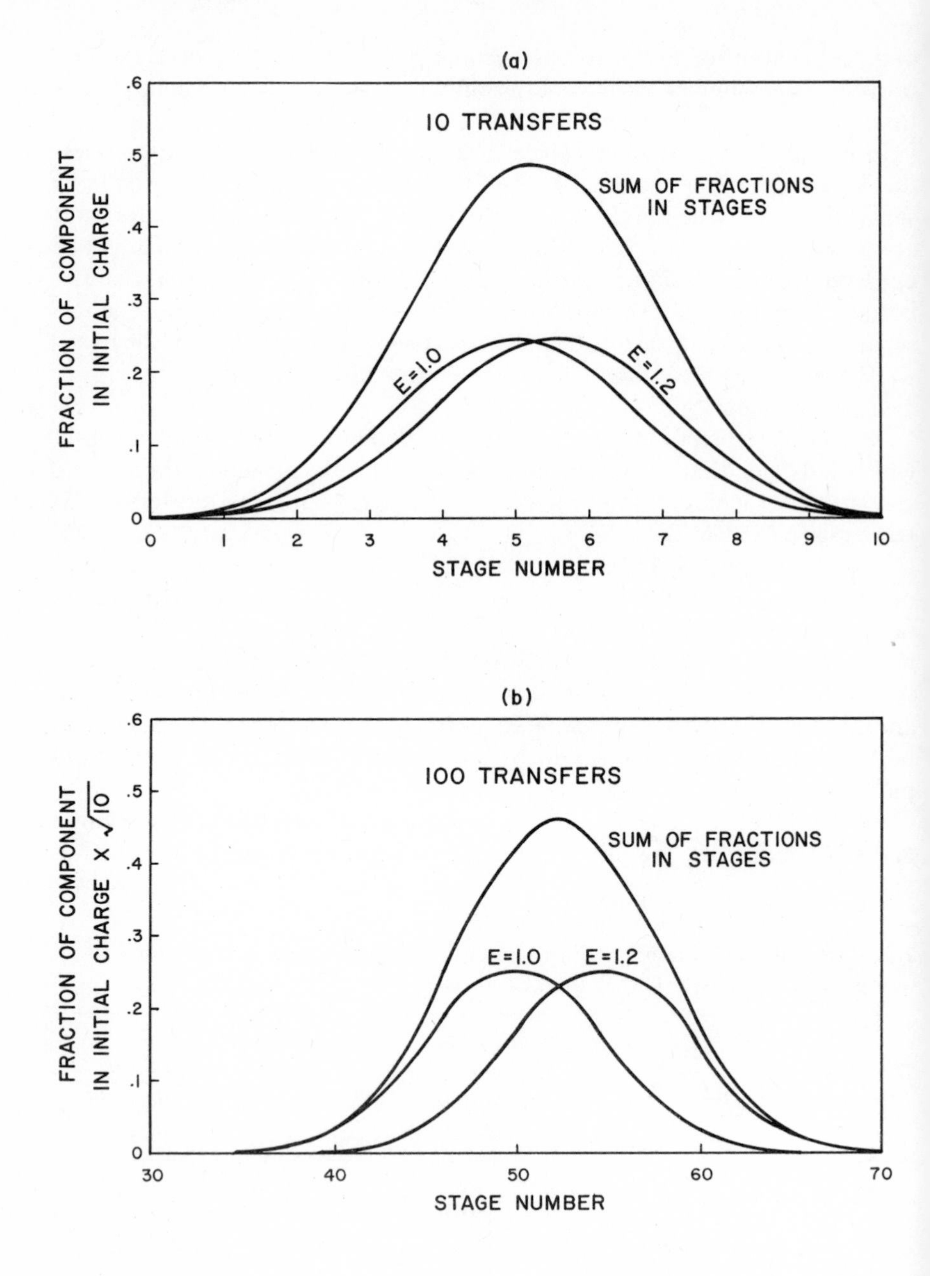
(a)
10 TRANSFERS
SUM OF FRACTIONS
IN STAGES
E=1.0
E=1.2
FRACTION OF COMPONENT
IN INITIAL CHARGE
STAGE NUMBER
(b)
100 TRANSFERS
SUM OF FRACTIONS
IN STAGES
E=1.0
E=1.2
FRACTION OF COMPONENT
IN INITIAL CHARGE X √10
STAGE NUMBER

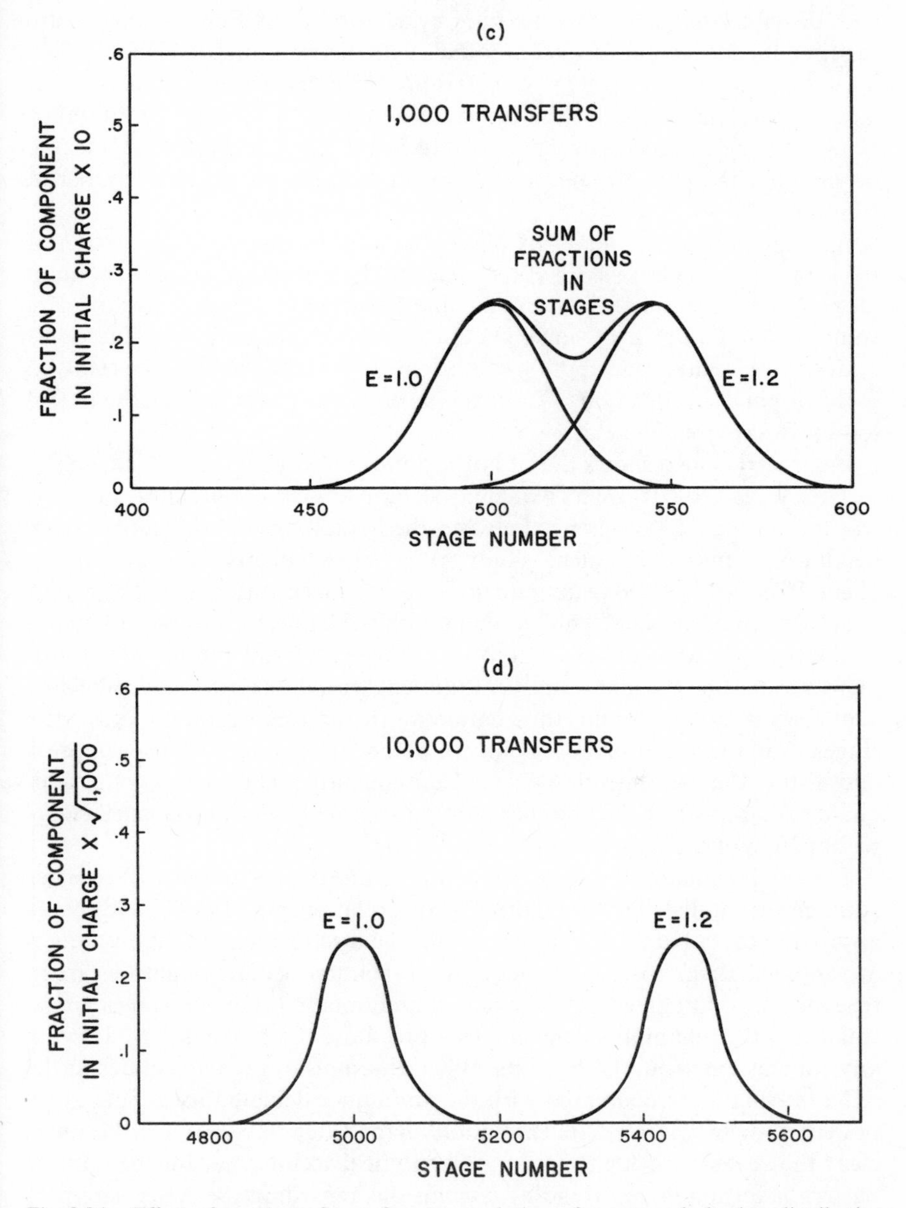

Fig. 3.24. Effect of number of transfers on resolution of compounds having distribution coefficients differing by 20%.

with distribution coefficients differing by a ratio of 1.2. Fewer transfers are required for the resolution of compounds with distribution coefficients differing by greater ratios. Similar calculations for compounds with extraction factors of 1.0 and 1.1 will show a peak difference of 238 stages at a total of 10,000 transfers and $r_m - r$ values of 116.5 and 116.3, respectively; thus the original purity and yield specifications would be achieved in this number of transfers.

Examination of the curves of Fig. 3.24 indicates that a major portion of the total stages on both sides of the peaks that are resolved does not contain measurable amounts of the desired solutes. Substantially all the solutes are contained in a number of stages equal to about 2.5 peak widths. Thus compounds with relative distribution coefficients of 1.1 could actually be resolved to the original specification by utilizing the proper 600 stages out of the 10,000 total previously calculated.

Figure 3.24*b* also shows that if both solutes responded to the same analytical technique, 100 transfers would not demonstrate that the charge material was a mixture. The difference between the actual and the theoretical peak heights for a pure component is only 10 %. When a probability curve is calculated for the observed maximum in the total concentration, it will be within a few percent of the data. This would be within the normal accuracy of analytical techniques and could be readily attributed to a small amount of entrainment during phase separation. Entrainment would have the same effect of suppressing the maximum, thus flattening the concentration curve for the stages. And the additional possibility of nonideal behavior will also suppress and distort the concentration curve. Consequently 100 transfers is not adequate to demonstrate the presence of impurities with distribution coefficients within 20 % of the major component.

Table 3.2 compares the calculation of the same two solutes in a 10-transfer countercurrent distribution by the rigorous relationship of (3.43) and by the approximate probability equation (3.44). The discrepancies at the end stages are appreciable in absolute values, but all the intermediate quantities are in reasonably good agreement. The actual amounts in these end stages are so small that the reliability of the analyses precludes the possibility of drawing any conclusions from these points. When the sums of the two solutes in the different stages are compared with the amounts calculated for a pure component with an intermediate extraction factor value of $E = 1.1$, it becomes clear that even with highly accurate analytical techniques, the 10-transfer pattern is incapable of providing conclusions regarding the composition of the initial material if the distribution coefficients are within 20 %.

The last column in Table 3.2 was calculated by the iterative technique of Fig. 3.22 based on a 10 % solute concentration in the initial feed stage and a linear variation in the distribution coefficient such that the extraction factor

Table 3.2 Comparison of Different Approaches to the Interpretation of a 10-Transfer Countercurrent Distribution for a Mixture of Equal Parts of Two Components with Extraction Factors of 1.0 and 1.2, Respectively

Stage No.	*Fractional Amount of Initial Charge in Each Stage*				*Sum of Solutes in Each Stage*		*Total Calculated for $E = 1.1$ from Total Initial Charge*		*Total Calculated for Nonideal System,*
	Component $E = 1.0$		*Component $E = 1.2$*						
	(3.43)	(3.44)	(3.43)	(3.44)	(3.43)	(3.44)	(3.43)	(3.44)	(3.43)
0	0.000977	0.001700	0.000377	0.000628	0.001354	0.002328	0.001199	0.002065	0.001628
1	0.009766	0.010285	0.004518	0.004633	0.014284	0.014918	0.013190	0.013797	0.016290
2	0.043945	0.041707	0.024398	0.022832	0.058343	0.064539	0.065288	0.061753	0.073868
3	0.117188	0.113372	0.078073	0.075175	0.195261	0.189547	0.191518	0.185094	0.202128
4	0.205078	0.206577	0.163952	0.165365	0.369030	0.371942	0.368661	0.371559	0.373632
5	0.246094	0.252313	0.236091	0.243023	0.482185	0.495336	0.486632	0.499491	0.486662
6	0.205078	0.206577	0.236091	0.238608	0.441169	0.445185	0.446080	0.449704	0.443198
7	0.117188	0.113372	0.161891	0.156515	0.279079	0.269887	0.280393	0.271152	0.271166
8	0.043945	0.041707	0.072851	0.068591	0.116796	0.110298	0.115662	0.109493	0.105374
9	0.009766	0.010285	0.019427	0.020082	0.029293	0.030367	0.028273	0.029611	0.023684
10	0.000977	0.001700	0.002331	0.003928	0.003308	0.005628	0.003110	0.005363	0.002372

varies from 1.2 at 10 % solute concentration to 1.0 at zero concentration. The calculations for a single component in a nonideal system also agree closely enough with the total solute concentrations that even if one evaluated precisely the distribution coefficient of a solute at infinite dilution, it would not be possible to interpret a 10-transfer distribution pattern and ascertain the existence of two components in the initial feed. Under all circumstances, 1000 or more transfers would be necessary to confirm this conclusion.

Inspection of the different curves in Fig. 3.24 reveals that it is not necessary to provide a thousand or more actual stages for the resolution of compounds with very nearly identical distribution properties. For example, if the proper 150 stages are selected to develop a 1000-transfer pattern, or the proper 400 stages in a 10,000-transfer pattern, the distribution curves in Fig. 3.24 could be attained within the limits of experimental accuracy. Craig designated the procedure of truncating the ends of the pattern as the alternate withdrawal technique and demonstrated its utility [22].

Figure 3.25 illustrates the application to a 1000-transfer operation to sepa-

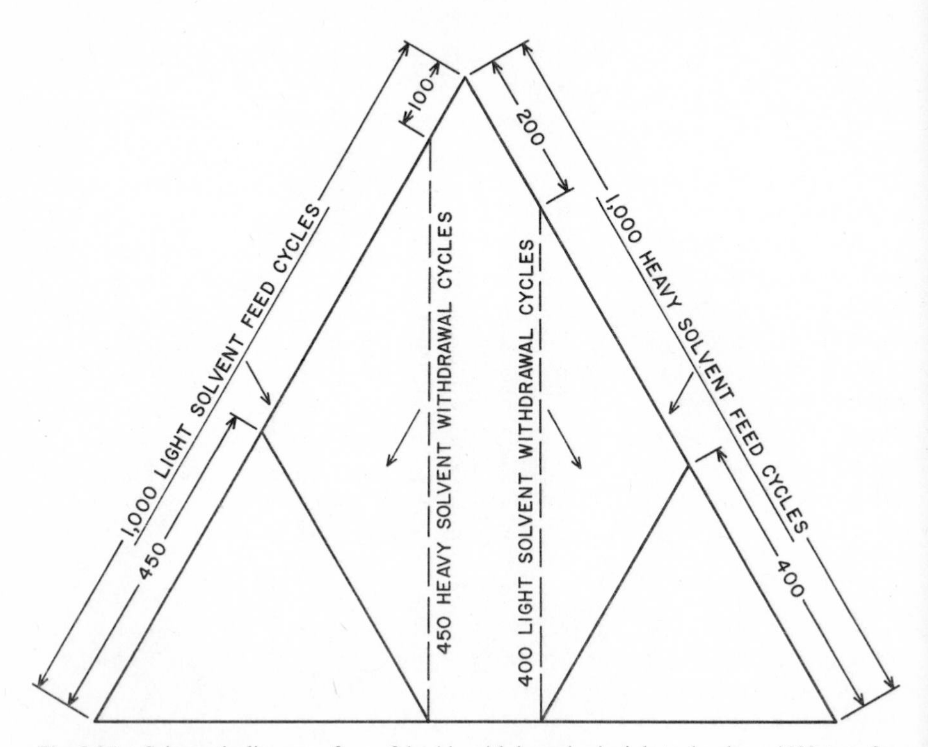

Fig. 3.25. Schematic diagram of use of double-withdrawal principle to develop a 1000-transfer separation in 150 stages.

rate the previous mixture according to the theoretical calculations for Fig. 3.24*c*. The 400 heavy solvent feed cycles at the lower right-hand side of the triangle and the 450 light solvent feed cycles at the lower left never reach the 150 center stages of interest and therefore may be omitted. Furthermore, the stages outside the vertical dashed lines contain only negligible amounts of either solute; thus by carrying out 400 light solvent withdrawals along the right-hand line and 450 heavy solvent withdrawals along the left-hand line, a separation equivalent to 1000 transfers can be achieved using only a total of 150 actual stages. This would be the theoretical minimum, and it would require frequent monitoring of the distribution obtained in the stages, to ensure that the concentration peaks were developing precisely according to the calculations on the basis of preliminary assumptions. Withdrawal procedures might have to be modified as the pattern developed. Alternatively, a larger number of stages such as 200 or 250 might be used, with only one or two monitoring patterns to confirm that the patterns are developing properly without excessive losses on either side. Furthermore, if 200 stages are utilized and the peaks develop at the theoretical locations, the retention of solutes in the total stages will be in excess of 99 %. Integrated values of the probability curve in Fig. 3.23 give a constant of 3.65 in (3.48) for the bandwidth at 99 % yield.

The solvent streams withdrawn from the system contain so little solute that they may be recycled to the opposite side of the extraction pattern, permitting the final separation to be achieved with no more solvent than that necessary to fill the number of tubes actually used.

This procedure also requires sensitive analytical techniques because the peak concentration in the stages of a 1000-transfer pattern is less than one-thirtieth the concentration in the initial feed. To minimize this problem, Craig [23] suggested adding the feed to a number of tubes not exceeding 5 % of the final total. When solute solubility in solvents is a limiting factor, this will permit a significant increase in the actual concentrations of the final stages. He noted that after a large number of transfers, this divided feed did not affect the shape of the concentration curves eventually developed. The effect will be within the normal analytical accuracy for determining the pattern.

Continuous Countercurrent Fractionation with Two Solvents

The withdrawal and multiple-feed techniques proposed by Craig to improve the resolution of a mixture and the accuracy of final results approach a continuous operation. When feed mixture is added to the initial feed stage at every transfer cycle to maintain the maximum allowable concentration in the stage, and the alternate withdrawal is continued, the compositions in the stages accumulate to a steady state condition where all the feed added is removed in the withdrawal streams. If the number of stages is adequate and the solvent ratio is properly selected, an essentially complete separation of the feed may be achieved. The possibilities can be readily recognized from the curves of Fig. 3.24. It was noted that the existence of two different components could not be ascertained from a total solute analysis in the 100-transfer pattern in Fig. 3.24*b*. If the solutions were withdrawn from the stages below the thirty-eighth and above the sixty-sixth, the solutes in these solutions would be essentially pure compounds. Then if a quantity of feed equal to that removed is added to the fifty-second stage at every withdrawal cycle and the countercurrent flows are continued, it will be possible to maintain the concentration pattern shown and ultimately obtain quantities of pure products. This is the basic mechanism of the continuous fractional liquid extraction process.

Detailed examination of the dual-withdrawal technique discloses that the withdrawals are made at every other transfer cycle and the number of countercurrent stages through which the solvents have passed is actually $2n - 1$, where n is the number of stages between and including the stages from which the solvents are removed from the system. Thus the separation obtained between the thirty-eighth and sixty-sixth stages has actually been achieved in 57 continuous countercurrent stages.

The mathematical relationships for continuous operation can also be derived from steady state considerations. Figure 3.26 shows the quantities of a given solute in the different streams passing between the equilibrium stages; H and L are now the flow rates of the heavy and light solvents, respectively, per unit time rather than per transfer. The extraction factor has the same definition, and the fractional distribution of solute between the phases in each stage will also be the same as previously derived. The functional relationships in all the streams will be the same for all components. Subscripts for the extraction factors will identify the different components of the mixture in Fig. 3.26.

Component 1 is more soluble in the light solvent and component 2 is more soluble in the heavy solvent. The column may be considered as two removal columns, since in the section below the center feed component 1 is extracted from the heavy solution, leaving component 2 in the solution. In the upper section component 2 is extracted from the light solution, leaving component 1 in this solution.

The derivation for the fractionation of ideal mixtures is similar to that for the simple extraction for removal in that the expressions are developed from both top and bottom of the column and equated at the feed stage. The calculations appear in Fig. 3.26. If the feed stage is the nth stage from the top and the mth stage from the bottom, the quantity of solute in equilibrium with the heavy liquid leaving the stage must be equal to the quantity leaving in the

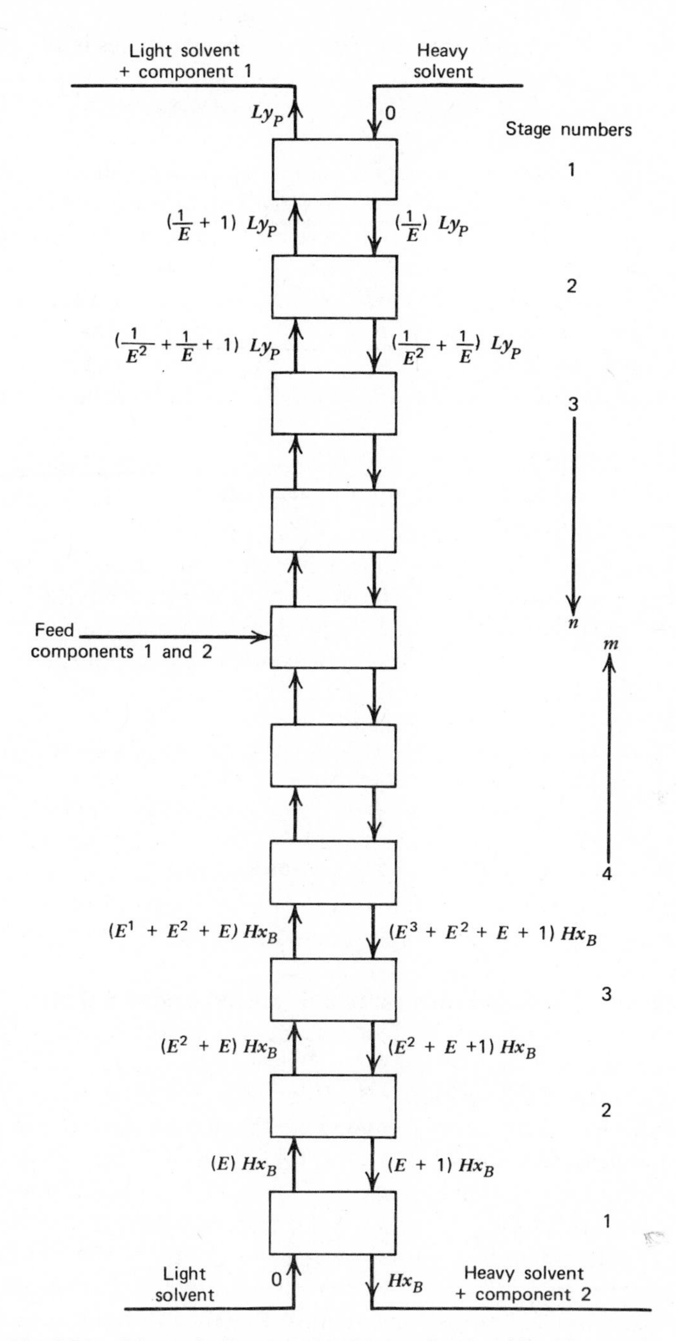

Fig. 3.26. Schematic diagram of multistage fractional liquid extraction.

light solvent from this stage and we have

$$E\frac{E^m - 1}{E - 1} Hx_B = \frac{(1/E^n) - 1}{(1/E) - 1} Ly_P \tag{3.49}$$

The ratio between the quantity of a component leaving the column in the light solvent to that leaving in the heavy solvent, which is called the "rejection ratio" R_1, has been derived by Bartels and Kleiman [24] as follows:

$$R_1 = \frac{E_1^{n+m} - E_1^n}{E_1^n - 1} \tag{3.50}$$

and

$$R_2 = \frac{E_2^{n+m} - E_2^n}{E_2^n - 1} \tag{3.51}$$

where n is the number of stages above and including the feed stage and m is the number of stages below and including the feed stage. Since both numbers include the feed stage, the total number of stages in the column is $n + m - 1$.

In the foregoing equations the value of R_2 is always less than unity because the column is operated to remove most of component 2 from the bottom of the column. For this component it is preferable to consider a "retention ratio," which is the ratio of the quantity of a component leaving at the bottom of the column to the quantity leaving at the top. This ratio is the reciprocal of the rejection ratio and is designated R_2'.

When the feed is introduced into the center of the column such that $m = n$, (3.50) reduces to

$$R_1 = E_1^n \tag{3.52}$$

and in terms of retention ratio, (3.51) becomes

$$R_2' = \frac{1}{R_2} = \frac{1}{E_2^n} \tag{3.53}$$

From the definition of extraction factor, it can be shown that

$$E_1 = \beta E_2 \tag{3.54}$$

where β = relative distribution = D_1/D_2.

If the fraction of component 1 passing overhead is equal to the fraction of component 2 passing out in the bottoms, the values of E_1 and E_2 are related such that

$$E_1 = \frac{1}{E_2} \tag{3.55}$$

$$E_1 = \sqrt{\beta} \quad \text{and} \quad E_2 = \frac{1}{\sqrt{\beta}} \tag{3.56}$$

When a given separation has been specified so that R_1 and R'_2 are fixed, it is possible to select a range of solvent ratios or extraction factors that will effect this separation. To obtain a reasonable separation of components into products of a high degree of purity at an appreciable yield of both, this results in both R_1 and R'_2 being greater than unity. The higher the purity desired, the greater the values of R_1 and R'_2.

Rearranging (3.50) gives

$$R_1 = E_1^n \frac{E_1^m - 1}{E_1^n - 1} \tag{3.57}$$

and to obtain values of R_1 appreciably greater than unity, E_1 must be greater than unity. Similarly, for R'_2 to be appreciably greater than unity, E_2 must be less than unity. This establishes the limits for the operating range of solvent ratios and indicates that a practical separation can be obtained only when

$$D_2 < \frac{H}{L} < D_1$$

Figure 3.27 shows the general shape of the curves obtained by stagewise calculations at different solvent ratios. Rigorous consideration of (3.57)

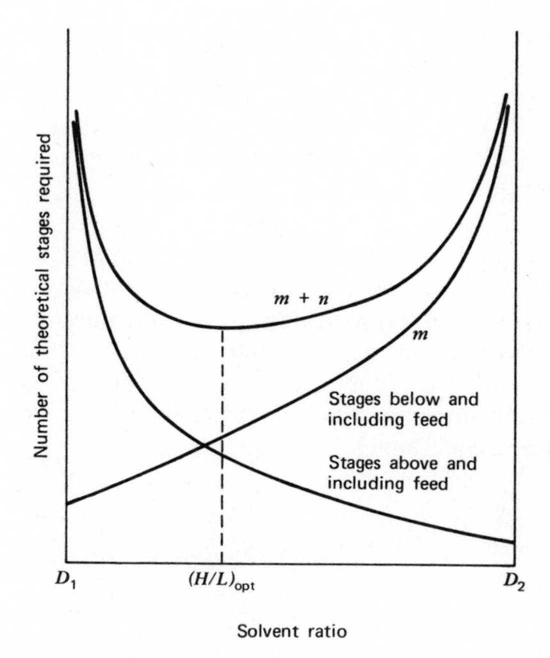

Fig. 3.27. Effect of solvent ratio on stages required in continuous fractional liquid extraction.

indicates that if m is appreciably greater than n, it is possible to obtain a value of R_1 greater than unity even if E_1 is slightly less than unity. Thus the curve of total stages is asymptotic to lines somewhat outside the values of D_2 and D_1, depending on the magnitude of the values of R_1 and R'_2 for the desired separation.

The curve of total stages passes through a minimum, and the solvent ratio giving this minimum is called the *optimum solvent ratio*. Van Dijck and Schaafsma [25] suggested operating at the geometric mean of the two limiting values, namely:

$$\frac{H}{L} = \sqrt{D_1 D_2} \tag{3.58}$$

and Stene [26] showed that this optimum solvent ratio holds for $R_1 = R'_2$ and at this solvent ratio $n = m$. This situation has been called the *symmetrical system*, and the first attempts to rationalize the performance of a fractional liquid extraction were based on systems of this type. Reference to the countercurrent patterns of Fig. 3.24 derived from (3.43) and (3.44) serves to demonstrate that the solvent ratio of (3.58) locates the maximum concentrations for the two components equidistant from the center stage.

Klinkenberg, Lauwerier, and Reman [27] developed algebraic equations to calculate the number of theoretical stages required above and below the feed to obtain a given separation at any solvent ratio. The equations could not be solved directly for n and m for all solvent ratios. However certain particular solutions were observed which provided the data necessary to plot the curves in Fig. 3.27, and the values of n and m could then be interpolated at any solvent ratio.

Klinkenberg [28] studied a large number of systems and developed an empirical correlation for the optimum solvent ratio and theoretical stages required above and below the feed, as a function of the rejection ratios of the two components. The correlations were given as families of curves.

Scheibel [29] developed empirical equations to express the relationships of Klinkenberg's curves. The optimum solvent ratio can be derived from (3.50), (3.51), and (3.54) by introducing approximations holding when n and m are large. The resulting equation is

$$\log E_1 = \frac{\log \beta}{1 + (\log R'_2/\log R_1)^{1/2}} \tag{3.59}$$

and this equation reproduces Klinkenberg's correlation over the range of usual separations.

The total number of theoretical stages required at the optimum solvent ratio is given as follows:

$$m + n = \frac{2 \log R_1 R_2'}{\log \beta} \left(1 - 0.04 \left| \log \frac{R_1}{R_2'} \right| \right) \tag{3.60}$$

This equation, without the term in parentheses on the right, was observed by Klinkenberg, Lauwerier, and Reman [27] for $m = n$, and the term in parentheses was evaluated empirically to correct for nonsymmetrical extraction systems. The deviations increase as the ratio between R_1 and R_2' becomes greater, but at a ratio of 1000 : 1 the error in the equation is only 3 %. This condition is considerably outside the range of usual separations.

The feed stage at optimum solvent ratio can be located from the relations of (3.61) and (3.62):

$$\frac{n}{m} = \left(\frac{\log R_2'}{\log R_1}\right)^{1/2} \left[1 + \frac{\log(R_1/R_2')}{R_2' \log R_1}\right] \tag{3.61}$$

where $R_2' < R_1$, and

$$\frac{m}{n} = \left(\frac{\log R_1}{\log R_2'}\right)^{1/2} \left[1 + \frac{\log (R_2'/R_1)}{R_1' \log R_2'}\right] \tag{3.62}$$

when $R_1 < R_2'$.

These equations, without the second terms in parentheses on the right, have also been derived by Klinkenberg and co-workers [27] for the symmetrical case, and the terms in parentheses have been developed empirically to fit the nonsymmetrical cases. The equations agree with the calculated distribution of stages above and below the feed at the optimum solvent ratio from (3.59) for the range of usual separations also covered by Klinkenberg's curves.

It is interesting to compare these equations with the estimate of requirements for a continuous extraction process deduced from the Craig transfer pattern of Fig. 3.24 previously discussed. When the extraction pattern developed by withdrawal of product streams from the thirty-eighth and sixty-sixth tubes of a 100-stage transfer is examined, it will be seen that the countercurrent solvents have been contacted with each other 57 times before leaving the system. Thus $m + n = 58$, which is twice the number of actual tubes used to carry out the separation. The previous calculations were based on a feed of equal parts of two components with distribution coefficients in the ratio of 1.2 : 1.0. If the actual values are 2.5 and 3.0, and product purities are to be equal so that rejection and retention ratios for components 1 and 2 are identical, the optimum extraction factor for component 1 according to (3.59) will be 1.095. Thus the ratio of light to heavy solvent required will be 1/(1.095 × 2.5) = 0.365. Since $R_1 = R'_2$, the rejection and retention ratios can be evaluated for the given value of $m + n$ from (3.60) as 14.1, indicating purities of 93.4 % for each product at steady state.

The equations are normally used to evaluate the number of theoretical stages and feed stage location in the design of a fractional liquid extraction process. As an illustration, consider a mixture of two components A and B, with distribution coefficients of 3 and 2, respectively. Component A is to be recovered at a 90 % yield and 99 % purity from 100 g of a mixture containing 30 g of A. Thus the following material balance is desired on the column:

	Feed	Light Solvent Product		Heavy Solvent Product	
Component	grams	grams	percent	grams	percent
A	30.0	27	99.0	3.0	4.1
B	70.0	0.27	1.0	69.73	95.9
	100	27.27		72.73	

From this table the value of R_1 is 27/3 or 9, and the value of R'_2 is 69.73/0.27 or 258. Thus the optimum solvent ratio is calculated from (3.54):

$$\log E_1 = \frac{\log 1.5}{1 + (\log 258/\log 9)^{1/2}}$$

whence $H/L = 0.39$. The sum of the stages between the feed stage and the ends of the column is calculated from (3.55):

$$n + m = \frac{2 \log(9 \times 258)}{\log 1.5}\left(1 - 0.04 \log \frac{258}{9}\right) = 36$$

and the total number of theoretical stages required is 35. The feed stage location according to (3.57) is such that

$$\frac{m}{n} = \left(\frac{\log 9}{\log 258}\right)^{1/2}\left[1 + \frac{\log(258/9)}{9 \log 258}\right] = 0.67$$

from which $n = 21.6$, $m = 14.4$, and the feed stage should be located above the bottom of the column about 40 % of the total column height.

A graphical technique for the stagewise calculations in a fractional liquid extraction process was first applied by Martin and Synge [30] to an ideal system. The method consists of constructing the operating lines for the sections above and below the feed stage on the equilibrium diagrams for each of the components. If the fresh solvents are free of both components being separated, the operating lines start at the axes as shown in Fig. 3.28. The

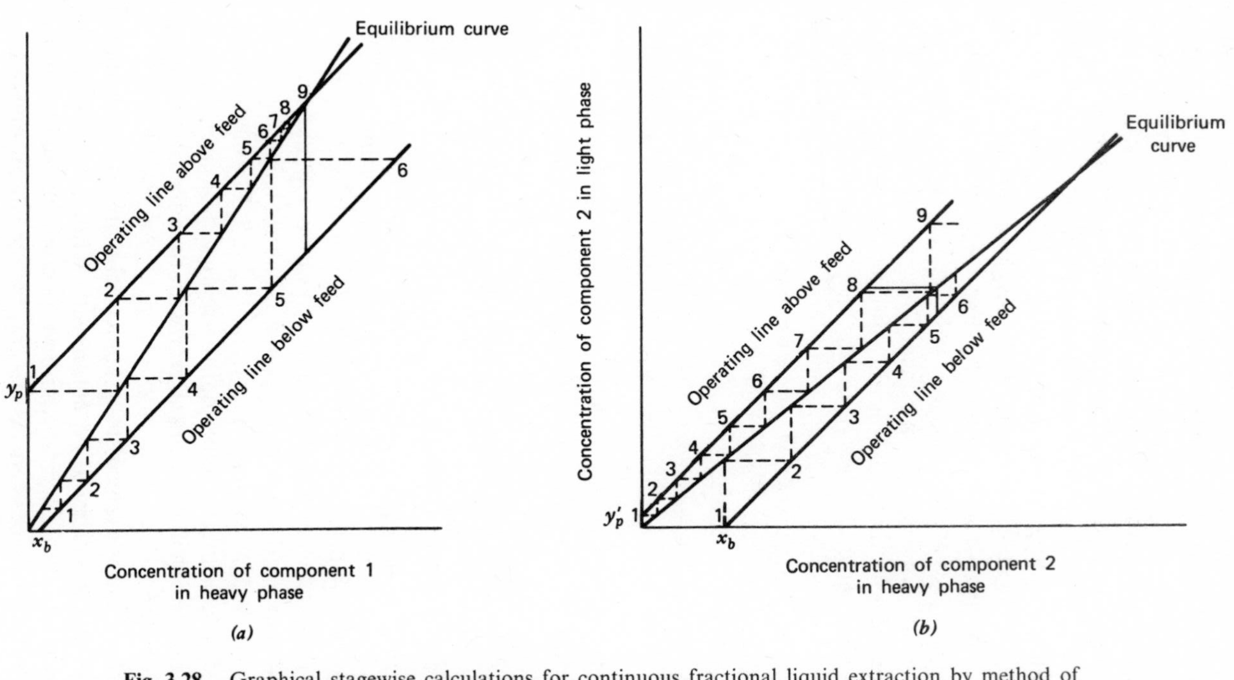

Fig. 3.28. Graphical stagewise calculations for continuous fractional liquid extraction by method of Martin and Synge. (*a*) For component 1. (*b*) For component 2.

operating lines are drawn through these points located from the overall material balance with slopes equal to the solvent ratio H/L for each of the sections of the column and for both components as given in Figs. 3.28*a* and 3.28*b*. The feed stage is the stage that matches the compositions at the same number of stages above and below the feed for each of the components, and it must be located by a trial-and-error technique. However the method is fairly direct because of the existence of the pinches. Thus for component 1 there is very little change in the stages required below the feed for more than six stages above, since the stages below increase only from about 5.1 to 5.4 for stages above numbering from 6 to infinity. It is, therefore, immediately apparent that about 5 stages are required below the feed; and when these are stepped off on the other section for component 2, it may be seen that they require about 8 stages above the feed. It is also apparent from Fig. 3.28 that a large increase in the number of stages below the feed from 5 to infinity only increases the stages required above from about 8 to 10. Thus the region of the match has been accurately defined, and alternate reference to the two parts of the figure rapidly establishes the complete match at 8.2 stages above the feed and 5.3 stages below. Since both these numbers include the identical feed stage, the total number of stages required for the separation is 12.5.

Compere and Ryland [31] applied this graphic technique to a nonideal system in which the equilibrium lines were curved but the equilibrium curves for each of the two components were unaffected by the presence of the other, allowing construction of the curves in Fig. 3.28. They demonstrated the method by determining the separation obtainable in a given number of equilibrium stages and compared the calculated values with their experimentally determined data on the given system.

Scheibel [32] proposed a method for matching components at the feed stage by plotting the concentrations of each of the components in one of the phases against the stage numbers above and below the feed, as in Fig. 3.29. This technique was also demonstrated on completely nonideal systems in which the distribution coefficient of each component varied with its own concentration and with that of the other component, making it impossible to represent the equilibrium data by single lines as in Fig. 3.28. The methods for prediction of nonideal distribution data and the representation of the data for convenient stagewise calculations have also been described by Scheibel [32, 33].

Asselin and Comings [34] applied the graphic technique of Martin and Synge to the study of fractional liquid extraction with reflux, as a later section discusses. They matched the concentration and located the feed stage by plotting the concentrations obtained in the graphic stepwise calculations against stage number as described by Scheibel [32].

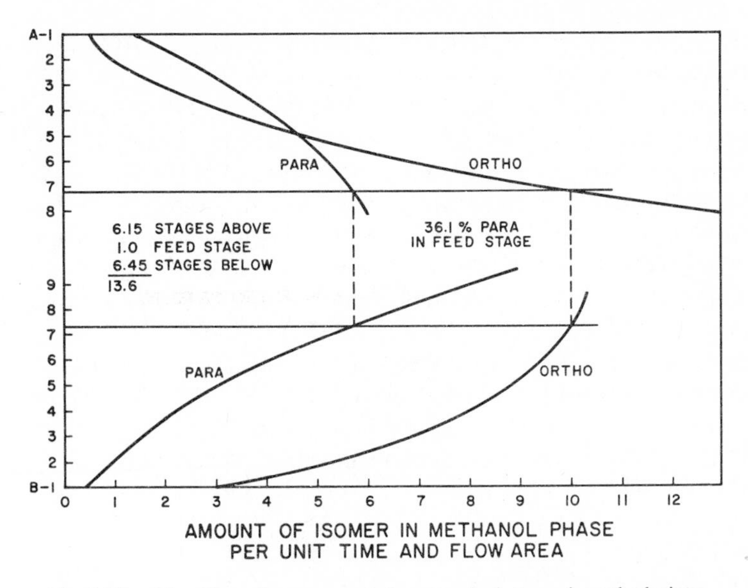

Fig. 3.29. Matching of components in numerical stagewise calculations.

Fractionation with Single Solvent and Reflux

Fractionation by distribution between two solvents is convenient for laboratory studies because the solutions of the purified solutes can be accumulated and the solute recovered as required. Solute concentrations are low and decrease as the distribution coefficients approach each other and more stages are needed for the separation. This makes fractionation by liquid extraction costly and commercially interesting only for separations that cannot be achieved by ordinary fractional distillation or by any of the special distillation techniques such as azeotropic and extractive distillation. Of course it is most applicable to the purification of compounds that cannot be distilled, either because of instability at boiling temperatures under available pressures or because of exceptionally low volatility requiring excessively high temperatures. Because of the higher cost, fractional liquid extraction is economically attractive, primarily to purifications involving valuable materials, particularly those for which chromatographic purification processes may be considered.

When the compounds to be separated are liquid and only partially soluble in a solvent, they may be subjected to countercurrent extraction to selectively remove the compound more soluble in the solvent. This compound is then separated from the solvent, which is recycled to the extractor. A portion of the extracted compound must be recycled to the extractor analogous to reflux in a distillation column; this serves to extract the other compound from the solvent phase. Figure 3.30 is the flow sheet for this process. The solvent separation step must be carried out simultaneously with the extraction if the purity of the extracted product is to be enhanced to desired specifications. Thus this process does not lend itself readily to batchwise laboratory operations. It is preferably carried out in a continuous pilot plant. Since only one solution must be distilled to recover the pure product, the lower operating cost generally makes it the most economic process for commercial applications.

Figure 3.31 presents an application of the Maloney–Schubert graphical technique for evaluating the number of theoretical stages required by this process. The illustration is based on the data of Varteressian and Fenske [35], who studied the separation of *n*-heptane and methylcyclohexane using aniline as the solvent. They derived the geometry for carrying out the stagewise calculations on a ternary diagram using experimentally determined

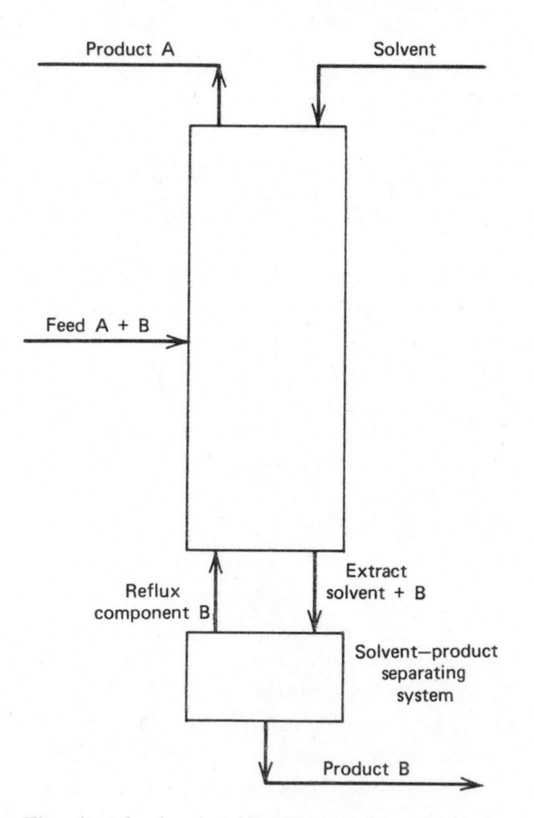

Fig. 3.30. Flowsheet for fractional liquid extraction using a single solvent.

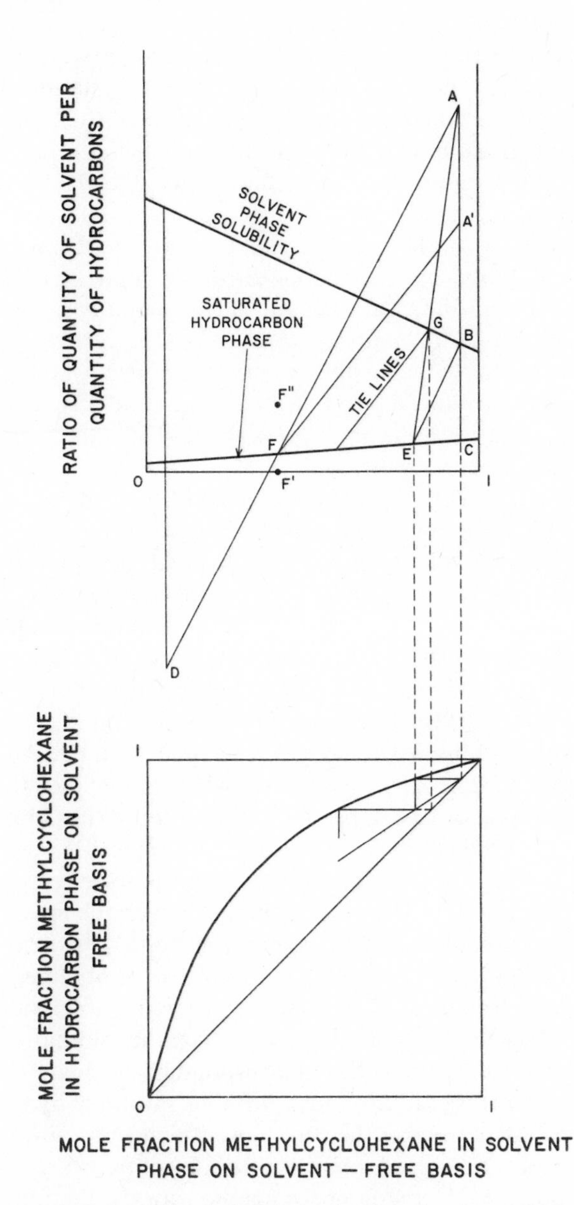

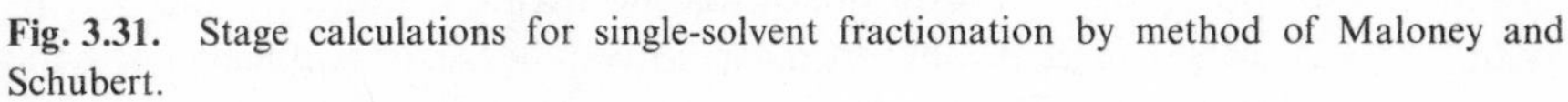

Fig. 3.31. Stage calculations for single-solvent fractionation by method of Maloney and Schubert.

tie lines. This method is rigorous, but it retains the inherent uncertainty, previously mentioned, in interpolating tie line data on a ternary diagram. This makes it less desirable than the approach of Maloney and Schubert [36], which utilizes the Ponchon method in distillation calculations but replaces the heat with solvent quantity. The upper section of the figure shows the amount of solvent per unit quantity of mixture being separated—in this case, methylcyclohexane and *n*-heptane—against the fraction of methylcyclohexane on a solvent-free basis. If point A is located on the vertical line such that AB/BC is the given reflux ratio, all lines passing this point intersect the solubility curves at concentrations that satisfy the solvent balance around the bottom of the column and any point below the feed stage. Also, if the line from A through F is extended to the vertical line at D, any line through D intersects the solubility curves at concentrations that satisfy the solvent balance around the top of the column and any point above the feed stage. These relationships can be derived similar to the heat balances on the Ponchon diagram.

Equilibrium conditions are indicated by tie lines in the upper section of Fig. 3.31, and the difficulty of interpolating these tie lines has been recognized. Maloney and Schubert have recommended obtaining the necessary equilibrium data from a curve such as that in the lower section of Fig. 3.31. From this equilibrium curve, the lower end of the tie line through B is located, and the line through EA locates the point on the operating line at G, which is then plotted on the equilibrium diagram. One apparent modification of this technique is to construct the operating line by a random selection of points, then step off the stages, as suggested by Varteressian and Fenske [32]. When the tie line crosses the line AD, the material balances are taken through point D, since the feed stage has been passed at this point in the column.

Figure 3.31 also indicates an interesting technique for determining the minimum reflux ratio in this type of extraction. Similar to the distillation operation, the minimum reflux ratio is obtained when the line AD coincides with the tie line through F. At this condition point A reaches the lowest possible position at which the desired separation can be obtained with infinite stages. At any lines below this position the separation is impossible, even with infinite stages. Thus the ratio of AB to BC is then the minimum reflux ratio.

In Fig. 3.31 the feed composition corresponding to point F is saturated with solvent. For purposes of this illustration the saturated hydrocarbon line has been shown farther above the zero solvent concentration line than the actual data indicate. If the feed mixture were solvent free, it would lie on this datum line at point F'. In the Ponchon method for distillation, this corresponds to subcooled feed. If some excess solvent were introduced with the saturated hydrocarbon feed, the feed point would be located at F'', which is analogous to partially vaporized feed. It can be seen that in the latter case the tie line through the feed point would be above that for the saturated feed only, and the minimum ratio would be higher. Conversely, at any actual operating reflux ratio, the number of stages required when excess solvent is introduced with the feed is greater than when all the necessary solvent is introduced at the end of the column. The same effect is observed in a two-solvent fractional liquid extraction when either or both solvents are introduced with the feed, but considerable calculation is necessary to prove it in the latter case, whereas it can be clearly seen from the Ponchon diagram for single-solvent fractionation.

Thus far in the discussion of fractionation by solvent extraction with reflux, no attention has been given to the method of recovering the component in the solvent stream so that it can be returned to the column for reflux. The obvious method is by fractional distillation in which the two-phase minimum boiling azeotrope is separated, the nonsolvent phase is removed, and the necessary amount is returned to the extraction column as reflux. The pure solvent from the bottom of the distillation column is recycled to the extraction column. It should be recognized that the overhead product from the distillation column is saturated with solvent, and a further distillation is necessary to remove this solvent to give a pure, solvent-free product. Similarly, the other product from the extraction column is saturated with solvent that must be removed. Figure 3.32 gives the complete process flowsheet for this separation as applied to the methylcyclohexane–*n*-heptane fractionation with aniline. The complexity of the process makes it difficult to simulate on a laboratory scale. It is best carried out in pilot plant operation.

Dual-Solvent Fractionation with Reflux

In fractionation with a single solvent as described in the previous section, the role of reflux is readily recognized. Frequently the equilibrium curve in Fig. 3.31 can be correlated to a constant selectivity equation analogous to the constant relative volatility in distillation

$$y = \frac{\beta x}{1 + (\beta - 1)x} \tag{3.63}$$

where β is solvent selectivity.

Under these conditions the number of theoretical stages required at total reflux can be derived as follows:

$$n + m - 1 = \frac{\log R_1 R_2'}{\log \beta} \tag{3.64}$$

where n and m have the same significance as in (3.60) and the total is 1 more than the total number of stages because the previous numbering system includes the feed stage twice.

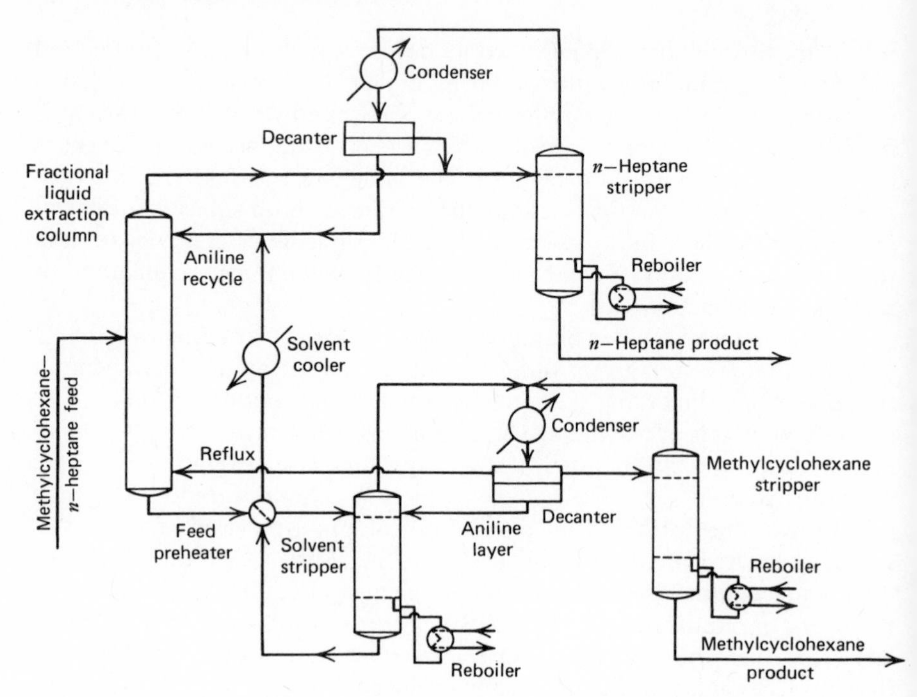

Fig. 3.32. Process flowsheet for fractionation of methylcyclohexane and heptane using aniline as the solvent.

The same equation can also be derived for a dual-solvent system, and comparison with (3.60) indicates that the number of theoretical stages required in a given separation may be reduced to about half through the use of a large reflux ratio. Since solubility in the solvent is always a limiting factor, however, the solvent flow rates must be increased to dissolve the reflux quantities in the solvents at the ends of the column. Thus reflux in liquid extraction increases the processing cost just as it does in distillation, and there is some optimum condition at which the cost saving in stages is offset by the cost of increased column diameter and increased utilities. In fractional liquid extraction the cost of isolating the purified solutes from the solvent is usually so large that an appreciable increase to reduce the number of stages is not economically justified.

However the calculations given for the batchwise fractionation in Figure 3.24 and for the continuous fractionation in Fig. 3.29 indicate that the total solute concentration is a maximum at the feed and a minimum at the ends of the extraction system. The use of a reflux rate at each end, which gives solute concentrations that do not exceed the concentration at the feed stage, decreases the number of stages without increasing the solvent recovery costs. Figure 3.33 illustrates the effect of reflux ratio on the concentrations in the stages of a fractional liquid extraction column separating a mixture of equal parts of *o*- and *p*-nitrochlorobenzenes with Skellysolve C and aqueous methanol ($\beta = 1.62$). The use of reflux increases the solute concentration at the feed stage only slightly, as shown. At reflux ratios above 9, the solute

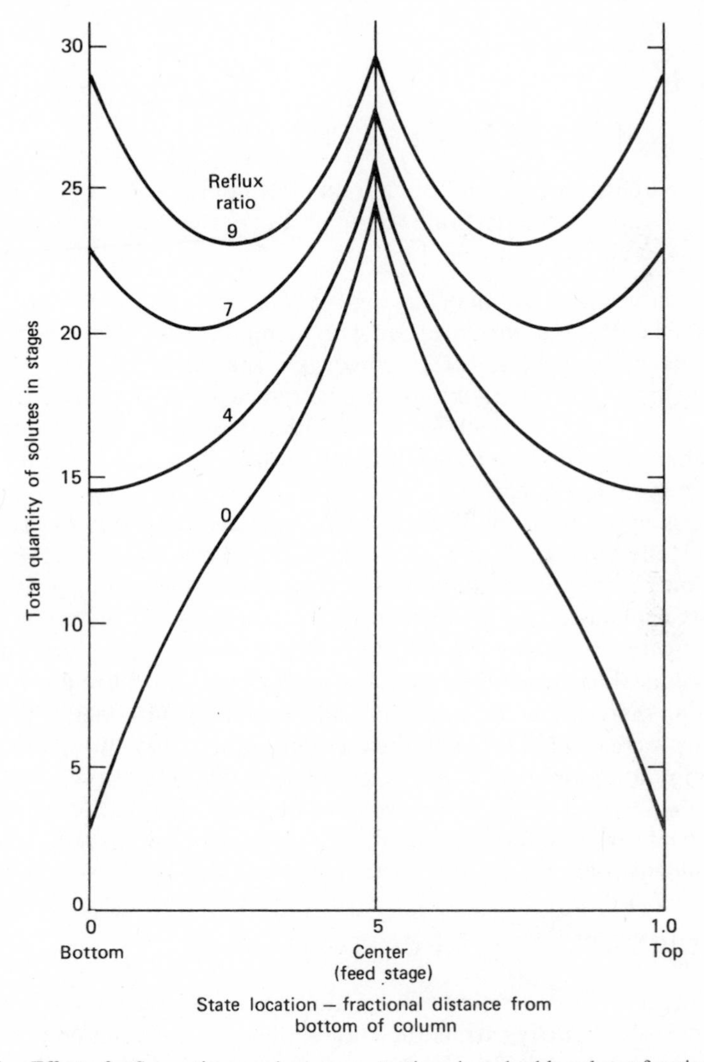

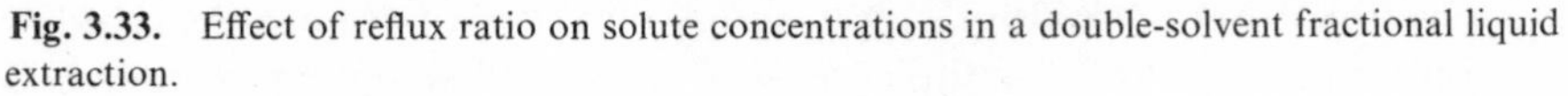

Fig. 3.33. Effect of reflux ratio on solute concentrations in a double-solvent fractional liquid extraction.

quantities in the end stages are greater than in the feed stage, and the solvent flow rates are then controlled by the reflux quantities. This condition would be uneconomic, and Scheibel [37] has defined an optimum reflux ratio at each end of the column such that the solute concentrations in both end stages are identical to the concentration in the feed stage. This is the highest reflux ratio that can be tolerated without adding proportionately to the solvent recovery costs. By making simplifying assumptions, the following equation was derived for estimating the optimum reflux ratio at each end of the column:

$$r_p = \frac{\sqrt{\beta}}{\sqrt{\beta} - 1} \frac{R_1 R_2' + 1}{(R_1 + 1)(R_2' + 1)} \times \frac{F}{P} - 1 \tag{3.65}$$

where r_p = reflux ratio at end of column where pure product P is removed
F/P = ratio of the feed rate to the product rate at the same end of the column

This equation is based on achieving the same end stage concentrations at reflux as the feed stage concentration with no reflux. Since the feed stage concentrations increase slightly with reflux ratio, the true optimum reflux ratio according to the previous definition will be somewhat greater than calculated from (3.65), and when calculated reflux ratios exceed 5, it may be desirable to omit the final numerical term in the equation to obtain a reflux ratio closer to the optimum as defined.

Nonideal systems normally exhibit reduced selectivity at high concentrations. This effect also limits the maximum solute concentrations in the column in the same manner as the solubility limits previously mentioned. When the selectivity is independent of concentration, the number of theoretical stages continually decreases with increasing reflux ratio to the value given by (3.64). In the nonideal case where selectivity decreases with increasing solute concentration, the number of stages initially decreases with reflux ratio and eventually increases [37]. Under these circumstances the reflux ratio requiring the minimum number of theoretical stages could be less than that calculated from (3.64) based on an average value of β, but the equation can provide the first trial for an iterative evaluation of the true optimum reflux ratio for such nonideal systems.

4 EQUIPMENT AND PROCEDURES

The chemical literature contains many descriptions of glass apparatus proposed for laboratory extraction studies. Some of the more popular items are available from laboratory supply companies. Patent literature also contains a wealth of ideas for the construction of multistage countercurrent extraction equipment, but relatively few designs have been commercialized. This section stresses the items that can be purchased as more or less standard equipment. Where they are not stock items, they can be fabricated to order according to standard designs. In some cases equipment is described that was formerly available as standard items and may still be used in some laboratories.

Analytical Applications

All the extraction stages considered in Section 3 could be provided by the classical technique of shaking the immiscible phases in a separatory funnel. The prospect of handling thousands of such funnels to resolve a mixture into pure components would deter a laboratory investigator from adopting this procedure for analysis. Thus countercurrent liquid distribution did not become popular for the analysis of compounds until Craig described his apparatus for the simultaneous mixing of the liquids and transfer of the light phases to the next stage.

Figure 3.34 shows his original design [38], which contained only 25 stages and quickly became obsolete as the value of this technique was recognized and it began to be applied to more difficult separation problems. The manipulation of the different operations is similar to that of the later models and can be easily explained on this simple device. It consists of a lower section *A*, containing 25 holes equally spaced around the diameter, all bored to the same depth; an upper section *B*, with matching holes bored through; and a cover *C*, with a single hole and plug. This first unit was machined from a solid stainless steel cylinder 6 in. in diameter, and the mating surfaces were ground flat to be leakproof without requiring a lubricant. Some of the center core of the upper and lower sections was also machined out to provide maximum pressure on the outer rim. This pressure is exerted on both pairs of surfaces by the spring *G* under the wing nut *F*. A sufficient portion of the center core of part *B* is retained to provide a bearing for rotation with respect to the bottom section about the shaft *E*. The bottom section is fixed to two arms *D* on each side, both containing pins *L*, about which the whole unit can be rotated by a crank on *L*. Another arm *I* is also fixed to the bottom section to hold a spring-loaded pin *J*, which serves to position the holes as the top section is rotated by snapping into indexing holes *K*, around the periphery of the upper section.

The apparatus is so designed that the capacity of the upper tube is twice that of the lower hole. This provides mixing space and also facilitates charging the heavy phase. The heavy phase does not have to be measured individually into each tube. A slight excess of the total amount required can be added to a sufficient number of consecutive tubes through the charge hole in the cover. By rotating the center section through a complete revolution, the

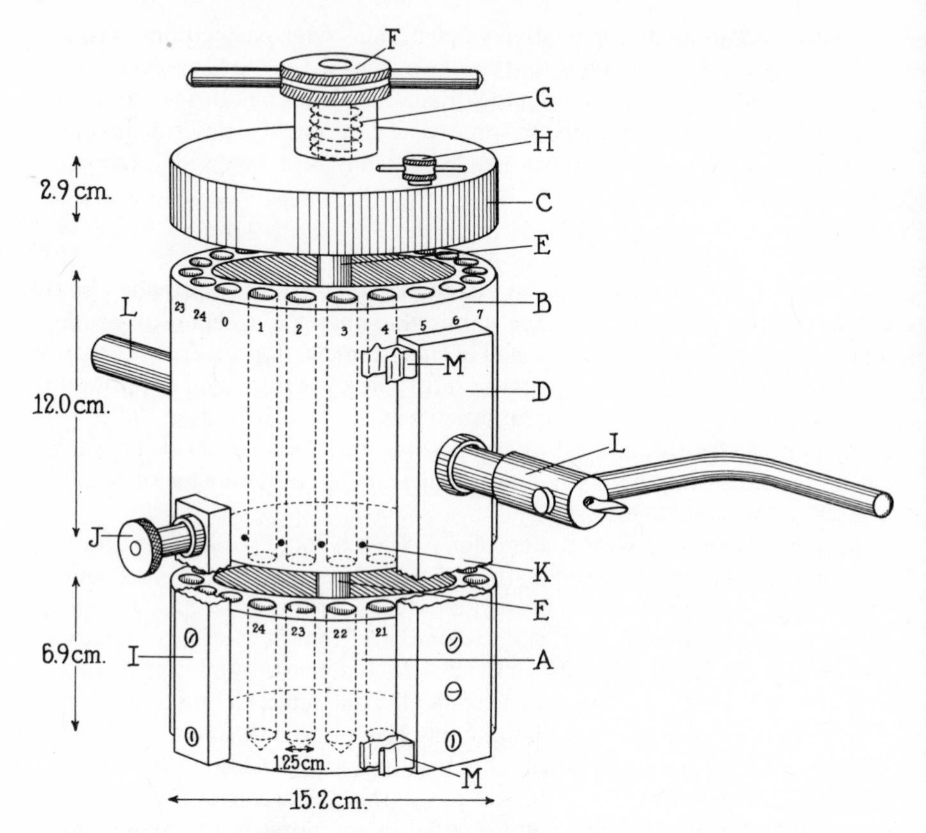

Fig. 3.34. A 25-tube countercurrent distribution apparatus for multiple transfers; labeled parts identified in text. Courtesy of Rockefeller University.

excess solvent can be removed from the last hole filled. If a quantity of light solvent is then added to the few tubes following the heavy phase charge and rotated through a complete circuit behind the heavy solvent charge with mixing at each transfer, the heavy phases in the tubes will be preequilibrated with the light phase and the excess light phase can be removed with excess heavy phase. The light solvent can also be saturated by shaking with an excess of heavy phase. After settling, a measured volume of preequilibrated light phase is added to all the tubes to give the desired solvent ratio. The geometry of this unit requires that the ratio of light to heavy solvent be less than 2, and this ratio may be increased by adding an appropriate volume of glass beads to the bottom chamber. The feed to be analyzed is introduced into one tube and distributed between the solvent phases by slowly rotating the unit with the crank about L. Then after allowing to stand in the upright position for a predetermined time, all the light phases are transferred to the adjacent stage by rotating the upper section B to the next position. The unit is again rotated about L to distribute the solutes between phases in the first two stages, and the procedure is repeated until one complete revolution of the upper section has been achieved. Solute analyses in each of the stages then give the 25-transfer countercurrent distribution pattern for the initial charge, and this can be interpreted by techniques previously discussed.

The 54-tube unit in Fig. 3.35 was fabricated with separate tubes [39] because the use of a solid 12-in. cylinder was impractical. It was also apparent that larger numbers of transfers would require a different design. The stainless steel unit also restricted the solutions that could be handled.

Fig. 3.35. Countercurrent distribution apparatus containing 54 tubes; labeled parts identified in text. Courtesy of Rockefeller University.

Craig [39] also devised the glass tube design illustrated in Fig. 3.36. Mixing of the stages is achieved by rocking the liquids in chamber A through a total angle of about 70°. When equilibrium is attained between the two phases, the tube is rotated 90° so that all the light phase drains into chamber C through tube B. Upon returning the tube to the original rocking position, the liquid in this chamber passes through tube D to the next stage. These tubes are located next to each other, normally in 10-tube units that can be manifolded into longer sections. They can also be placed on multiple tiers to reduce the overall length. Similar designs are used in all the countercurrent distribution machines currently available. These units can be provided with robot operators, permitting all mixing and transfer operations to be carried on automatically. Solvent feed and solution withdrawals may also be programmed automatically.

The glass tubes allow use of a broader range of solvents, but the linear arrangement eliminates one very interesting use of the original design which

Fig. 3.36. Glass countercurrent distribution machine. Courtesy of Rockefeller University.

was not fully exploited. The calculations in Fig. 3.24 show that only 150 stages of a 1000-stage pattern contain measurable solute when the components in the feed have distribution coefficients differing by 20 %. Consequently, with a 150-tube circular unit, the upper section could be rotated continuously until 1000 transfers were obtained and the concentration pattern would be essentially as shown in Fig. 3.24*c*. Since countercurrent distribution found its first important applications in verifying the purity of different products, the initial charge could be resolved by continual rotation until the amount of solute on one side of the peak overlapped that on the other side in the circular arrangement. The original 54-tube unit, which was the largest of this type fabricated, could be rotated nearly 10 times before the sides of the peak met around the circular arrangement of tubes if the initial charge were pure. This is equivalent to more than 500 transfers, and if the peak failed to reveal significant horizontal distortion, one might conclude that the distribution coefficients of impurities, if such were actually present, were probably within 10 % of the distribution coefficient of the pure compound. This continuous circular resolution is most effective if the extraction factor is close to unity, and the usual practice in the study of ionizing solutes is to buffer the aqueous solution to the proper pH to give such distribution coefficients.

The geometry of all the standard tubes for the Craig apparatus limits the ratio of light to heavy solvent to less than 2.0. Entrainment of heavy phase in the light phase will also create a sufficient distortion of the distribution pattern to render the results inconclusive if the light solvent is less than 0.2 times the heavy solvent. It is thus essential that the distribution coefficient for the compound under investigation be within the range of 5 to 0.5. In the case of nonionizing solvents this generally calls for preliminary investigation of different solvent pairs and use of a mixed solvent for at least one of the phases.

Nonideality also contributes to the distortion of the distribution pattern and initial charge quantities are normally kept to the lowest amounts that will provide meaningful analytical data on the final solutions. The use of radioactive tracers for analyses will permit operation at the lowest possible concentrations, thus providing the most reliable data for interpretation by the countercurrent distribution technique.

At present, countercurrent distribution equipment based on the Craig tube design is available in standard capacities from 1 to 3 ml size to the 200 to 300 ml size in cell units from 10 to a maximum of 240. They can be supplied for manual, semiautomatic, and automatic operation by Spectrum Medical Industries, Inc. Figure 3.37 shows the 1000-cell automatic countercurrent distribution instrument used by Craig at Rockefeller University. Automatic fraction collectors are also available for these units. E-C Apparatus Corporation supplies similar designs utilizing a modification of the original Craig tube.

Fig. 3.37. The 1000-tube countercurrent distribution instrument used by Craig at Rockefeller University.

Simulation of Continuous Processes

The continuous multistage fractional liquid extraction illustrated in Fig. 3.26 can be simulated by a stagewise extract pattern with double withdrawal if fresh feed is added to every feed stage, as in Fig. 3.38. The concentrations of the different components of the feed then accumulate at the feed stage to a peak, at which the withdrawal streams remove from the system the total amount of each component in the feed. Stene [40] recognized that the amount at any product withdrawal cycle of the continual feed system appearing in Fig. 3.38 is the sum of all the terms in the single feed relationship above the given cycle. The sum of all the terms of the series converges to the quantity in each stage at steady state.

Scheibel [41] noted that these terms approached a geometric series. The sum of such a series must equal the steady stage quantities obtained from (3.49), and on this basis he derived a simple relationship for estimating the

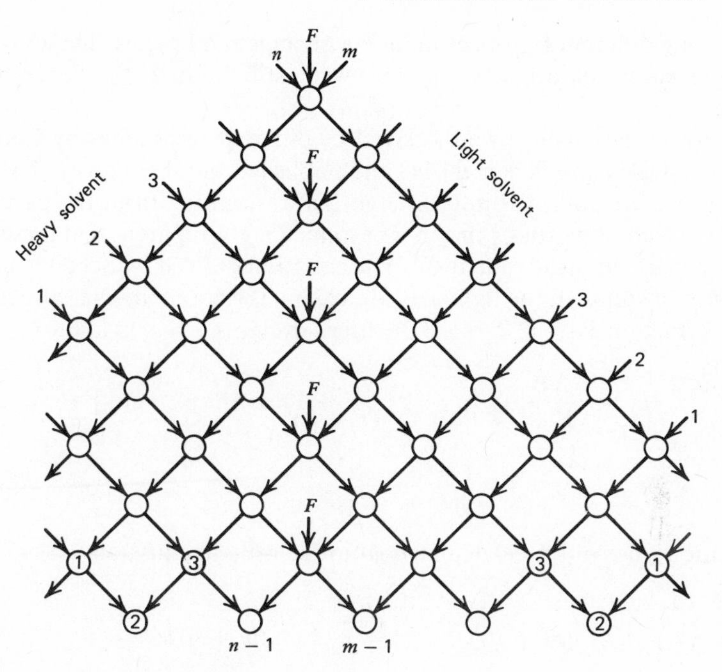

Fig. 3.38. Development of steady-state conditions by continual feed in the batchwise technique.

deviation of any given product cycle from steady state conditions. In a symmetrical extraction system (such that $n = m$), the fractional deviation of any product cycle t from the steady state condition is designated δ and is given as

$$\delta = \left[\frac{4E(1 - 1/2^{n-1})}{(E + 1)^2}\right]^t \qquad (3.66)$$

Curves can be calculated to give the number of cycles required for any given system to reach a specified approach to steady state. This simple relationship showed reasonably good agreement up to about 15 stages, but the discrepancies from the rigorous calculations increased as the extraction factor approached unity and as the number of cycles increased to attain a closer approach to steady state conditions.

The discrepancies resulted because the first few terms of the exact series differed appreciably from the geometric series. Since these comprised a major portion of sum, the limiting ratio of the terms in the actual series was significantly different from the ratio derived in (3.65), which is based on reaching steady state at a constant ratio. Peppard and Peppard [42] developed two sets of empirical equations to represent the deviation from steady state, each set covering different ranges of numbers of stages and cycles. These covered a greater range of applicability but were still limited by the empirical approach.

Rigorous relationships were derived by different techniques by Compere and Ryland [43] and Scheibel [44] and published simultaneously. The individual terms are collected into different series resulting from the particular derivation technique, and it may not be immediately apparent that these solutions are mathematically identical. The equation derived by Scheibel for the fraction of a solute in the light solvent leaving the countercurrent fractional liquid extraction system in Fig. 3.38 after t cycles is given as follows:

$$P_t = \left(\frac{E}{E+1}\right)^n \left\{ \varepsilon_1 + \varepsilon_2 \frac{E}{(E+1)^2} + \varepsilon_3 \left(\frac{E}{(E+1)^2}\right)^2 + \cdots + \varepsilon_t \left[\frac{E}{(E+1)^2}\right]^{t-1} \right\} \tag{3.67}$$

where the values of ε are calculated from the general equation

$$\varepsilon_t = \sum_{i=0}^{\infty} \binom{n+2t-3}{t-1-i(m+n)} - \sum_{i=0}^{\infty} \binom{n+2t-3}{t-2-i(m+n)} \\ \sum_{i=0}^{\infty} \binom{n+2t-3}{t-m-1-i(m+n)} + \sum_{i=0}^{\infty} \binom{n+2t-3}{t-m-2-i(m+n)} \tag{3.68}$$

in which the terms inside the summation designate the binomial coefficients such that

$$\binom{n}{r} = \frac{n!}{r!(n-r)!} \quad \text{and} \quad \binom{n}{o} = 1$$

and for values of r greater than n and less than zero, the terms do not exist. This requirement thus limits the summations to relatively few terms for the smaller values of t. Larger values of t approach so closely to the steady state conditions that it would not be possible to measure the deviation experimentally. The fractional approach to steady state is obtained by dividing the calculated value of P_t by the steady state quantities derived from (3.50) and (3.51), noting that the fraction of a feed component in the light solvent product stream is $R/(R+1)$ and the fraction of the component in the heavy solvent product stream is $R'/(R'+1)$.

This equation can best be applied through the computer calculation of curves similar to Fig. 3.39 to determine the numbers of cycles necessary to achieve different fractions of steady state. Appreciation of the large numbers of cycles required to approach within 10% of steady stage conditions at large numbers of stages is important in the interpretation of experimental data on

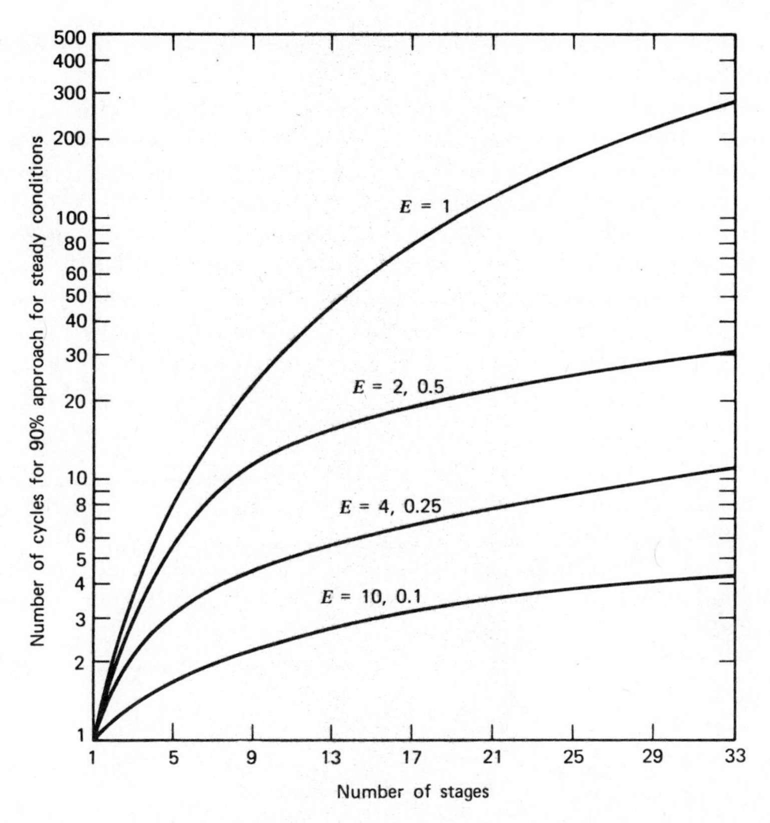

Fig. 3.39. Cycles required for given approach to steady state for different extraction factors and numbers of stages with uniform continual feed.

continuous multistage columns. The equations have been derived for stagewise operation, but if one considers operating a 30-stage column on a relatively easy separation with a selectivity of 2, so that $E_1 = \sqrt{2}$ at the optimum solvent ratio, about 100 turnovers of the complete contents of each stage are required. This is equivalent to passing sufficient solvents through the 30-stage column to replace the contents 3.3 times. Since the changes actually occur by dilution rather than by discrete steps, and the ratio of the solvent phases in the liquid contents of the column will not necessarily be in the same ratio as their feed rates, however, it becomes apparent that the total solvents passed through a 30-stage column must be of the order of 10 times the contents of the column to approach within 10 % of steady state. And twice this number will be required to approach to within 1 % of steady state. Calculations of this type will give an indication of the time that must be allowed after startup before one can expect meaningful data to evaluate the performance of continuous equipment.

Because of the slow approach to steady state conditions with a unit feed quantity in all the feed stages in Fig. 3.38, Scheibel has proposed adding sufficient feed to each of the feed stages to maintain the steady state concentrations at each feed cycle [45]. Rigorous calculation of the additional feed required over the steady state quantity at each feed cycle is complex, but Scheibel noted that this quantity also approached a geometric series at increasing cycles. The ratio is approximately $2E/(E + 1)^2$.

The additional amount of each component required in the first feed cycle over the steady state amount may be calculated as follows:

$$2\frac{E^n - E}{E - 1}F_s \tag{3.69}$$

where F_s is the amount of the given component in the steady state feed and applying the previous ratio to this term will give feed rates about 10% above that necessary to maintain steady state conditions in feed stages at the second cycle, and subsequent cycles will approach closer to the theoretical concentrations.

The amount of additional feed calculated to maintain steady state concentrations in the second feed stage can be derived as

$$2\frac{E^n - E^2}{(E - 1)(E + 1)} \tag{3.70}$$

and when this value is diminished by the ratio of $E/(E + 1)^2$ in successive feed stages, the total amount of this solute will not exceed the steady state condition at any succeeding cycles. Since the numerical values of all the previous terms are identical for values of E and $1/E$, operation at the optimum solvent ratio for a symmetrical extraction pattern ($n = m$) will give the same ratio for both components and will thus apply to the total amount of a binary feed mixture.

By adopting this technique of adding excess feed quantities at the different feed cycles, the steady state concentrations can be approximated within 10% after a number of product cycles equal to the number of stages. The approach will be closer for larger values of E with this number of feed cycles. For extraction factors approaching unity, the number of feed cycles may be reduced to about one-tenth those required with a uniform feed quantity. This empirical equation errs on the low side; thus an analysis of the feed stage concentration after a number of feed cycles will enable one to increase the excess feed quantity to bring the stages closer to steady state in a given number of feed cycles.

All the previous theoretical derivations are based on a constant distribution coefficient. This condition may be approached in analytical applications by maintaining dilute concentrations. When the objective is the preparation of samples of pure products or the simulation of process conditions for the design of a separation plant, solute concentrations must be as high as possible. Under such conditions the distribution coefficient varies with concentration, and a sound theoretical treatment calls for a knowledge of the variation to permit the making of iterative computer calculations on the extraction pattern to determine the best procedure for varying the feed quantity. Lacking such detailed information, the most practical approach would be the use of the previous empirical relationships with periodic monitoring of the feed stage composition for readjustment of the excess feed rate. In this way the maximum yield of pure product will be obtained, and the closest approach to steady state can be attained in a minimum number of feed cycles.

Operation at maximum solute concentrations also precludes the use of any device in which the phase separation is effected based on a constant volume of one phase as in the Craig apparatus. Individual phase volumes frequently vary by more than 50 % over the different stages in a separation at high solute concentrations. This poses no problem in a multistage unit that directs the solvent flows based on the density difference. In batchwise simulation of continuous operation, each phase separation requires individual attention. These manual operations may be conveniently carried out in a funnel rack described by Scheibel [46]. The unit represented in Fig. 3.40 consists of a two-tier rack, holding 10 separatory funnels on one row and 11 on the other; thus a countercurrent process of up to 21 theoretical stages can be studied. Funnels must be removed and shaken manually. After phase disengagement, the members of the row of funnels containing the solutions are moved to the upper position and all are tilted to drain into the lower tier. When drained to the interface, they are all tilted in the other direction and the light phase is drained into the adjacent funnels in the lower level. These funnels are then shaken, and the operations are repeated.

When steady state conditions have been approached in this unit, it is possible to sample both light and heavy phases to evaluate the distribution coefficients over the range of concentrations expected in the commercial process. This device requires far more attention than the Craig apparatus, but it has no limitations on solvent ratio or solute concentration. There is generally some increase in the liquid volume in the feed stage, and when the 1-liter capacity funnels are used, the total volume of the feed and the two solvents used in each cycle must be less than 800 ml. When the total volume or the volume of either phase in the feed funnel is found to increase excessively, the feed quantity can be reduced in the subsequent cycles to prevent the entire mixture from becoming homogeneous in this stage. In a mixer-settler or a

Fig. 3.40. Funnel rack for countercurrent extraction studies of up to 21 stages. Courtesy of Suntech, Inc.

multistage column, the operator cannot recognize this effect until the center section of the unit has become homogeneous. It is then necessary to repeat the run at a lower feed-to-solvent ratio. By observing the variation in the interface level and the total level in the different funnels, the operator can recognize immediately the development of an abnormal situation and can take corrective action before the interface in the feed stage disappears.

A funnel rack of this type can be used to obtain data on liquid extraction processes and, in some cases, to prepare small samples of purified product. Analysis of the different phases provides design data at high solute concentrations in the region of the feed stage. If the 21 stages are not adequate for the desired product purity, solute concentrations at the ends of the extraction system will be low enough to allow extrapolation of distribution coefficients to zero concentration. Thus the data can be reliably extrapolated to higher purities for the design of the commercial process.

Preparation of Small-Scale Samples

The Craig countercurrent distribution apparatus is suitable for the isolation of microgram quantities of purified products. The use of separatory funnels in a double-withdrawal pattern can provide gram-size samples of purified products from a fractional liquid extraction operation. Funnels can also provide liter volumes of solutions when applied in a solvent extraction process for removal. The funnel rack design previously described makes extractions of both these types more convenient, but the most practical approach to the preparation of small samples is through the operation of a continuous multistage extractor. Equipment for such a process can provide any given number of extraction stages and requires only monitoring the solvent flow rates and, in fractionation processes, the solute feed rate as well.

Most development scientists have different concepts of the boundary between laboratory and pilot plant production. Equipment manufacturers normally refer to their smallest units as "laboratory models" even though capacities may be an order of magnitude larger than the capacities of other so-called laboratory units. Rather than attempt to define the nebulous line that separates laboratory and pilot plant equipment, the following discussion of the different types of extraction equipment includes the capacity ratings, allowing the reader to decide for himself whether they agree with his personal concepts of what constitutes laboratory equipment. Terminology is that of the equipment manufacturer, to facilitate identification of the items.

Some items are described for their historical value and impact on future developments, but prime emphasis is on equipment that is now or has been in the past commercially available as standard items.

Mixer-Settler Devices

The first continuous multistage extraction units consisted of alternate mixer-settler combinations (Fig. 3.41). The similarity to the batchwise operation in discrete stage is obvious. However a theoretical stage is achieved only if the mixer is operated at sufficient speed to obtain substantial equilibrium between the phases and the settling unit provides enough time for complete phase separation. A greater degree of agitation in the mixer requires a larger settler, and in all cases the operation of these units must be a compromise between low agitation with complete phase separating but incomplete equilibrium, and excessive agitation resulting in entrainment in the phases leaving the settler. Both cases give less than a theoretical stage for each mixer-settler combination, but by the proper selection of settler size and agitator speed, each unit can be operated to provide very nearly one equilibrium stage.

Mixer-settler units are generally assembled for specific applications. One type is the mixer-settler unit of Coplan, Davidson, and Zebroski [47, 48],

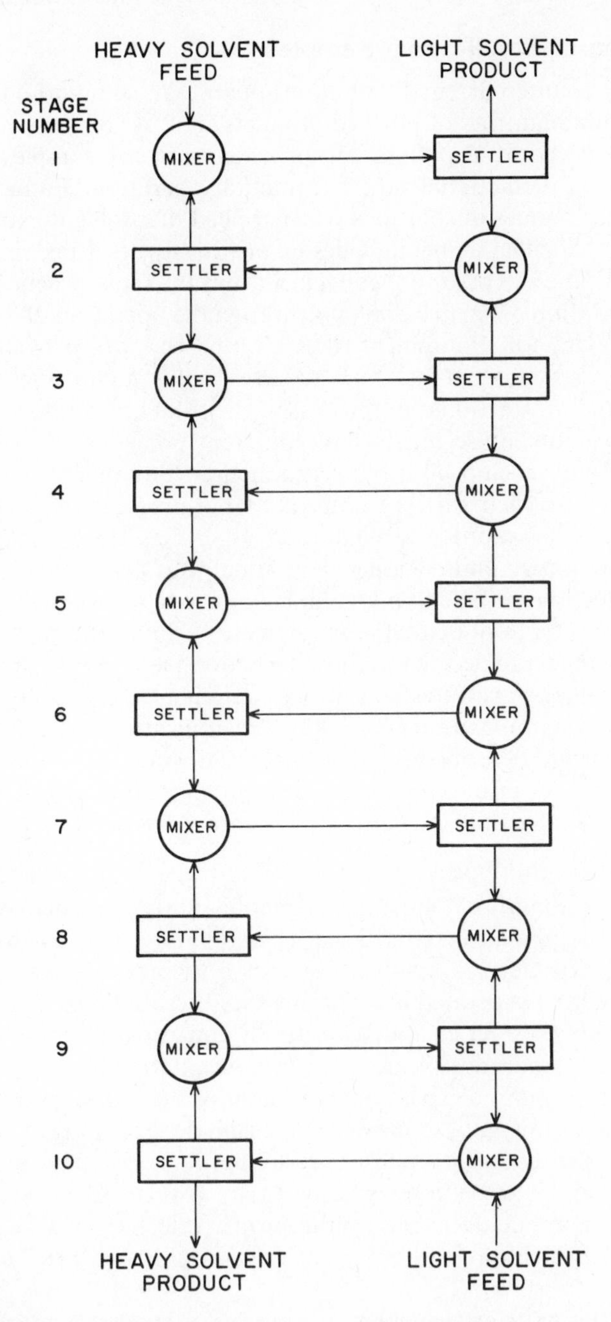

Fig. 3.41. Schematic diagram of 10-stage mixer-settler extraction unit.

which utilizes a pump impeller to draw up heavy phase from the bottom of a mixing chamber and allows the mixture to pass between baffles to the settling chamber where ports at different levels allow the appropriate phases to flow to the proper adjacent mixing chambers. In another type, developed by Fenske and Long [49, 50], agitation is obtained from a pulsating perforated disc with all mixing stages stacked vertically and agitated with a common shaft. Adjacent settling chambers with appropriate traps allow the heavy phase to pass downward to the mixing stage and the light phase to pass to the mixing stage above. At the proper feed rate and pulsation rate of the agitator, this design gave stage efficiencies close to 100 %

Treybal [51, 52] also described a vertical arrangement of mixer-settlers in which the residence times in the mixing stages could be adjusted. Figure 3.42 presents one adaptation of his design available from Chemical Process Products Division, Norton Company, Akron, Ohio. Mixing is effected by

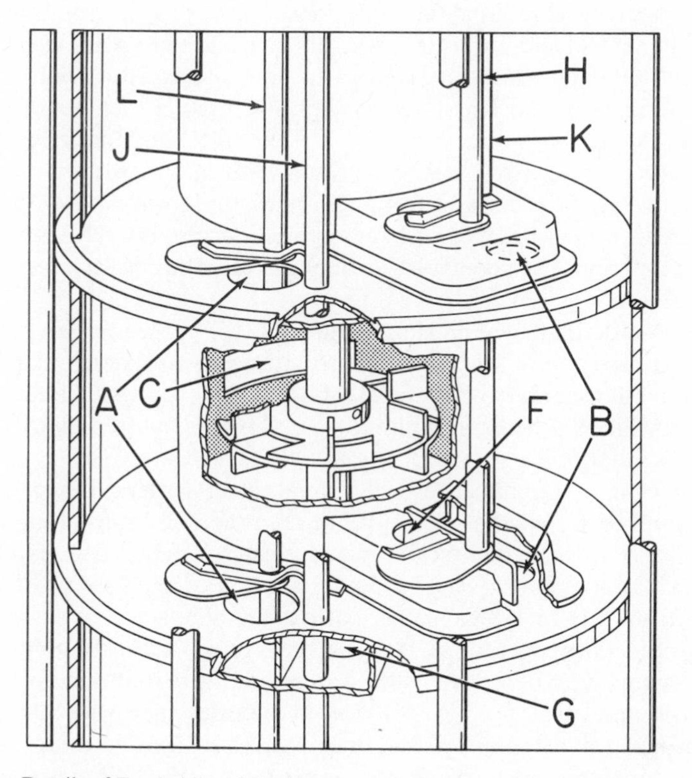

Fig. 3.42. Details of Treybal liquid-liquid contactor. Courtesy of Chemical Process Products Division, Norton Company.

flat-bladed turbine-type agitators mounted on a central shaft in a cylindrical mixing section within a larger column. The wall of this section contains exit ports for the mixture and settling occurs in the annular space. The stages are separated with horizontal baffles having ports A and B which, respectively, permit the heavy phase from a given stage to mix with the light phase recycle from the settling zone to the mixer of the next lower stage and the light phase to mix with the heavy phase recycle of the next upper stage. These flows are regulated by externally rotating rods L and K to vary the opening of ports A and B. The residence times of the heavy and light phases in the mixing section are also regulated externally through rods H and J which vary the opening of the recycle ports F and G respectively into the recycle ducts. In Figure 3-42 only control discs on rods L and H are shown, those on rods J and K may be assumed to be out of sight below the ports that they control. It is possible to externally adjust the residence times to maximize the stage efficiency for any given set of operating conditions. Stage efficiencies of 80% have been attained on difficult extraction systems.

Solvent systems exhibiting high interfacial tension or processes requiring widely different solvent flow rates generally give low efficiencies in extraction equipment relative to solvent systems with low interfacial tension and very nearly equal flow rates. The latter systems give high efficiencies in agitated columns and may even show reasonable efficiency in a packed or spray column. Acetic acid extraction from water with methyl isobutyl ketone is typical of the latter system, and acetic acid extraction from water with toluene or *o*-xylene is one of the more difficult extraction systems. The effect of the type of system must be taken into account in evaluating the relative efficiencies of different equipment designs.

A mixer-settler patented by Hazen and Cline [53] is fabricated by Denver Equipment Division of Joy Manufacturing Company. These units may be installed in a horizontal arrangement, and Fig. 3.43. shows their small two-stage unit, which has a mixing volume of 2 gallons and a throughput capacity of about 1 gallon/min.

More recently, Hazen has devised a small mixer-settler arrangement that can be supplied for smaller capacities in the range of 2 ml/minute up to 1 gallon/min. A 10-stage laboratory unit fabricated by Hazen Research, Inc., Golden, Colorado, appears in Fig. 3.44. Larger pilot plant sizes with mixing volumes from 1.18 to 128 gallons are also available.

Many other concepts for providing adjacent mixer-settler combinations in single columns were described in the patent literature, but none progressed beyond the initial prototype design. Most of these designs do not offer control of the residence times in the mixing zones; thus the stage efficiencies are functions of throughput and mixer speed. Unlike mixer-settler combinations that can provide essentially one theoretical stage over a wide range of operating

Fig. 4.43. Two-stage laboratory solvent extraction unit. Courtesy of Denver Equipment Division, Joy Manufacturing Company.

Fig. 3.44. Ten-stage laboratory mixer-settler unit. Courtesy of Hazen Research, Inc.

conditions and solvent properties, these units must be tested to evaluate the fraction of a theoretical stage obtained in each actual stage, and the efficiency may vary as the solvent properties change over the range of concentrations in the extraction process.

In 1948 Scheibel [54, 55] described a vertical column with alternate mixing sections and wire mesh packed sections in which the mixing was performed with flat-bladed turbine-type agitators on a common shaft (Fig. 3.45). A 1-in. column has a capacity of the order of 1 liter of total solvents per hour. The capacity of this column depends on the properties of the solvents (i.e., density difference, viscosity, and interfacial tension), as well as the solvent ratio. When flow rates differ appreciably, a high agitator speed is required to disperse the smaller liquid volume sufficiently in the mixing stage to approach equilibrium between the phases. This requires a larger mesh section for complete phase disengagement, and the standard design results in a lower stage efficiency because of incomplete phase separation. Stage efficiencies can also

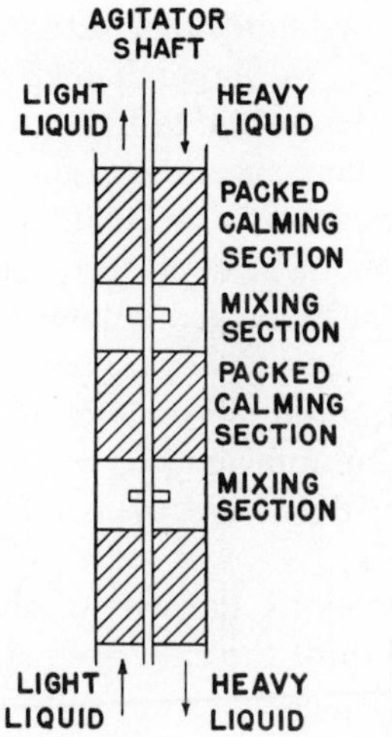

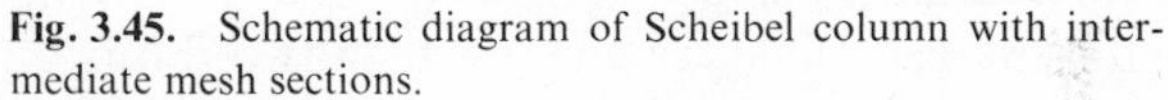

Fig. 3.45. Schematic diagram of Scheibel column with intermediate mesh sections.

be greater than 100 % because of mass transfer that occurs between some solvents during the countercurrent settling in the mesh section. This is typical of solvent pairs having low interfacial tension. Stage efficiencies in the standard design average about 80 % in difficult extraction systems, and when low efficiencies are obtained, an increase in the height of the intermediate mesh will provide phase separation at a higher degree of mixing, thus increasing the efficiency of a given stage [56]. Because of the smaller number of actual stages in a given column height, however, such modifications do not always increase the total number of theoretical stages.

Columns of this design up to 36 in. in diameter have been installed commercially. The height of the wire mesh required between stages for settling increases with column diameter and makes large columns uneconomic relative to other designs. Present applications are primarily in smaller diameters such as pilot plant and particularly laboratory columns, since the most economic designs for larger columns cannot be scaled down to 1 in. diameter.

Agitated Columns

The previous section mentioned several columns that consisted of discrete mixing and settling zones provided at the same level, as in the Fenske-Long and the Treybal designs, and at alternate levels, as in the Scheibel column, which avoids the need for interface control arrangements in the settling or calming zones. Economics of fabrication have limited the columns utilizing wire mesh calming zones for phase separation to the smaller commercial sizes.

Experimental studies have shown that the most effective use of the column height is achieved in large-diameter columns with adjacent mixing zones, eliminating the intermediate phase separation. In agitated multistage extraction columns, it is difficult to achieve the same geometry as the larger diameter mixing stages. Compromises that must be made in the design and fabrication of the small units result in substantial reductions in efficiency and capacity. This factor was not immediately recognized, and one can only speculate on different designs that were perhaps studied on too small a scale in the past and rejected because of the poor performance. Ordinary packed columns, which were used for gas-liquid contacting in distillation and absorption operations, were found to be relatively ineffective for liquid-liquid contacting. In 1941 Ney and Lochte [57] reported data on a small column agitated with a central spinning tube which indicated poorer efficiency than the same height of small Berl saddle packing.

The 1-in. diameter column with alternate mixing sections and wire mesh settling zones previously described has less than one-third the capacity per unit cross-sectional area and less than two-thirds the stage efficiency of the larger diameter columns. This was achieved only after numerous trials with mesh of different physical characteristics. It was fortunate that the first column of this design was 4 in. in diameter and gave more than 100 % stage efficiency because without this knowledge the extraction studies on the smaller column could have been abandoned in the early stages. The approach adopted by Scheibel for the development of new designs of agitated extraction columns consists of constructing the first prototype 12 in. in diameter because this is the smallest size that will give reasonable scale-up data. Only after the design parameters of such a column have been established can one proceed with confidence to determine the smallest diameter column that will give reasonable efficiencies and capacities for laboratory investigations of liquid extraction processes. Furthermore, scale-up factors for the small columns must be derived empirically from previous experiences on similar extraction systems.

In 1952 Oldshue and Rushton [58] reported stage efficiencies for the acetic acid–methyl isobutyl ketone–water system up to 86 % on a 6-in. column in which the agitators were separated by a horizontal annular baffle and the mixing stages also contained vertical baffles. Figure 3.46 is a schematic diagram of this design, and Mixing Equipment Company, Inc. (Mixco), Rochester, New York, has supplied columns up to 10 ft in diameter for commercial applications. The 6-in. unit in Fig. 3.47 is the smallest available and maximum capacity on the previous system was about 100 gallons/hour, but maximum efficiency was obtained at about half this throughput.

A schematic diagram of the rotating disc contactor (RDC) developed by Reman [59, 60] appears in Fig. 3.48. This unit depends on the shearing action of rotating discs to promote mixing of the countercurrently flowing phases. Stator rings are located on the column wall midway between the rotating

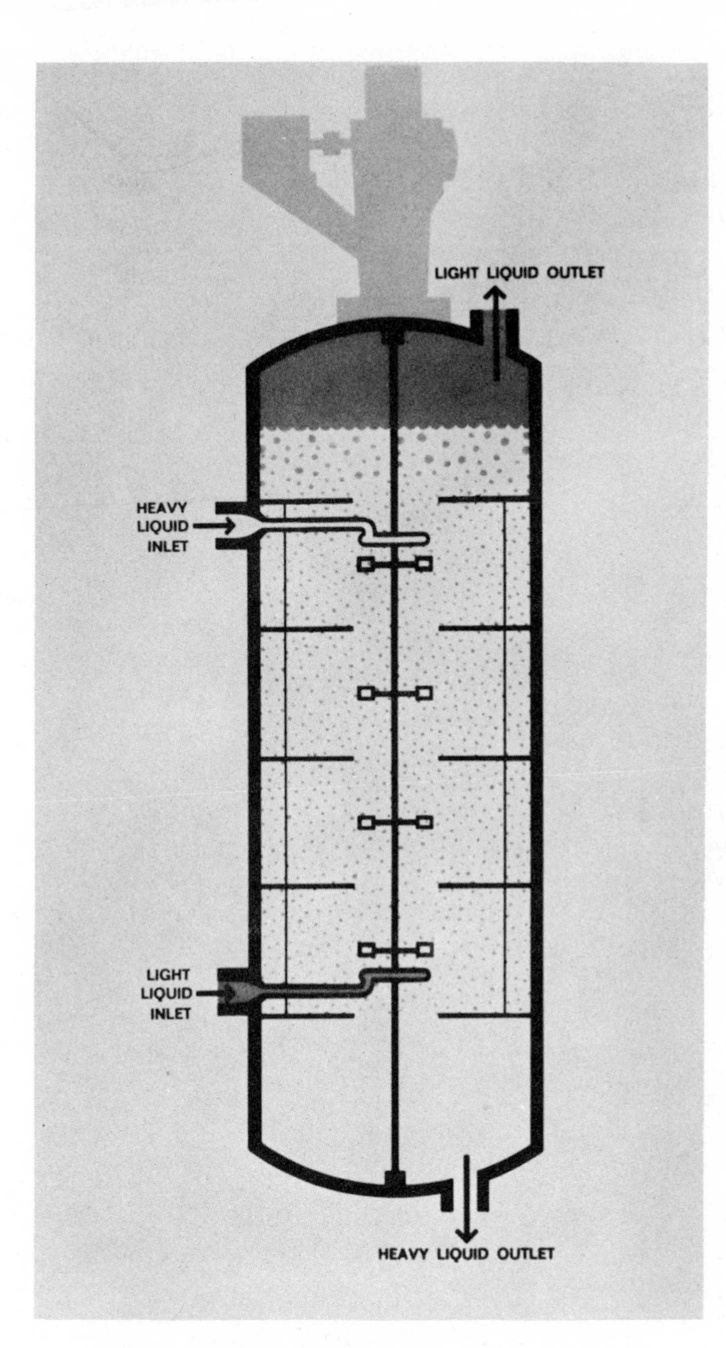

Fig. 3.46. Schematic diagram of Mixco extraction column.

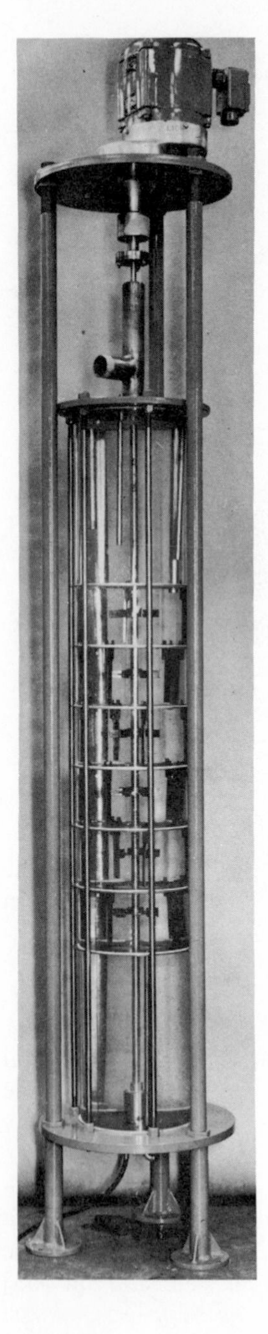

Fig. 3.47. A 6-in. Mixco column containing five-stages.

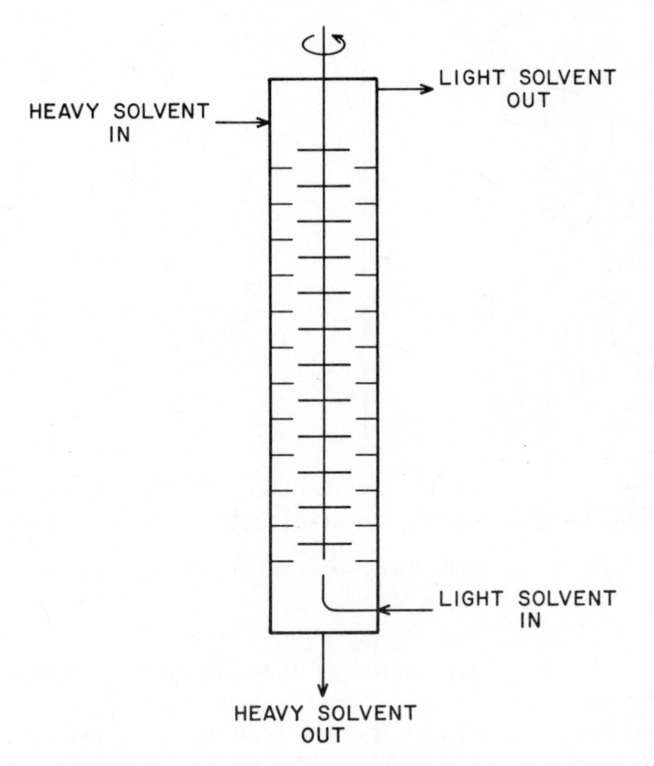

Fig. 3.48. Schematic diagram of rotating disc contactor.

discs. The flow pattern in the stage derives from the frictional drag of the rotating disc, and the vertical baffling required in the Oldshue-Rushton column is not used. Stage efficiencies in this unit are less than in the previous designs and range from 15 % on difficult extraction systems up to 40 % on easy extraction systems such as acetic acid–methyl isobutyl ketone–water. Subsequently in 1956 Reman [61] patented a modification in which annular baffles were installed above and below the annular disc to promote horizontal flow. Since the outer diameters of these discs were smaller than the inside diameter of the annular baffles on the column wall, the entire mixing assembly could be removed from the column.

Various other modifications of the rotating disc contactor have been proposed. Misek [62] has installed the rotating disc assembly eccentrically in a cylinderical column and provided calming zones at the opposite side of the mixing section as shown in Fig. 3.49. Ganapathi, Krishna, Raju, and Rao [63] have found increased stage efficiencies by providing radial grooves in the rotating disc which increase the circulation pattern in the column. The

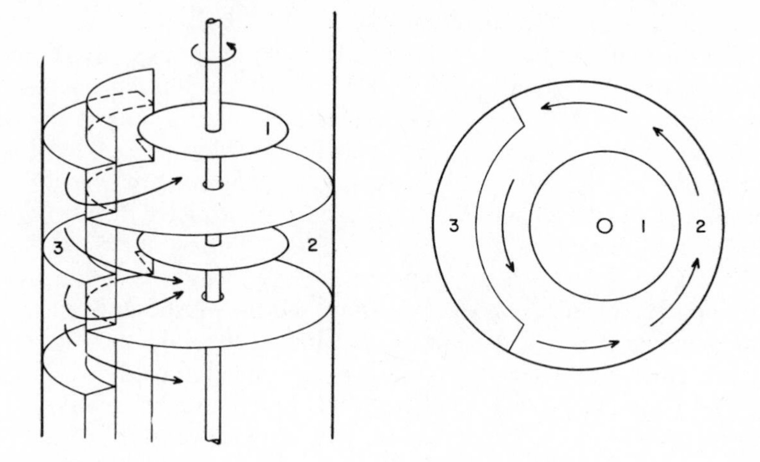

Fig. 3.49. Schematic diagram of asymmetric rotating disc extractor.

pumping action of this type of disc is intermediate between that of the flat disc and the turbine type agitator used in the Oldshue-Rushton column.

In 1956 Scheibel [64, 65] described an agitated extraction column (Fig. 3.50). It contains central horizontal baffles above and below the agitator to direct the horizontal flow of the two-phase mixture to the wall of the column and annular baffles at the wall to direct the flow back to the agitator

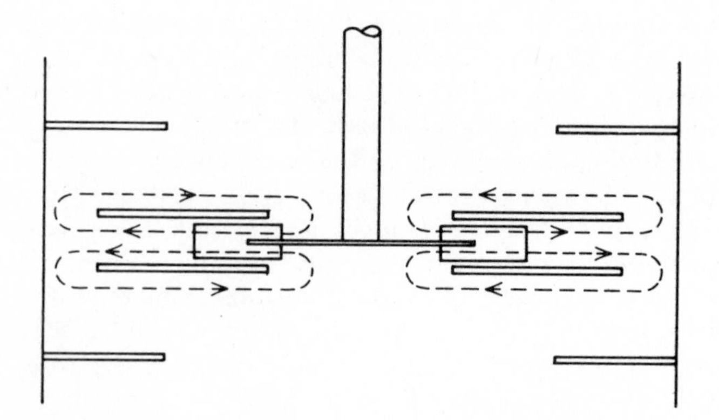

Fig. 3.50. Schematic diagram of Scheibel column with baffled mixing stages. Courtesy of Fluid Separation Design, Inc.

through the inlet of the center baffle. Partial phase separation occurs by impact at the wall so that the light phase rises to the stage above and the heavy phase drops to the stage below. The bulk of the phase mixture recycles to the agitator for additional dispersion and extraction. The initial design consisted of alternate mixing sections and wire mesh packing. The objective was to reduce the height of a mixing and the packed settling section of the original design to provide more economic scale-up to large diameter columns. The initial studies were made on a 12 in. column. The outer annular baffling was so effective that the mesh could be removed from the settling section without appreciable loss of stage efficiency, and the most effective use of the column height was achieved by complete elimination of the settling sections. This was particularly advantageous on the difficult extraction systems where relatively little extraction occurs in the countercurrent flow in the settling section. In easy extraction systems where transfer occurs in the settling zone there is very little difference in the overall performance of a column of a given height with the two designs. Columns of this type have been fabricated as small as 2 in. in diameter, but this size gave such a low efficiency and capacity that data obtained were not suitable for scale-up to larger sizes. Columns of this design are available in sizes starting at 3 in from York Process Equipment Corp., Parsippany, New Jersey, and Fluid Separation Design Inc., Fairfield, New Jersey. Figure 3.51 shows a 3-in. column containing 33 stages capable of handling up to 30 liter/hr of total solvents.

Previous agitated columns depend on rotation of a central shaft for mixing the phases. In 1959 Karr [66] presented performance data on a column that promoted mixing by reciprocating action of a series of perforated plates on a central shaft. Efficiencies of up to 30 % for each plate have been reported in 1-in. columns on the acetic acid–methyl isobutyl ketone–water systems [68]. This design has been tested in sizes up to 36 in. and laboratory columns in diameters starting at 1 in. are available from Chem-Pro Equipment Corporation, Fairfield, New Jersey. The 3-in. column containing 12 plates, used by Karr in his initial studies on this design at Hoffmann–La Roche, Inc., Nutley, New Jersey, is illustrated in Fig. 3.52.

Many other studies have been reported on the effect of pulsing perforated trays to contact two liquid phases. Recently Wellek et al. [69] have reported performance data on a column in which wire mesh packing, similar to that used for the settling section of the first Scheibel column, was pulsed in the column to promote mixing. The work was based on a column designed by Carr [70].

Pulse Columns

The concept of pulsing the contents of a column to obtain mixing was first suggested by Van Dijck in 1935 [71]. Early attempts at pulsing the liquids in

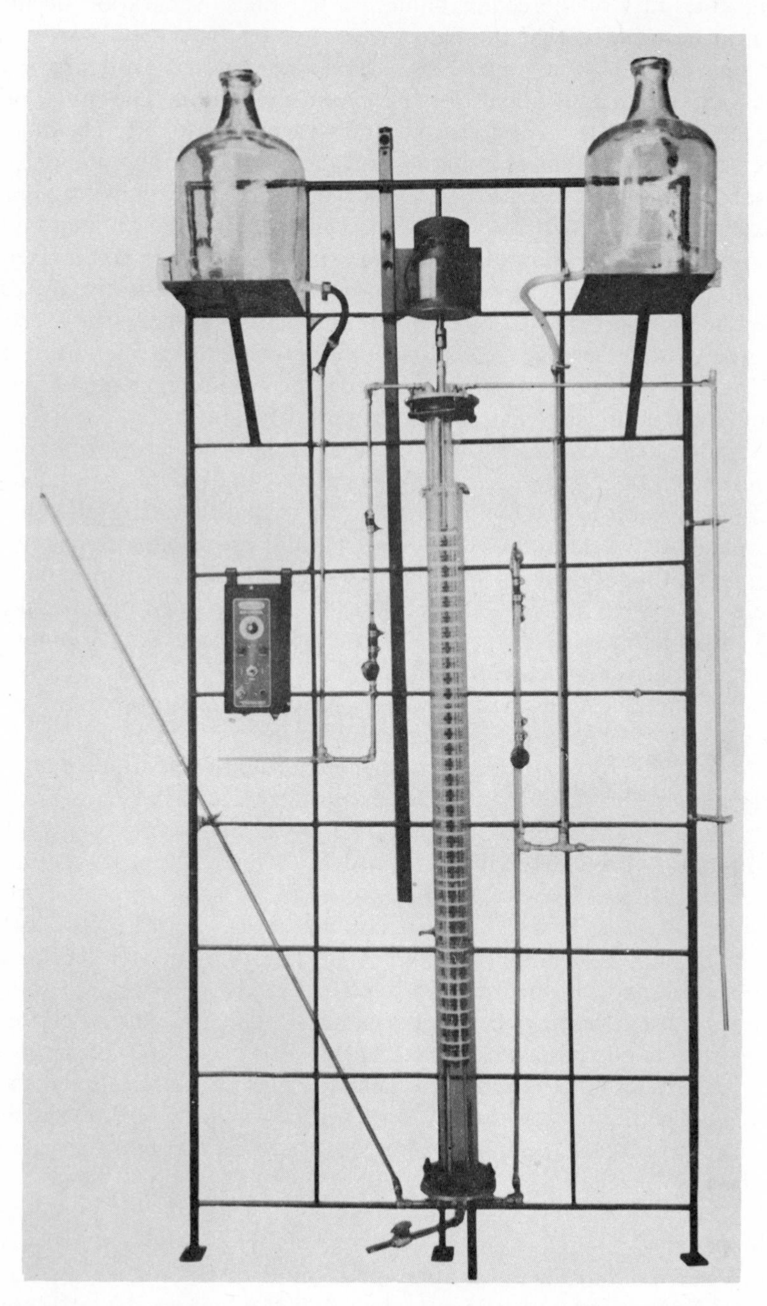

Fig. 3.51. A 33-stage baffled column. Courtesy of Fluid Separation Design, Inc.

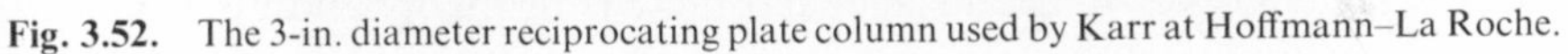

Fig. 3.52. The 3-in. diameter reciprocating plate column used by Karr at Hoffmann–La Roche.

packed columns did not yield significant improvement in the performance [72, 73]. Pulsing of ordinary perforated tray columns with downcomers shows very little increase in tray efficiency [74, 75]. Under normal gravity flow these columns have tray efficiencies in the range of 4 to 50 % depending on hole size, column diameter, and tray spacing, as well as the type of extraction system involved. A system with high interfacial tension requires larger holes with a corresponding decrease in tray efficiency. Best efficiencies are obtained in pulse columns using sieve trays—perforated trays sealed to the walls without separate passages for the countercurrent flow of the continuous phase in the column, containing holes so small that interfacial and surface tensions prevent gravity flow.

Figure 3.53 illustrates the effect of pulsing the contents of this column.

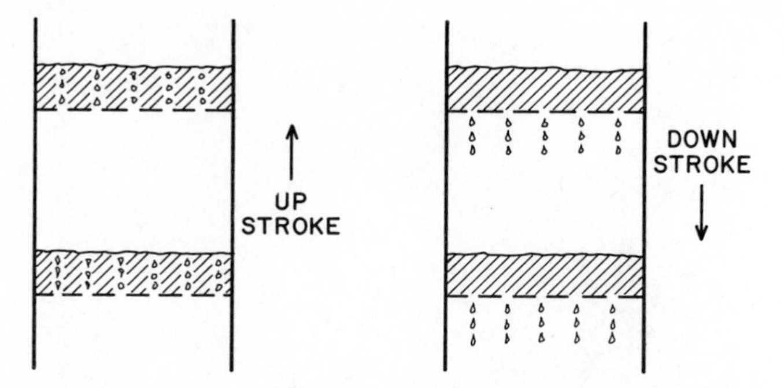

Fig. 3.53. Schematic representation of pulse column behavior.

In the upward portion of the pulse the light phase is dispersed in the heavy phase as it is expelled from the holes. On the reverse stroke the heavy phase is dispersed in the light phase. When pulsations are carried out with a sine curve variation, there is an optimum frequency and amplitude that allows the maximum amount of dispersed phase to settle away from the nozzle, preventing this phase from being trapped in the reverse flow. A small amplitude and excessive frequency could return most of the droplet phase back to the point of origin with very little net flow in the system. In these columns the total solvent flow is proportional to the total volumetric displacement of the pulse, and entrainment of the droplets in the reverse flow will reduce the proportionality constant to a small fraction.

Small holes result in smaller droplets, which provide more surface for mass transfer but are also more easily entrained in the reverse flow. Back-mixing reduces the efficiency of each stage, and there is probably an optimum hole size for every solvent system depending on density difference and interfacial tension. For every hole size the optimum pulse frequency and amplitude also depends on the properties of the solvent system, and the variables have not been fully explored. Under optimum pulsing conditions, however, stage efficiencies above 50% per perforated tray have been achieved. More studies are needed to evaluate the design parameters for different solvent systems.

In 1956 M. R. Cannon [76] proposed a controlled cycling technique to overcome some of these problems. Instead of the sine curve pulse imparted to the column contents, he proposed a short-period, high-pressure pulse to atomize the liquid as it was ejected through the holes of the perforated tray. He also provided a double-cone screen for impacting the small droplets to promote coalescence (Fig. 3.54). Most efficient mixing is obtained with a square pulse wave rather than the sine wave previously used, and this characteristic was obtained by controlling the length of time the solvent feed valve and the discharge valve for the solvent at the other end of the column were

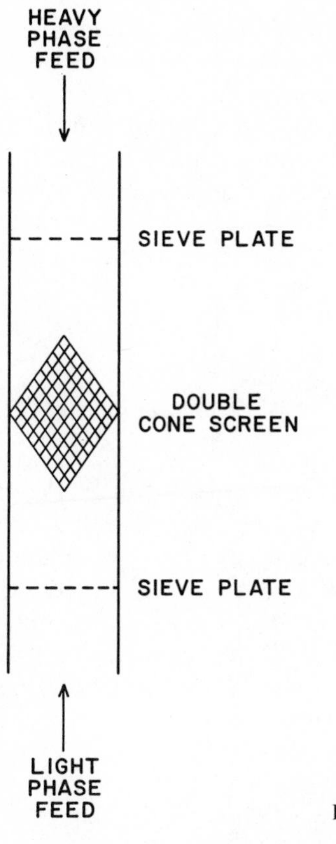

Fig. 3.54. Stage details of controlled cycle extractor.

open. The solvent ratio was controlled by regulating the lengths of time the valves for each solvent flow were open. By spacing the solvent cycles, it is possible to provide time for coalescence and phase separation in the column. Thus this design gives better phase contacting and flow rate regulation than had been obtained by previous pulsing techniques.

For this column to continue to operate properly, there must be no change in the volumes of the phases over the column. Szabo, Lloyd, Cannon, and Speaker [77] found that the mutual solubility change in the acetic acid–methyl isobutyl ketone–water system required an additional cycle in which the contents of the column were adjusted to maintain the interface level above the top stage. In this case it was necessary to withdraw heavy phase from the column and allow an equivalent volume of light phase to return. Thus controlled cycling requires five different cycles in sequence: namely, light solvent feed, phase separation, heavy solvent feed, phase separation, and product phase volume adjustment.

Figure 3.55 shows a 2-in. controlled cycle extractor formerly available from Cannon Instrument Company, College Park, Pennsylvania, in which the double-cone screen was eliminated. Stage efficiencies of this column reported on the previous system varied from 50 to 85 %, depending on solvent ratio, but were essentially independent of total throughput. Column capacities appeared to be larger than those of the agitated extraction columns on the same system.

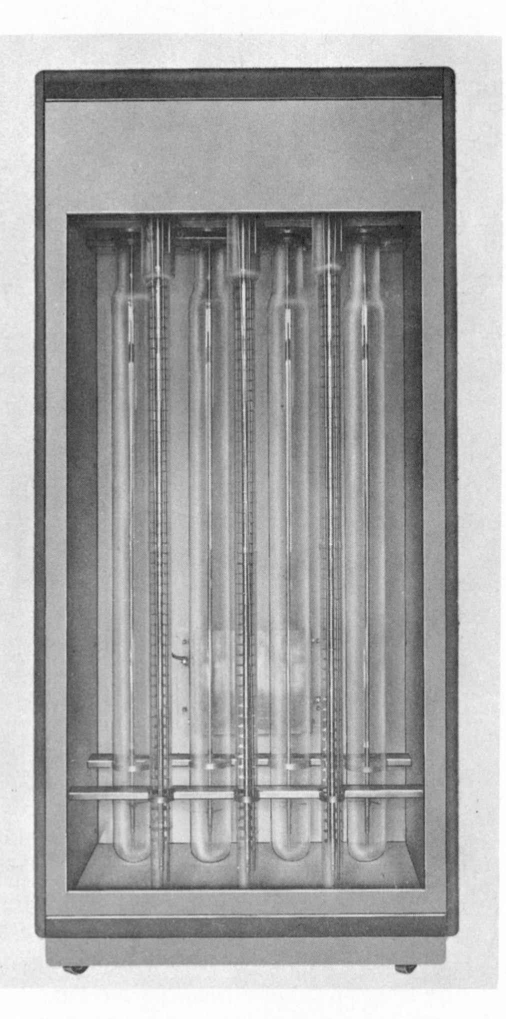

Fig. 3.55. Controlled cycle extractor. Courtesy of Cannon Instrument Company.

Centrifugal Extractors

All extractors discussed in the previous sections depend on gravity for phase separation. When density differences are small, solvent throughputs are limited. Agitation is usually reduced, ensuring that droplet sizes are large enough to provide phase separation in the settling zones, and this results in low stage efficiencies. Systems that emulsify readily also require operation at minimum agitation to avoid emulsion formation. Stable emulsions can be due to the presence of surface-active agents or to insoluble matter that collects at the interface and retards coalescence of the droplets. All these problems will be minimized in a stronger gravitational field such as obtained in a centrifuge.

Podbielniak [78] patented a centrifugal extractor in 1935 that consisted of a spiral passageway in a rotor, through which the heavy phase flowed outward along the outer wall and the light phase flowed to the center along the inner wall of the spiral. It was subsequently found that the surface renewal resulting from the passage of the heavy phase through louvers in the outer wall to the next turn of the spiral, and passage of the light phase through louvers in the opposite direction, produced more mass transfer than was obtained across the liquid interface in a single turn.

This concept did not gain acceptance until it was found to satisfy the particular requirements for the extraction of penicillin from fermentation broth. The broth had to be acidified to provide a favorable distribution coefficient for the biologically active compounds, and the free acid was unstable in the aqueous phase, requiring immediate transfer to the butyl acetate phase. Even though the broth was filtered before extraction, however, additional solids precipitated on acidification, and these tended to stabilize emulsions. The Podbielniak extractor was the ideal device to achieve rapid, efficient extraction of the penicillin acid from the acidified aqueous phase with a minimum residence time. It has also been used in the extraction of other antibiotics from fermentation broth and for subsequent purifications. The centrifugal extractor produces relatively clean liquids. The solids accumulate in the passages, eventually plugging the unit. Operating procedures include periodic shutdown and washing with caustic to dissolve solids deposited by the acidification of the broth. A centrifugal extractor capable of passing the solids was subsequently developed, but the new feature sacrificed some of the extraction efficiency for this capability.

In present Podbielniak extractors the spiral passageway has been replaced with a series of concentric cylinders with perforated walls, which thus resemble a perforated tray column under separating forces up to 10,000 g. Capacities range from 0.1 to 100 m^3/hr; Fig. 3.56 is a cutaway diagram of a typical extractor available from Baker-Perkins, Inc., Saginaw, Michigan.

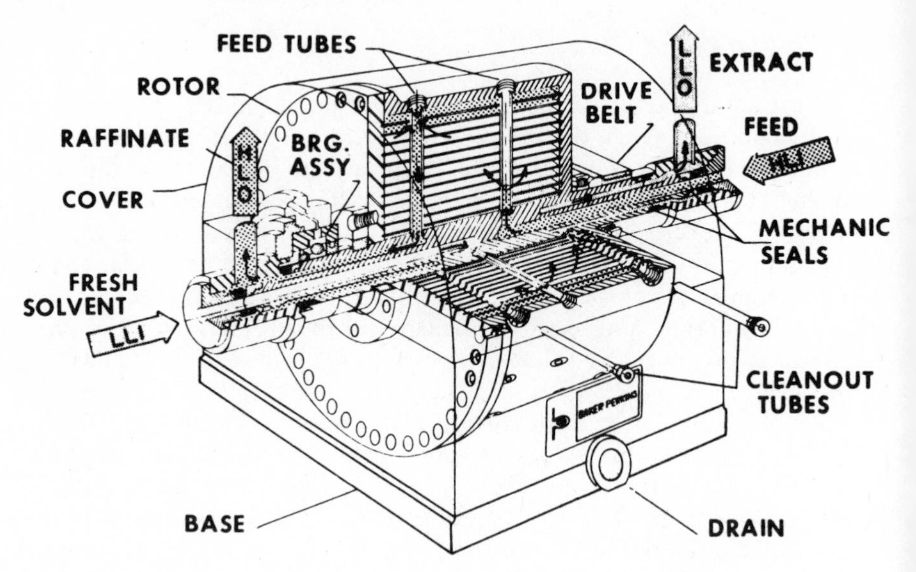

Fig. 3.56. Cutaway of Podbielniak centrifugal extractor. Courtesy of Baker–Perkins, Inc.

The light solvent enters the center of the shaft at one end and passes to the outer shell, which is analogous to the bottom of a gravity extraction column. Heavy solvent enters the other end of the shaft and passes to the inner annulus, analogous to the top of a gravity column. The solvents pass countercurrently through perforations in the walls, and the light phase exits through an annular space in the shaft at the same end as the heavy solvent feed, while the heavy phase product leaves the rotor through an annular space at the light solvent feed end of the rotor.

A small centrifugal extractor with an internal width of 25 mm and a total internal volume of 0.5 liter (Fig. 3.57) provides four or five theoretical stages at a total throughput of 1.3 liters/min on the acetic acid–methyl isobutyl ketone–water system [79]. Larger models of the Podbielniak extractor are available with capacities up to 2300 liters/min. Other centrifugal extractors with capacities in excess of 100 liters/min for large commercial applications are the Westfalia Extractor, available from Centrico, Inc., Northvale, New Jersey, and the DeLaval Extractor, fabricated by DeLaval Separator Company, Poughkeepsie, New York.

Centrifugal extractors of all sizes must be fabricated with precision, and although they may appear to be small relative to a gravity separating unit of the same capacity, they call for more maintenance, especially in the seals between the rotating shaft and the feed and product streams. And like all

Fig. 3.57. Pilot plant centrifugal extractor. Courtesy of Baker–Perkins, Inc.

large, high-speed equipment, they require more operator attention; thus their commercial applications have been primarily those requiring short residence times and separating forces greater than 1 g, such as the extraction of fermentation broths to recover biologically active compounds.

Operation of Continuous Countercurrent Extraction Equipment

Batchwise studies of liquid-liquid extraction processes involving alternate mixing and phase separation steps can serve as reference conditions for continuous extraction studies. Mixing may consist of dispersing the light phase in the heavy, or vice versa, and phase equilibrium concentrations are the same regardless of how the equilibrium is attained. However the ratio at which equilibrium is approached differs for the two conditions, and this kinetic parameter controls the stage efficiency of all continuous liquid-liquid contacting equipment.

When two liquids are mixed, one becomes the continuous phase and the other the dispersed phase. Dispersion of one phase as small droplets within larger droplets of a second phase dispersed in the original phase does not normally occur. Only if the dispersed phase is viscous enough to trap small droplets of the continuous phase for an extended period will a dispersion within the dispersed droplets be noticeable. The major portion of the transfer always occurs at the droplet interface, and stage efficiency has been found to differ depending on the direction of mass transfer as well as on the choice of the dispersed phase.

As a general rule, when two phases are shaken together or otherwise mixed with an agitator, the phase present as the larger volume becomes the dispersed phase. This also occurs when external agitation is provided. In mixer-settlers the relative volumes of the two phases are controlled by the locations of the light and heavy phase ports and cannot be changed without rebuilding the unit. On the other hand, all columns previously described may be operated by either dispersing the light phase in the heavy or the heavy phase in the light, depending on whether one controls the interface above the top stage or below the bottom stage, respectively.

When mixing is obtained by rotation of an agitator, centrifugal force causes the heavy phase to flow to the periphery and the light phase to flow to the center. This tendency exists even though vertical baffles are installed in the mixing zone to prevent rotation of the mixer contents. Thus it is always easier to maintain a dispersion of light phase in heavy phase. Dispersion of heavy phase in light requires a greater pumping action of the impeller, to circulate the heavy phase droplets to the center inlet against the centrifugal force, permitting them to be redispersed and eventually brought to equilibrium with the light phase. Agitator speeds required to disperse heavy phase in the light phase are always greater than those needed to disperse light phase in the heavy phase. The difference depends on the solvent properties (i.e., density, viscosity, and interfacial tension) and, in some cases, the required agitator speeds differ by a factor of more than 3.

Column throughput generally is maximized by dispersing the solvent flowing at the smaller rate; although when the rates differ by a factor of less than 2, either phase may be dispersed. Since dispersing the light phase generally gives higher stage efficiencies, this is the preferred method of operation in all cases in which it does not limit the overall capacity of the column.

Interface effects are also a factor in the capacity and efficiency of agitated columns. The direction of transfer of the solute has a significant effect on the performance characteristics of all agitated columns including the Karr reciprocating plate column [80]. This was first noted by Scheibel and Karr [81] on three different solvent systems, and the reason is not entirely clear. It could be related to thermal effects at the interface due to the difference in the heats of solution of the solute in the two solvents, or it could be attributable to the internal convection currents within the droplet, which are induced by the density difference as the solute is extracted at the interface.

The diffusion rate of a solute through a stationary liquid phase is very low. Major factors in the transfer of solute from one phase to another are surface renewal and convection currents set up in the continuous phase as the dispersed phase moves through it. However this can only bring the surface of the droplet to equilibrium with the continuous phase. Transfer within the dispersed phase also depends on the internal eddy currents induced by the viscous drag of the surface of the continuous phase as the droplet moves through it. Thermal effects and density differences at the interface resulting from the transfer of solute may either oppose or supplement this internal circulation, depending on the direction of transfer. Since the mechanism is not fully understood, performance data on a particular solvent system are necessary for confident scale-up to commercial plants.

The geometry of the scale-up process results in a longer residence time in the larger columns. This permits lower degrees of agitation and a corresponding increase in capacity. These factors can be taken into account based on prior experience on other systems, but the empirical nature of the scale-up design dictates a conservative approach. All large agitated columns should be provided with a variable speed drive to permit optimization of the column performance, to achieve maximum purity at design throughput, maximum throughput at design purity, or some intermediate set of conditions. And if the scale-up has not been sufficiently conservative, it may be necessary to sacrifice one of the design parameters. Thus in either case a variable speed drive is essential for the most effective operation of agitated extraction columns of any given size.

The preferred mode of dispersion must be evaluated on smaller columns, either laboratory or pilot plant size. Agitator drives on the larger columns rarely have a range of speed and power sufficient to permit operation in both modes. Agitator designs based on dispersing the light phase frequently are incapable of handling the power input needed to disperse the heavy phase, so that changing the external drive mechanism is not always possible.

The recommended procedure for starting up an agitated column is to fill all mixing stages with the continuous phase. The dispersed phase flow is then started with the agitator running, and the flow rate of the continuous phase is controlled to maintain this phase at the desired interface level. When the dispersed phase appears at this level, the continuous phase flow is started at the desired rate, maintaining the interface level by adjusting the product flows from the column.

Dispersed phase holdup in the column increases as the maximum or flooding capacity is approached. At these conditions it is essential that the interface level controller respond slowly to avoid excessive cycling, since the amount of dispersed phase holdup is sensitive to the flow of the continuous phase. A sudden large change in a product flow rate to maintain an interface will have an opposite initial effect, thus will induce cycling in a proportional controller. Consider, for example, a column operating in the light phase dispersed mode with an interface level below the desired control point. This calls for a decrease in the rate of heavy phase drawoff, which reduces the net downflow rate of the heavy phase. The Stoke's law effect of the latter change releases some of the dispersed light phase holdup in the column, which can cause the interface at the top of the column to fall even further below the control point. A proportional controller will continue to reduce the heavy phase drawoff until it is finally shut off. Then, when the bulk of the holdup is released from the column and replaced with heavy phase, the interface level will move up toward the control point at a rate equivalent to the combined flows of both phases. This will cause the heavy phase drawoff valve to open and the dispersed phase holdup in the column will begin to accumulate again. Consequently, there will be no change in the rate at which the interface level is rising, and the heavy phase product controller may proceed to the full open position. Under these conditions the column may give the appearance of flooding (i.e., light phase carryover in the heavy phase product). When the dispersed phase holdup reaches its maximum amount and the light phase starts to accumulate at the top of the column, the reduction of the heavy phase drawoff releases this holdup and the cycle repeats.

Clearly this type of cycling reduces both efficiency and capacity of the column. In the small laboratory columns 3 to 4 ft high the dispersed phase holdup is so small that the cycling effect usually is unnoticed. In pilot plant columns 10 to 15 ft high it is more noticeable but does not cause the operator undue concern and is rarely reported. In columns more than 30 ft high, however, the dispersed phase holdup may be equivalent to more than one hour of the flow rate, and proper interface level control is essential for maximum efficiency and capacity. It is also critical in pilot plant columns containing large numbers of stages so that the height is more than 20 ft. Scheibel has found the most practical control system for these large columns to be a wide band proportional flow controller reset by a liquid level controller with the widest band available (300 to 500%), and the latter instrument must respond over such a long period that no automatic reset can be tolerated. Precise interface level control becomes less important for columns operating well below their flooding rates, but is essential if the maximum capacity and efficiency is to be attained under any given circumstances.

The different agitator speeds required to disperse the light and heavy phases also make it impractical to attempt to control the interface at an intermediate point in the column. Under these conditions the agitator speed required to disperse the heavy phase in the upper section of the column will emulsify the lower section and produce flooding, while the agitator speed required to disperse the light phase would provide almost no efficiency in the upper section of the column. Theoretically one could design the agitators to achieve proper mixing above and below the interface in the column, but it would then be necessary to control the interface precisely between the stages having the different agitator designs.

Pilot Plant Extraction Studies

The previous section described the details for operation of agitated extraction columns. The development of an extraction process from basic chemical concepts to commercial fruition generally necessitates a pilot plant study after the initial laboratory investigations. The ideal pilot plant is one that provides all the necessary chemical engineering data for scale-up as well as data on accumulation of feed impurities in the solvents and in the equipment so that proper steps can be taken to control these problems in the large unit. The pilot plant should furnish all this information while producing small quantities of finished product for test marketing to evaluate consumer acceptance.

These requirements are mutually exclusive, although they may be fulfilled over different periods on the same unit with minor modifications. A pilot plant operated to obtain performance data for scale-up will yield very little useful product. Conditions will be varied to explore the operating characteristics, and product of specified quality may be obtained at times, but not for any sustained period. Equipment modifications will preclude continuous operation for prolonged periods. A study of this type is more essential for the newer equipment designs, particularly those which have not yet been applied to commercial operations. It is of less importance in the case of the more widely used equipment.

On the other hand, solvent extraction processes recycle relatively large quantities of solvents that will, in time, accumulate impurities. Thus before the commercial plant can be installed with complete confidence, one must establish the effect of prolonged operation on the solvents and evaluate processes for recovering them when they can no longer be recycled. This will provide the basis for suitable purges to maintain the solvent impurities at safe levels for continuous operation.

The problem of solids accumulation at the interface of the extractor must be considered. Prolonged operation is necessary to determine whether these solids will be removed from the column in the liquid phase, leaving that end where the interface is controlled, or whether a continuous purge must be taken, filtered, and returned to the column to control these solids. It is also possible for solids to accumulate within the extractor, affecting its performance, and this may be particularly detrimental in columns containing wire mesh packing, which provides a large surface for deposition of solids and may eventually become plugged. Changes in the operating characteristics of any extractor must always be viewed with suspicion. An increase in efficiency with time could be due to partial plugging, increasing dispersed phase holdup, and perhaps ultimately necessitating a reduction in column throughput.

Pilot plant operation of limited duration can ascertain only that the full-scale plant will operate for this length of time without detectable problems. Operation for purposes of detecting impurity accumulations is compatible with small-scale production for market evaluation and can also provide useful data on corrosion and mechanical failure of equipment, but may not necessarily yield basic data for optimization of plant design.

5 APPLICATIONS OF LIQUID EXTRACTION

Liquid extraction may be used simply to transfer a component from one liquid phase to another, or advantage may be taken of the difference in the distribution coefficients of two or more compounds between a pair of immiscible solvents to effect a separation. These processes have different applications and are considered separately. In all applications they compete with distillation, chromatography, and chemical separations by way of the formation of other compounds.

Extraction for Removal

Distillation must always be given first consideration for solvent recovery, and if azeotropes are encountered, the case for extractive distillation should be reviewed. If distillation is inapplicable for any reason or if dilute solutions are involved, liquid extraction and adsorption should be given next consideration. Adsorption may be economic if concentrations are less than a few hundred parts per million. Normally when concentrations are of the order of several percent or more, extraction will probably be the most economic. The exception occurs when the solute is higher boiling than the rest of the solution, with evaporation therefore requiring considerable heat input. In this case extraction of the solute into a solvent of lower volatility than the solute may be the most economic process, since only the small amount of solute need be distilled away from the extract solution to recycle the solvent.

Liquid extraction is ideally suited to remove pollutants or other contaminants from large volumes of solution in the concentration range from 100 ppm up to 5 % or more. Aqueous phenol solutions make an interesting case study for liquid extraction because they represent a common waste disposal problem. At ordinary temperatures a saturated phenol solution in water contains about 9 % phenol by weight. Since this is very close to the azeotropic composition at atmospheric pressure, the azeotrope does not readily separate into two immiscible phases as in most azeotropic distillations. Thus phenol can be a serious problem in wastewater discharged into a river or waterway from which drinking water is drawn by other downstream communities. Concentrations of less than 1 ppm can be detected in the taste of the water, and removal to less than 100 ppb in total plant effluent is preferable.

Many solvents have been used for the extraction of phenol from water. The preferred solvents are those whose solubility in water can be tolerated, eliminating the need for further purification steps. Since phenol is acidic, it can be recovered from the extract solution by a second extraction with an aqueous caustic solution at concentration sufficient to cause liquid phenol to separate out on acidification of the caustic solution. This process thus recovers the phenol in a reusable form, after further purification by distillation. A process of this type consumes chemicals approximately equal in value to the phenol recovered; thus the operating cost must be borne by the process producing the polluted aqueous stream.

Methyl isobutyl ketone has also been used to extract phenol from water [82]. This solvent is soluble in water to about 2 % and is removed by distillation. The amount of solvent required to extract the phenol is small enough to be distilled from the extract, leaving a phenol residue. This process can remove phenols profitably from aqueous solutions containing more than 1500 ppm. Below 1000 ppm extraction costs with this solvent become comparable to extraction followed by caustic wash to remove the phenol to permit recycling of the solvent.

Adsorption processes are also competitive at concentrations below 1000 ppm, and at lower concentrations (in the range of 20 ppm), biodegradation becomes a viable choice. In some cases a two-step purification may be considered, recovering the bulk of the phenol by extraction and finishing with either the adsorption or biological process for final cleanup to less than 1 ppm. In most cases, however, the incremental cost of reducing a 20-ppm effluent to 1 ppm is less than the cost of a second processing unit. Thus adsorption and biodegradation present an advantage over liquid extraction only if the initial phenol concentrations are less than a few hundred parts per million.

Solvent extraction has also received considerable attention as a general solution to pollution control problems because it does not require temperature changes in the water. Because of the large volumes involved, thermal effects can be a major operating cost, and attempts to minimize them necessitate large heat exchangers. Chloroform and other impurities can be removed by solvent extraction. The best solvents for removal of such compounds usually have a high solubility in water that cannot be tolerated, and solvents having an acceptably low solubility in water do not have sufficiently favorable distribution coefficients. The most practical approach may be to extract the undesirable impurities with the most effective solvent, then to extract the dissolved solvent with second solvent having an acceptably low solubility in water and a boiling point sufficiently high that the first solvent can be easily stripped off. For this purpose, the preferred solvent for the first extraction must be volatile enough to be economically removed by steam stripping. It must also be readily separated from the impurities it removes from the water solution. Methyl isobutyl ketone exhibits strongly negative deviations from ideality with chloroform [83] and can be steam stripped from the solution as the heterogeneous azeotrope with water. Methyl isobutyl ketone can be extracted with a high-boiling hydrocarbon, from which it also can be steam stripped as the heterogeneous azeotrope.

Processes of this type offer interesting economic possibilities in the general area of pollution control, and applications have not been fully explored. Economics of solvent extraction processes are most favorable at the higher concentrations; therefore the most efficient means of control is always at the source of the pollution, where it is possible to select the most suitable solvent to control a particular contaminant. There is very little chance of finding a single solvent to extract the myriad of contaminants found in river water, to purify it for drinking purposes. Several different solvents would probably be required, posing solvent recovery and purification problems. The relatively low concentrations of contaminants in the usual river water make for economics more favorable to chemical treatment methods and adsorption as currently used in water purification.

Liquid extraction processes are most effective when directed toward the removal of single compounds or a group of specific compounds closely related, thus having comparable distribution properties between solvents. This is the basis for present applications of liquid extraction for removal, as distinguished from fractionation by liquid extraction.

Fractionation Processes

As has been previously noted, fractional liquid extraction is theoretically applicable to all separation problems. Equipment designs are also available for implementing this separation and purification process. Limitations are entirely economic. Where fractional or azeotropic distillation is a viable process, liquid extraction cannot compete. In most cases it cannot compete with extractive distillation, which is usually somewhat more costly than ordinary fractional distillation. The exception is in the separation of a wide-boiling mixture of one type of compound from a similar wide-boiling mixture of another type. Extractive distillation is best applied to a mixture with narrow boiling range, and where pure compounds are not required, the prefractionation cost can be prohibitive. The classical example is the separation of aromatics from paraffins in petroleum mixtures. The original demand for toluene was met during World War II by extractive distillation of a narrow-boiling petroleum fraction, using phenol for the extractive solvent. At present the entire benzene-toluene-xylene mixture is extracted from a wide-boiling reformate fraction with diethylene glycol or sulfolane to obtain an aromatic fraction and a paraffinic raffinate. The aromatic fraction is readily separated in pure compounds by fractional distillation, since relative volatilities for adjacent members of this homologous series are about 2.0. The paraffinic raffinate from the solvent extraction is not resolved into pure compounds.

Thus liquid extraction is the most economic process for separating one group of similar compounds, such as isomers or members of a homologous series, from another group, because it is applicable to a mixture with a wide boiling range. The advantages are most significant when there is no value in separating the groups into individual compounds, although it is much easier to achieve the fractional distillation of pure compounds from the groups than from the original mixture. This is the only condition for which fractional liquid has been found to have an economic advantage over fractional distillation and extractive distillation.

Chromatography, like fractional liquid extraction, is also a potentially universal separation technique. Almost any mixture can be resolved into pure components through the selection of the proper column. Chromatographic columns are smaller and simpler to operate; they provide separations in a much shorter operating time and require much smaller feed samples. In addition, chromatographic instruments have undergone so much development over the past two decades that they cost less than the equipment for countercurrent distribution. Thus chromatography has replaced liquid extraction as an analytical tool.

However the small charge sample that constitutes one of the advantages of chromatography for analytical measurements is the major disadvantage in commercial separation processes. Large diameter columns are required, and effluent streams are usually very dilute, making recovery of the desired compound expensive.

Fractional liquid extraction, on the other hand, gives higher concentrations of the desired component in solution, thereby reducing the recovery costs. Theoretically, the process represented in Fig. 3.58, where reflux at each end of the column is provided by extraction of the effluent streams, might approach a saturated solution in the two product streams. However the size of the stripping extractors in such cases would be excessive, and the economic optimum occurs when the incremental cost of evaporation equals the incremental cost of the stripping extractors required to achieve the same increase in concentration. When suitable solvents are found, fractional

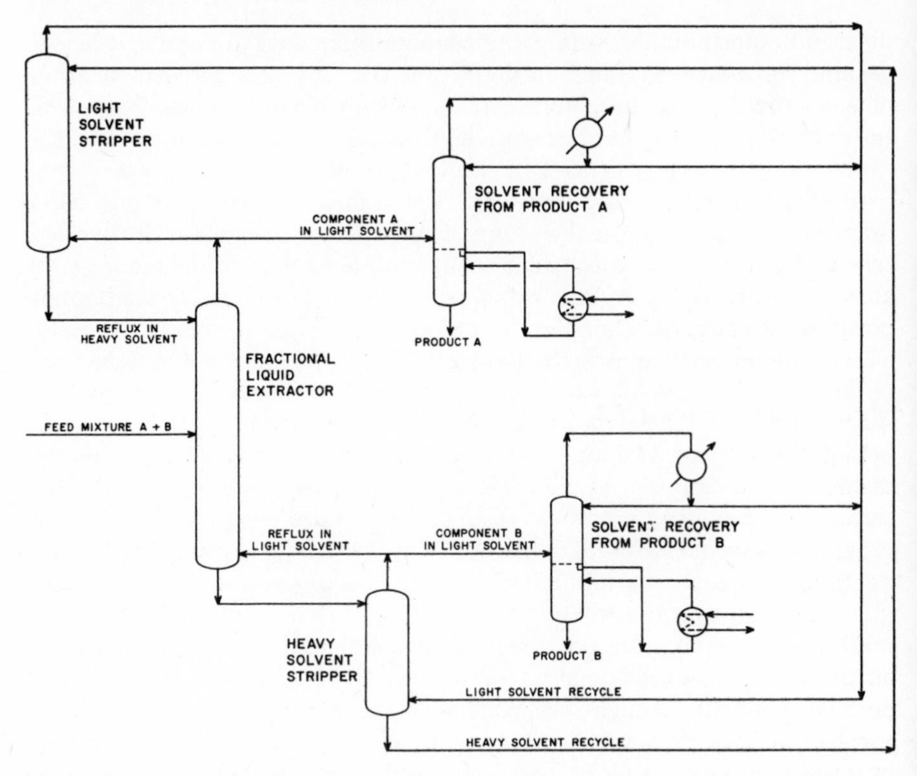

Fig. 3.58. Fractional liquid extraction process with reflux supplied by stripping extraction and product recovery from light solvent streams.

liquid extraction offers a more practical, economic process than liquid chromatography for separating large quantities.

At present the Craig countercurrent distribution equipment is primarily useful in the purification of samples larger than are obtainable from chromatographic equipment and in the application of the data thereby generated to the process design of fractional liquid extraction pilot plants and larger installations. Mixer-settler units can simulate countercurrent liquid extraction because the effectiveness of the different steps can be enhanced by reducing throughput, which, in effect, increases residence times. By operating at a sufficiently low throughput, one theoretical stage may be obtained from each unit, and the time required for a given approach to steady state will be inversely proportional to the throughput. Thus the most practical approach involves selecting the largest throughput at which one can confidently expect substantially one theoretical stage for each mixer-settler combination.

Agitated columns, on the other hand, exhibit different efficiencies on different systems at different operating conditions and are not reliable for developing phase equilibria data. Equilibria data are best obtained by other means, then applied to the interpretation of performance data on the given system in a particular continuous multistage column for design of a full-scale unit.

Fractional liquid extraction is most likely to be economic in the separation of nondistillable compounds. Two major fields of application are the refining of metals and the isolation and purification of biologically active compounds, particularly those which are subject to loss of activity or decomposition at elevated temperatures. These fields are considered separately, although the principles involved in the different areas are very similar.

Metallurgical Processes

Liquid extraction has been extensively used for the separation and purification of heavy metals, particularly the rare earths. Chemical methods of separation normally segregate the metals according to group in the periodic system. Isolation and purifications of the individual components within the group are frequently solved by liquid extraction techniques.

Metal salts that are complexed with bulky organic molecules become soluble in organic solvents, and the distribution coefficients vary depending on the particular metal in the original salt. The organic compound may bond to either the negative or positive ion in aqueous solution, or it may complex with the neutral salt only in an organic solvent, thus remaining essentially insoluble in the aqueous phase. The neutral coordination compound must have sufficient solubility in organic solvents to provide a viable fractional liquid extraction process.

The extraction of uranium in the form of uranyl nitrate with tributyl phosphate (TBP) supplies a well-known illustration of this process. The uranyl nitrate–TBP complex in the solvent consists of two TBP molecules per molecule of uranyl nitrate, and simplifying assumptions are generally made to develop the equilibrium phase relationship. To illustrate a fundamental technique that will be applicable to all systems of complex formation, however, all chemical equilibria relationships must be considered. Thus the association reactions

$$UO_2(NO_3)_2 + TBP \rightleftharpoons UO_2(NO_3)_2 \cdot TBP$$

$$UO_2(NO_3)_2 \cdot TBP + TBP \rightleftharpoons UO_2(NO_3)_2 \cdot 2TBP$$

exist in both liquid phases, and the respective equilibrium constants in the aqueous phase may be expressed as

$$K_{A1} = \frac{[UO_2(NO_3)_2 \cdot TBP]}{[UO_2(NO_3)_2][TBP]} \tag{3.71}$$

and

$$K_{A2} = \frac{[UO_2(NO_3)_2 \cdot 2TBP]}{[UO_2(NO_3)_2 \cdot TBP][TBP]} \tag{3.72}$$

The corresponding equilibrium constants in the solvent phase are designated with primes (i.e., K'_{A1} and K'_{A2}, respectively). In these relationships concentrations are given in moles per liter. If the same equations were expressed in terms of the more fundamental thermodynamic concept of activities, the equilibrium constants thus defined would be identical in both phases because the activities of a component are identical in two phases that are in equilibrium with each other. The same condition applies to ionization equilibria. Thus the ionization constants may be expressed as follows:

$$K_1 = \frac{[UO_2(NO_3)^+][NO_3^-]}{[UO_2(NO_3)_2]} \tag{3.73}$$

$$K_2 = \frac{[UO_2^{2+}][NO_3^-]}{[UO_2(NO_3)^+]} \tag{3.74}$$

If concentrations are replaced with activities to obtain the thermodynamic equilibrium constant, the value must be identical in both phases. However when the ionization constants are calculated from conventional concentrations in moles per liter, the activity coefficients for the ions in the solvent phase are so large that ionic equilibria may be neglected.

When all equilibria conditions are considered in both phases, the apparent distribution coefficient may be derived by the technique described in the last part of Section 2 as

$$D = D' \frac{1 + K'_{A1}[TBP](1 + K'_{A2}[TBP]) + K'_1/[NO_3^-](1 + K'_2/[NO_3^-]}{1 + K_{A1}[TBP](1 + K_{A2}[TBP]) + K_1/[NO_3^-](1 + K_2/[NO_3^-]} \tag{3.75}$$

The form of this equation is completely rigorous, but the designated equilibrium constants will be functions of concentration. They will be constant only when the concentration terms are replaced with activities. Thus these equilibrium constants are related to the true thermodynamic equilibrium constants κ as follows

$$K_{A1} = \kappa_{A1} \frac{\gamma_{UO_2(NO_3)_2}\gamma_{TBP}}{\gamma_{UO_2(NO_3)2 \cdot TBP}} \tag{3.76}$$

$$K_{2A} = \kappa_{A2} \frac{\gamma_{UO_2(NO_3)_2 \cdot TBP} \cdot \gamma_{TBP}}{\gamma_{UO_2(NO_3)_2 \cdot TBP}} \tag{3.77}$$

$$K_1 = \kappa_1 \frac{\gamma_{UO_2(NO_3)_2}}{\gamma_{UO_2(NO_3)^+} \gamma_{NO_3^-}} \tag{3.78}$$

$$K_2 = \kappa_2 \frac{\gamma_{UO_2(NO_3)^+}}{\gamma_{UO_2^{2+}} \gamma_{NO_3^-}} \tag{3.79}$$

where γ represents the activity coefficients for the particular specie in the given phase.

If activity coefficients are very nearly constant over the concentration range involved, equilibrium constants defined in conventional terms of moles per liter will also be very nearly constant.

Since activity coefficients are inversely proportional to solubility, (3.76) through (3.79) indicate that association in the aqueous phase may be neglected when the solubilities of the complex compounds are negligible in this phase. Similarly, ionization in the solvent phase may be neglected when ion solubilities are low. In the case of the uranyl nitrate–TBP mixtures, the equimolar complex is also assumed to be essentially nonexistent, and its activity coefficient is thereby so large that K_{A1} is negligible. Similarly the monovalent $UO_2(NO_3)^+$ ion may be assumed to be essentially nonexistent in both phases. Thus (3.75) may be simplified to

$$D = D' \frac{1 + K'_{A1} K'_{A2}[\mathrm{TBP}]^2}{1 + K_1 K_2/[NO_3^-]^2} \tag{3.80}$$

and K'_{A2} and K_2 are then assumed to be large enough to render the first terms in the numerator and denominator negligible with respect to the second terms. Thus we have

$$D = \frac{D' K'_{A1} K'_{A2}}{K_1 K_2} [\mathrm{TBP}]^2 [NO_3^-]^2 \tag{3.81}$$

and this relationship gives a reasonable correlation of the observed distribution data.

Similar assumptions may be made to interpret distribution data for the formation of other complex compounds. The same theoretical techniques may be applied to the formation of chelate complexes in which the metal forms coordination bonds with two or more basic groups of the organic compound. Tartaric acid, 8-quinolinol, *N*-methoxysalicyaldoxime, dimethyl glyoxime, and various β-diketones have been used as chelating agents for different metals. Maximum selectivities between different metals can sometimes be achieved through the use of both a chelating agent for the one metal and a masking agent that prevents the formation of a chelate of the other metal. Ethylenediaminetetraacetic acid has been found to be an excellent masking agent for some separations.

Possibilities for liquid extraction processes in the separation of metals are virtually unlimited, particularly in the case of rarer metals where processing costs for obtaining higher yields or purities can be readily justified on an economic basis.

In all solvent processes for metal refining, the organic compounds must be large enough to encapsulate the metal salt, either on a molecular scale or through the formation of colloidal aggregates of sufficient size to dissolve in an organic solvent. Reaction kinetics involving such large molecules normally are controlling in the extraction. Larger residence times than are normally used to transfer a given solute from one phase to the other are required. Since mixer-settler combinations can provide any desired residence time, they have been preferred in the past. Agitated extractors do not provide sufficient flexibility in this respect because when solvent flow rates are considerably different, an attempt to increase the residence time of the smaller flow by increased agitation may result in emulsion formation. For this reason data on the small laboratory columns have not been encouraging, and larger columns with corresponding greater residence times have not been adequately investigated. Controlled cycling or pulse columns with small hole sizes have also been found to be applicable because the residence times can be partially controlled by the pulse frequency and magnitude.

In the general area of metal refining, solvent extraction provides many interesting possibilities for the development of new processes and equipment designs for particular applications.

Biologically Active Compounds

The importance of liquid extraction in the isolation of antibiotic compounds was discussed earlier. This method is also widely used in the isolation and purification of alkaloids, which contain at least one basic nitrogen atom and can be extracted with acidic solutions. Solubilities in organic solvents are also functions of the particular acid group used in the extraction, as well as pH of the aqueous phase. It thus becomes possible to prepare pure compounds from the plant extracts by first removing all water-insoluble compounds by extraction of an acid solution with an organic solvent, usually an ether. After this so-called defatting extraction, the desired alkaloid is recovered from the aqueous solution by extraction procedures, using suitable acids at appropriate pH values. Distribution coefficients can be calculated from equations similar to those in Section 2 for the selection of optimum conditions.

Chemical structures of these compounds can be determined from X-ray diffraction patterns. Distribution data at different pH can provide supportive evidence of the structure. In this respect, some of the most interesting distribution curves are obtained with amphoteric compounds that contain acidic groups, such as phenolic OH, in addition to the basic nitrogens. Figure 3.59 illustrates the type of distribution coefficient variation that might be expected for a compound containing one such acidic group and two basic nitrogens. The theoretical equation for the distribution coefficient would be a combination of (3.23) and (3.31) with appropriate terms in the denominator:

$$D = \frac{D'}{1 + K_1/[\mathrm{H}^+] + K_{1\mathrm{B}}/[\mathrm{OH}^-] + K_{1\mathrm{B}}K_{2\mathrm{B}}/[\mathrm{OH}^-]^2} \tag{3.82}$$

where the terms have the same significance as in the previous equations, and the first and second ionization constants for the basic nitrogen groups have the additional B subscript. At low pH the last term in the denominator dominates, and at high pH the second term provides the major contribution to the value of the denominator. Thus the slope of the curve approaches 2.0 at low pH. The curve passes through a maximum, then approaches a slope of -1.0. Such a behavior could be used to confirm the chemical structure. If the ionization constant for the second nitrogen is smaller than 10^{-14}, however, the slope at low pH would not be sufficiently close to 2.0 to be conclusive.

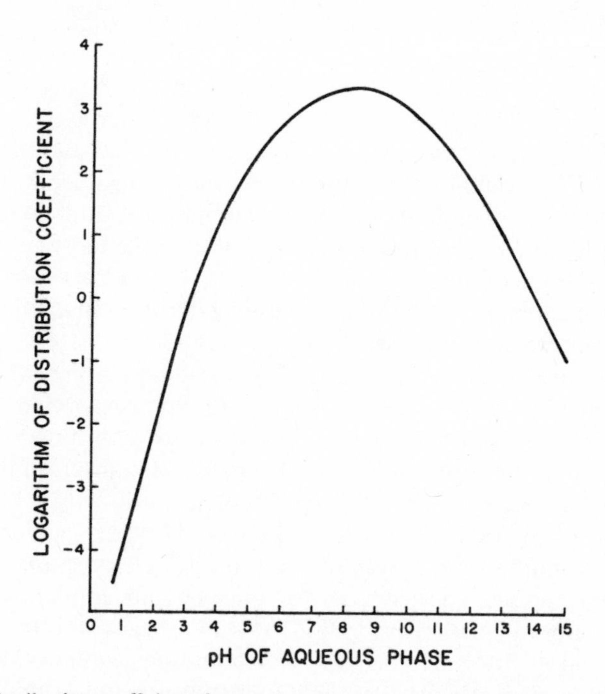

Fig. 3.59. Distribution coefficients for amphoteric compound containing one acidic group and two basic groups.

This technique has been used to postulate the molecular formula of metal chelates [84] in the extraction of lanthanides and is applicable to all compounds having one or more acidic or basic groups. In addition to the theoretical implications, such data provide the basis for determination of the optimum conditions for purifying the particular compound.

Current work on the isolation and purification of anticarcinogens furnishes many opportunities for the use of liquid extraction techniques. Stabilities of the compounds are not known, and molecular structures are so complex that they are not distillable. Isolation techniques must be based on either liquid extraction or chromatography, or a combination of both. Because of the simplicity of operation, chromatography is preferable for the exploratory work. Liquid extraction can then produce quantities needed for more extensive testing, but the processes must be developed.

Extraction techniques similar to those described by Scheibel and Karr [85] for the purification of vitamin A esters are applicable. The fractional liquid extraction column was first operated at a ratio of light to heavy solvent low enough to allow the vitamin A ester to pass out of the column in the heavy phase, while all impurities less soluble than the vitamin A ester were removed in the light phase. The product was recovered from the heavy phase and introduced into the extractor a second time, using a higher light-to-heavy solvent ratio sufficient to separate the vitamin A ester in the light phase from all impurities that were more soluble in the heavy phase. The heavy solvent was distilled away from the impurities and recycled to the extractor. The light solvent extract was concentrated and cooled to crystallize out pure vitamin A ester. The same principles apply to all purification problems. The effect is essentially the same as achieved by Craig's batchwise countercurrent distribution technique, which separates the less soluble and the more soluble impurities on opposite sides of the distribution pattern and collects the desired compound in the center. Fractionation by liquid extraction in a column permits continuous operation to prepare samples of any desired size.

Since vitamin A ester is unstable at elevated temperatures and is decomposed in the presence of water, it presents more problems than most of the anticarcinogenic compounds now being studied. Solvents must be sufficiently low boiling to permit evaporation at reasonable vacuum from the solutions and recycling to the extractor. The requirement of nonaqueous immiscible solvent systems greatly restricts the selection. The light solvent may be a paraffinic or naphthenic hydrocarbon such as hexane or heptane with limited amounts of aromatics. The heavy solvent must be sufficiently polar to be only partially miscible in the hydrocarbon. Possible heavy solvents are acetonitrile, methanol, nitromethane, methyl Cellosolve, aniline, nitrobenzene, ethylene glycol, and any mixtures of these solvents. Only the first two are

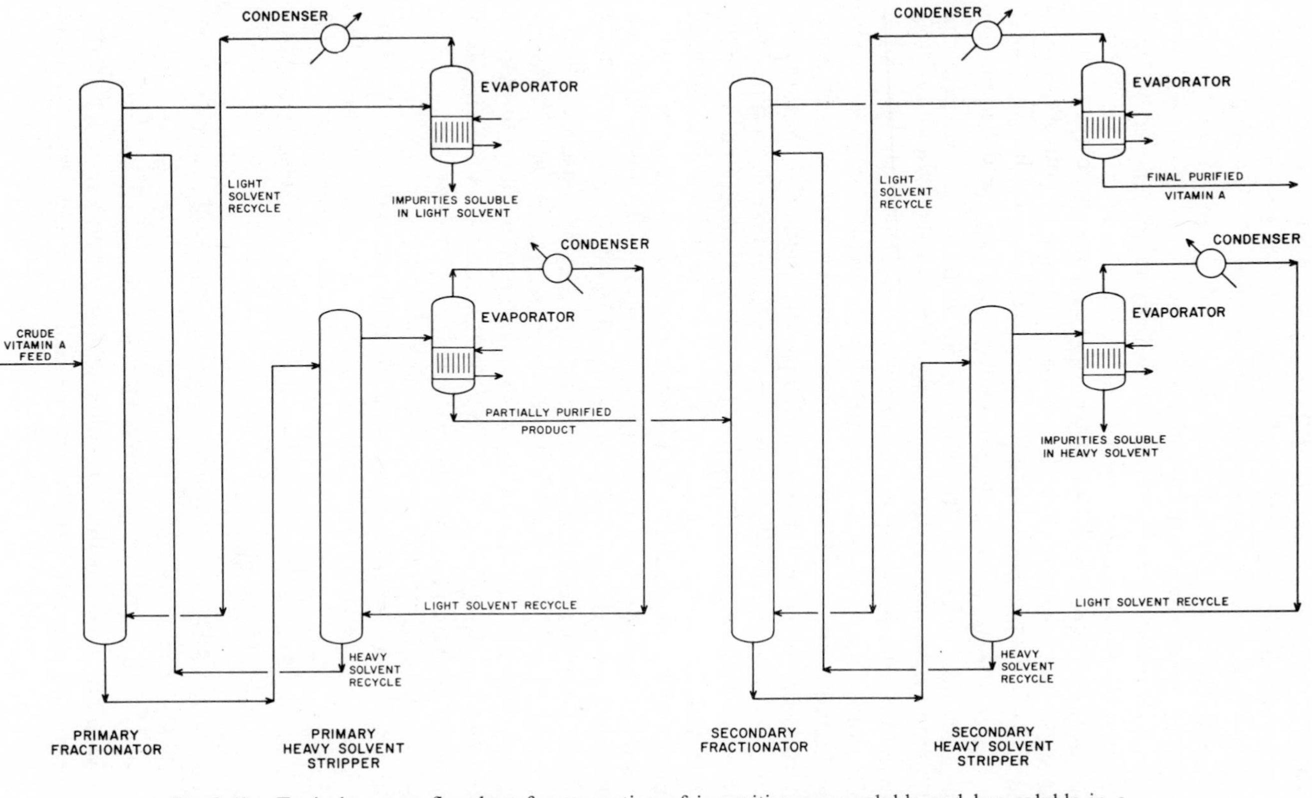

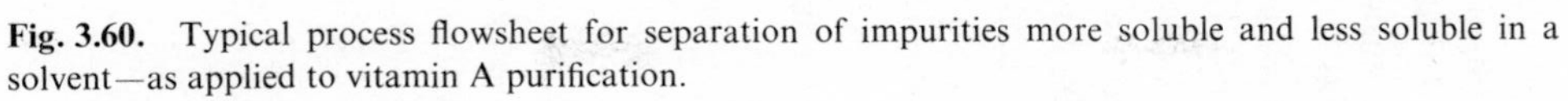
Fig. 3.60. Typical process flowsheet for separation of impurities more soluble and less soluble in a solvent—as applied to vitamin A purification.

volatile enough to be distilled away from the vitamin A fraction without decomposing it. The latter solvents and their mixtures can be utilized by extracting the heavy phase solutions from the fractionation column with a sufficient volume of light solvent to remove the dissolved solutes. The solvent can then be recycled to the extractor. The second extract is evaporated to recover the solvent for recycle and to isolate the solutes, which will contain the vitamin A ester after the first fractionation step and the impurities after the second fractionation step. The process flowsheet for the purification of vitamin A using high-boiling solvents for the heavy phase appears in Fig. 3.60.

Most of the anticarcinogens presently identified contain a basic nitrogen atom, which means that a fractionation at low pH can also effect a purification. In this case the first liquid extraction can be carried out at a pH lower than the acid ionization constant, defined as the p*K* for water divided by the ionization constant for the base. This process will remove all weaker bases and separate the desired compound with all stronger basic components. A second fractional liquid extraction at a pH higher than the acid p*K* will separate the desired compound from the stronger bases; if other impurities are present with similar basic properties, further purification can be effected using the solvent systems previously mentioned for the vitamin A purification. Since there is almost no probability that two different compounds will have nearly identical distribution coefficients and ionization constants, this dual-extraction system with two different solvent pairs should yield pure products. Of course the sequence of the extraction processes may be reversed, or the four different fractionation steps may be interchanged in any manner, and in many cases one or more of the individual steps may be omitted.

This approach offers intriguing possibilities in the purification of all compounds, but because of the operating cost, it is most practical for fine organics that command a higher price. It is ideally suited to the isolation and purification of pharmaceutical compounds of all types.

NOMENCLATURE

- a — ionic radius (Å)
- C — ion concentration (moles/liter)
- D — apparent distribution coefficient = concentration in light phase/concentration in heavy phase
- D' — distribution coefficient of free acid or base
- D_1 — distribution coefficient of component 1 more soluble in light phase
- D_2 — distribution coefficient of component 2 more soluble in heavy phase
- E — extraction factor

E_1	extraction factor of component 1
E_2	extraction factor of component 2
F	feed rate in continuous countercurrent extraction
F_s	amount of solute in steady state feed quantities in batchwise simulation
H	flow rate of heavy solvent in continuous extraction; also
H_B	quantity of heavy solvent per cycle in batchwise extraction
I	ionic strength
K	equilibrium or ionization constant for acids or bases (normally based on concentrations in moles/liter)
L	flow rate of light solvent in continuous extraction; quantity of light solvent per cycle in batchwise extraction
m	feed stage number counting from bottom of column
n	feed stage number counting from top of column
p	fraction of a component in light solvent at equilibrium contacting between phases $= E/(E + 1)$
P_t	fraction of a solute leaving batchwise extraction system in light phase after t extraction cycles
q	fraction of a component in heavy solvent at equilibrium contacting between phases $= 1/(E + 1)$
r	tube number in batchwise pattern
R	rejection ratio = quantity in light product phase/quantity in heavy product phase
R_1	rejection ratio of component more soluble in light phase
R'	retention ratio = quantity in heavy product phase/quantity in light product phase
R'_2	retention ratio of component more soluble in heavy phase
t	number of cycles in batchwise extraction
T	total amount of a given solute in a stage
x	concentration of solute in heavy solvent phase
x_B	concentration in heavy phase product stream in continuous extraction
y	concentration of solute in light solvent phase
y_p	concentration in light phase product stream in continuous extraction
Z	ion valence
β	relative distribution $= D_1/D_2$
γ	activity coefficient

δ fractional deviation from steady state concentrations
ε constants in (3.67) for deviation from steady state
κ thermodynamic equilibrium constant in terms of activities

References

1. M. Frenc, *Z. Angew. Chem.*, **38**, 323 (1925).
2. E. Jantzen, *Dechema Monogr.*, **5** (48), 81 (1932); Verlag Chemie, Berlin.
3. M. H. Tuttle, U.S. Patent 1,912,349 (May 30, 1933) to Max B. Miller & Co.; reissue 19, 763 (November 19, 1935).
4. L. C. Craig, *J. Biol. Chem.*, **155**, 519 (1944).
5. L. C. Craig and O. Post, *Anal Chem.*, **21**, 500 (1949).
6. E. G. Scheibel, *Chem. Eng. Prog.*, **44**, 681–690, 771–782 (1948).
7. G. H. Reman, U.S. Patent 2,601,674 (1952).
8. J. Y. Oldshue and J. H. Rushton, *Chem. Eng. Prog.*, **48**, 297 (1952).
9. A. E. Karr, *Am. Inst. Chem. Eng. J.*, **5**, 446 (1959).
10. A. N. Campbell and A. J. R. Campbell, *J. Am. Chem. Soc.*, **59**, 2481 (1937).
11. J. D. Cox, *J. Chem. Soc.*, **1954**, 3183.
12. F. A. H. Schreinemakers, *Z. Phys. Chem.*, **33**, 78 (1900).
13. J. Kielland, *J. Am. Chem. Soc.*, **59**, 1675 (1937).
14. R. A. Robinson and R. H. Stokes, *Electrolytic Solutions*, Academic Press, London, 1959, pp. 492–493.
15. M. T. Bush and P. M. Densen, *Anal. Chem.*, **20**, 121 (1948).
16. J. B. Hunter and A. W. Nash, *Ind. Eng. Chem.*, **27**, 836 (1935).
17. K. A. Varteressian and M. R. Fenske, *Ind. Eng. Chem.*, **28**, 928 (1936).
18. M. Frenc, *Z. Angew. Chem.*, **38**, 323 (1925).
19. E. Jantzen, *Dechema Monogr.*, **5** (48), 81 (1932); Verlag Chemie, Berlin.
20. M. H. Tuttle, U.S. Patent 1,912,349 (May 30, 1933) to Max B. Miller & Co.; reissue 19, 763 (November 19, 1935).
21. S. Stene, *Ark. Kem., Miner. Geol.*, **A18** (18) (1944).
22. L. C. Craig, G. H. Hogeboom, F. H. Carpenter, and V. du Vigneand, *J. Biol. Chem.*, **168**, 665 (1947).
23. L. C. Craig and D. Craig, *Technique of Organic Chemistry*, Vol. 3, Part 1, 2nd ed., A. Weissberger, Ed., Wiley-Interscience, New York, 1956, p. 193.
24. C. R. Bartels and G. Kleiman, *Chem. Eng. Prog.*, **45**, 589 (1949).
25. W. J. D. Van Dijck and A. Schaafsma, U.S. Patent 2,245,945 (June 17, 1941).
26. S. Stene, *Ark. Kem., Miner. Geol.*, **A18** (18) (1944).
27. A. Klinkenberg, N. A. Lauwerier, and G. H. Reman, *Chem. Eng. Sci.*, **1**, 93 (1951).
28. A. Klinkenberg, *Ind. Eng. Chem.*, **45**, 653 (1953).
29. E. G. Scheibel, *Ind. Eng. Chem.*, **46**, 16 (1954).
30. A. J. P. Martin and R. L. M. Synge, *Biochem. J.*, **35**, 91–121 (1941).
31. E. L. Compere and A. Ryland, *Ind. Eng. Chem.*, **43**, 239 (1951).
32. E. G. Scheibel, *Chem. Eng. Prog.*, **44**, 681–690, 771–782 (1948).
33. E. G. Scheibel, *Ind. Eng. Chem.*, **42**, 1497 (1950).

34. G. F. Asselin and E. W. Comings, *Ind. Eng. Chem.*, **42**, 1198 (1950).
35. K. A. Varteressian and M. R. Fenske, *Ind. Eng. Chem.*, **29**, 270 (1937).
36. J. O. Maloney and A. E. Schubert, *Trans. Am. Inst. Chem. Eng.*, **36**, 741 (1940).
37. E. G. Scheibel, *Ind. Eng. Chem.*, **47**, 2290 (1955).
38. L. C. Craig, *J. Biol. Chem.*, **155**, 519 (1944).
39. L. C. Craig and O. Post, *Anal. Chem.*, **21**, 500 (1949).
40. S. Stene, *Ark. Kem., Miner. Geol.*, **A18** (18) (1944).
41. E. G. Scheibel, *Ind. Eng. Chem.*, **43**, 242 (1951).
42. D. F. Peppard and M. A. Peppard, *Ind. Eng. Chem.*, **46**, 34 (1954).
43. E. L. Compere and A. Ryland, *Ind. Eng. Chem.*, **46**, 24 (1954).
44. E. G. Scheibel, *Ind. Eng. Chem.*, **46**, 43 (1954).
45. E. G. Scheibel, *Ind. Eng. Chem.*, **44**, 2942 (1952).
46. E. G. Scheibel, *Ind. Eng. Chem.*, **49**, 1679 (1957).
47. B. V. Coplan, J. K. Davidson, and E. L. Zebroski, *Chem. Eng. Prog.*, **50**, 403 (1954).
48. B. V. Coplan, J. K. Davidson, and E. L. Zebroski, *Chem. Eng.*, **61** (3), 132 (1954).
49. M. R. Fenske and R. B. Long, *Chem. Eng. Prog.*, **51**, 194 (1954).
50. M. R. Fenske and R. B. Long; *Ind. Eng. Chem.*, **53**, 791 (1961).
51. R. E. Treybal, *Chem. Eng. Prog.*, **60** (5), 77 (1964).
52. R. E. Treybal, U.S. Patent 3,325,255 (June 13, 1967).
53. W. C. Hazen and R. L. Clive, U.S. Patent 3,206,288 (September 14, 1965).
54. E. G. Scheibel, *Chem. Eng. Prog.*, **44**, 681 (1948).
55. E. G. Scheibel, U.S. Patent 2,493,265 (January 3, 1950).
56. E. G. Scheibel, *Ind. Eng. Chem.*, **42**, 1497 (1950).
57. W. O. Ney, Jr., and H. L. Lochte, *Ind. Eng. Chem.*, **33**, 825 (1941).
58. J. Y. Oldshue and J. H. Rushton, *Chem. Eng. Prog.*, **48**, 297 (1952).
59. G. H. Reman, *Proceedings of the Third World Petroleum Congress*, The Hague, 1951, Sect. III, p. 121.
60. G. H. Reman, U.S. Patent 2,601,674 (June 24, 1952).
61. G. H. Reman, U.S. Patent 2,729,545 (January 3, 1956).
62. T. Misek, "Rotating Disc Extractors and Their Calculation," State Publishing House of Technical Literature, Prague, 1964.
63. M. Ganapathi, M. S. Krishna, C. J. V. J. Raju, and C. V. Rao, *Indian J. Technol.*, **12**, 273 (July 1974).
64. E. G. Scheibel, *Am. Inst. Chem. Eng. J.*, **2**, 74 (1956).
65. E. G. Scheibel, U.S. Patent 2,850,362 (September 2, 1958).
66. A. E. Karr, *Am. Inst. Chem. Eng. J.*, **5**, 446 (1959).
67. A. E. Karr, U.S. Patent 3,527,650 (1970).
68. T. C. Lo and A. E. Karr, *I&EC Proc. Design Devel.*, **11**, 495 (1972).
69. R. M. Wellek, M. W. Ozsoy, J. J. Carr, D. Thompson, and T. V. Konkle, *I&EC Proc. Design Devel.*, **8**, 515 (1969).
70. J. J. Carr, M. S. thesis, University of Missouri, Rolla, 1963.
71. W. J. D. Van Dijck, U.S. Patent 2,011, 186 (August 13, 1935).
72. G. Feick and H. M. Anderson, *Ind. Eng. Chem.*, **44**, 404 (1952).
73. H. F. Wiegandt and R. L. Von Berg, *Chem. Eng.*, **61** (7), 183 (1954).
74. W. M. Goldberger and R. F. Benenati, *Ind. Eng. Chem.*, **51**, 641 (1959).
75. L. D. Smoot, B. W. Mar, and A. L. Babb; *Ind. Eng. Chem.*, **51**, 1005 (1959).

76. M. R. Cannon, *Oil Gas J.*, **54**, January 23, 1956, p. 68.
77. J. J. Szabo, W. A. Lloyd, M. R. Cannon, and S. S. Speaker, *Chem. Eng. Prog.*, **60**, 66 (1964).
78. W. J. Podbielniak, U.S. Patent 2,003,308 (June 4, 1935).
79. D. B. Todd and R. D. Gordon, *Proceedings of the International Solvent Extraction Conference, Lyon*, **3**, 2380 (1974).
80. T. C. Lo and A. E. Karr, *I&EC Proc. Design Devel.*, **11**, 495 (1972).
81. E. G. Scheibel and A. E. Karr, *Ind. Eng. Chem.*, **42**, 1048 (1950).
82. F. C. Lauer, E. J. Littlewood, and J. J. Butler, *Iron Steel Eng.*, **46**, 99 (May 1969).
83. A. E. Karr, E. G. Scheibel, W. M. Bowes, and D. F. Othmer, *Ind. Eng. Chem.*, **43**, 961 (1951).
84. T. R. Sweet and H. W. Parlett, *Anal. Chem.*, **40**, 1885 (1968).
85. E. G. Scheibel and A. E. Karr, U.S. Patent 2,676,903 (April 27, 1954).

Chapter **IV**

ION–EXCHANGE CHROMATOGRAPHY

Phyllis R. Brown
Anté M. Krstulovic

---

1 Introduction 199

2 Ion-Exchange Terminology 199

3 Organic Ion Exchangers 201

Ion-Exchange Resins 201
Physical Properties 201
Particle Size and Form 201
Cross-Linkage 202
Swelling and Porosity 203
Chemical Properties of Ion-Exchange Resins 204
Stoichiometry of Ion-Exchange Reactions 204
Breakthrough and Total Capacity 204
Affinity of the Resin for the Counterion 205
Stability of Ion-Exchange Resins 207
Nomenclature 207
Mechanism of Ion-Exchange Reactions 207
Ion-Exchange Equilibria 207
Ion-Exchange Kinetics 211
Ion-Exchange Celluloses 211
Ion-Exchange Dextrans and Polyacrylamide Gels 214

4 Inorganic Ion Exchangers 216

Hydrous Oxide Exchangers 216
Acid Salt Exchangers 216
Heteropoly Acid Salts 216
Metal Sulfides 217
Phosphate Gels 217

5 Miscellaneous 217

6 HPLC Packing Materials 217

Liquid Chromatography Columns 219
- The Unpacked Column 224
- Packing the Column 224

7 Ion-Exchange Techniques and Applications 226

Choice of Exchanger 226
Ion-Exchange Processes 227
- Batch Operation 227
- Column Process 227
- Continuous Process 230

Mixed-Bed Exchangers 231
Ion Exclusion 231
Ionophoresis 231
Ion-Exchange Membranes 231
Oxidation-Reduction Polymers 232
Chelating and Ligand-Exchange Resins 233
Ion-Exchangers as Catalysts 233
Ion Exchange in Nonaqueous Solvents 234
Ion-Exchange Chromatography 234
- Principles of Chromatography 234
  - Resolution in LC 236

Components of a Liquid Chromatograph 238
- Injection System 238
- Solvent Delivery System 239
- Detection Devices 239

Applications: Organic Substances 240
- Amino Acids, Peptides, and Proteins 240
- Carbohydrates and Carbohydrate Derivatives 241
- Organic Acids 241
- Amines 242
- Nucleic Acid Components 243

Applications: Inorganic Substances 247
- Cations 247
  - Alkali and Alkaline Earth Metals 247
  - Transition Metals 247
  - Rare Earth Metals 248
- Anions 248
  - Halide 248
  - Phosphorus Oxyanions 248

Supplementary Reading: Ion-Exchange Literature 255

1 INTRODUCTION

Even though the first ion exchangers were originally known and used more than a century ago, the field of ion exchange is still experiencing a phenomenal increase in its applications and importance. Because many types of synthetic ion-exchange materials are now being "tailored" for specific purposes, their applications have grown tremendously, and almost every branch of science makes use of them. Ultimate proof of their utility is the "promotion" of ion-exchange materials from the laboratory to industrial scale. This was a logical result of the replacement of the original low-capacity resins with modern synthetic organic polymers, which lend themselves to industrial use.

The use of exchange materials can be traced back to 1850, when H. T. Thompson and J. Spence discovered that soil can perform an ion-exchange reaction in which ammonia is displaced from a solution of its salts. Their observations aroused interest, and the result was a publication in collaboration with J. T. Way [1, 2]. In 1905 R. Gans [3, 4] used a fusion process to synthesize ion-exchange materials he called "Permutits." These sodium-aluminum silicates were capable of exchanging their constitutional sodium ions for some other ions (e.g., calcium and magnesium). After great initial enthusiasm for Gans' Permutits, it was found that they had undesirable properties. An improvement over Gans' patented method of preparation was the substitution for the fusion process of a precipitation method employing sodium silicate and sodium aluminate solutions. The result was a "gel" that was found to have many practical uses.

The event that marked the real birth of modern ion-exchange technology was a now classical paper published by B. A. Adams and E. L. Holmes [5] describing the preparation of a cation-exchanging resin using phenolsulfuric acid and formaldehyde. Adams and Holmes also prepared another ion-exchange resin, using a phenyldiamine type of amine, and formaldehyde.

Since then the list of commercially available synthetic resins has grown tremendously and it is now a formidable task to enumerate them all.

2 ION-EXCHANGE TERMINOLOGY

An ion exchanger can be defined as a substance that is capable of reversible exchanges of its constitutional ions with some other external ionic species bearing the same charge. Ion exchangers are insoluble in the solvents containing ions to be exchanged, and they are resistant to chemicals at temperatures below 100°C. Among their basic physical properties, hardness and resistance to abrasion are the most desirable.

All ion exchangers have the same basic structure. This fundamental framework or skeleton is a three-dimensional network that is fixed and chemically insensitive to the surrounding electrolytes. It acts like a semipermeable membrane through which charged species can diffuse. The "site of exchange" is attached to this inert matrix. The "site of exchange" is hydrophilic and contains functional groups that are ionizable and mobile counterions. These counterions can be replaced when brought in contact with an external solution containing ions of suitable charge and size. The fixed functional groups determine the nature of the exchange process. Thus ion-exchange materials are categorized as cation exchangers and anion exchangers. The cation exchangers are capable of exchanging cations and the anion exchanger can exchange anions with the surrounding electrolytes. Cation and anion exchangers are further subdivided into strong, moderately strong, and weak, depending on the nature of the functional groups. Table 4.1 lists some of the ionic substituents frequently used as ion exchangers.

Strongly acidic cation exchangers contain sulfonic acid groups and can be represented as follows:

$$RSO_3^-H^+$$

where R stands for the matrix, SO_3^- is the site of exchange, and H^+ is the mobile counterion that is exchanged.

Substituted phosphoric acids are used as moderately strong acidic ion exchangers. The carboxylic acid group is found in weakly acidic cation exchangers of the type

$$RCOOH \tag{4.2}$$

Strongly basic anion exchangers use tetraalkylammonium groups and can be represented as follows:

$$[RN(CH_3)_3]^+Cl^- \tag{4.3}$$

Weakly basic ion exchangers usually contain tertiary amine groups:

$$[RNH(CH_3)_2]^+Cl^- \tag{4.4a}$$

Table 4.1 Structure and Types of Some Resin Ion Exchangers

Type	*Matrix*	*Functional Group*
Strong acid	Cross-linked polystyrene	$-SO_3^-H^+$
Moderately strong acid	Cross-linked polystyrene	$-PO(OH)_2$
Weak acid	Polymerized acrylic acid	$-COO^-Na^+$
Strong base	Cross-linked polystyrene	$-(CH_2NR_3)^+Cl^-$
Weak base	Cross-linked polystyrene	$-(CH_2NHR_2)^+Cl^-$
		$-(CH_2NH_2R)^+Cl^-$

or secondary amine groups

$$[RNH_2CH_3]^+Cl^- \tag{4.4b}$$

Since ion-exchange reactions involve ionic equilibria, chemical reactions representing these processes can be written in the usual manner, for example,

$$RSO_3^-H^+ + K^+Cl^- \rightleftharpoons RSO_3^-K^+ + H^+ + Cl^- \tag{4.5}$$

or

$$2[RN(CH_3)_3]^+Cl^- + 2K^+ + SO_4^{2-} \rightleftharpoons [RN(CH_3)_3]_2^+SO_4^{2-} + 2K^+ + 2Cl^- \tag{4.6}$$

3 ORGANIC ION EXCHANGERS

Ion-Exchange Resins

Physical Properties

PARTICLE SIZE AND FORM

Most commercially available ion-exchange resins are spherical particles called beads. Figure 4.1 illustrates a cross section of a single cation-exchange

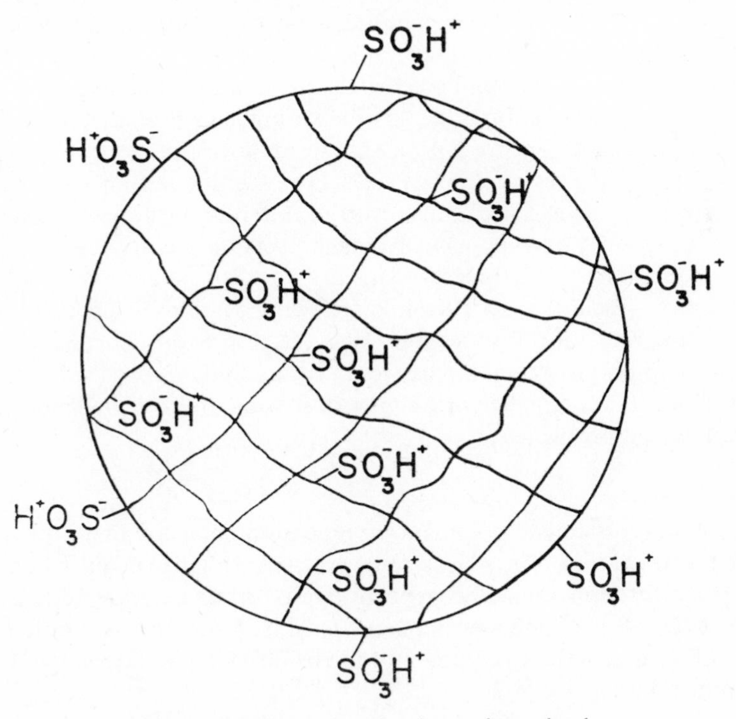

Fig. 4.1. Cross section of an ion-exchange bead.

Table 4.2 Particle Size Ranges for Resins[a]

Mesh Range Designation	*Screen Analysis*	*Diameter of Particles*		
		in.	*mm*	μm
16– 20	Wet	0.0460–0.0331	1.168–0.84	1168–840
20– 50	Wet	0.0331–0.0117	0.84 –0.297	840–297
50–100	Dry	0.0117–0.0059	0.297–0.149	297–149
100–200	Dry	0.0059–0.0029	0.149–0.074	149– 74
200–400	Dry	0.0029–0.0015	0.074–0.038	74– 38

[a] Reproduced (with changes) from *Dowex: Ion Exchange*, 1958, 1959, 1964, by permission of the Dow Chemical Company.

bead. When placed in water, the bead swells because water will diffuse into the interior. The organic network remains intact, while the hydrogen ions of the sulfonic functional group are completely dissociated and mobile. They can therefore be exchanged with an equivalent amount of ionic species of like charge.

These resin beads vary in size ranging from 1 μm up to 1 mm in diameter. Particle sizes are usually expressed in mesh sizes of the U.S. standard screens (Table 4.2). Resins with smaller particle sizes exhibit higher pressure gradients in column operation; they also have higher efficiency and a shorter equilibration period, and require a smaller quantity of resin for a specific reaction than do the resins of larger sizes. However the settling rate decreases with decreasing particle size; thus resins of different mesh sizes are used for different purposes: 200–400 mesh resins are used in gravity-feed chromatographic columns; 100–200 mesh resins are used for batch operations; very large beads (<100 mesh) are used in large-scale reactions where there is need for fast flow rates and fast settling rates. High-pressure chromatographic operations employ very small particles in the colloidal range (1–10 μm), and this provides extremely efficient systems for solving sophisticated biochemical and biomedical problems.

CROSS-LINKAGE

To obtain cross-linked polymers, styrene monomers are first polymerized to give linear chains of the polystyrene, which are then covalently bonded to each other through divinylbenzene bridges, forming a three-dimensional matrix. When this copolymer is reacted with sulfuric acid, sulfonic groups are introduced into the structure, and the resulting cation exchanger has the structure shown in Fig. 4.2.

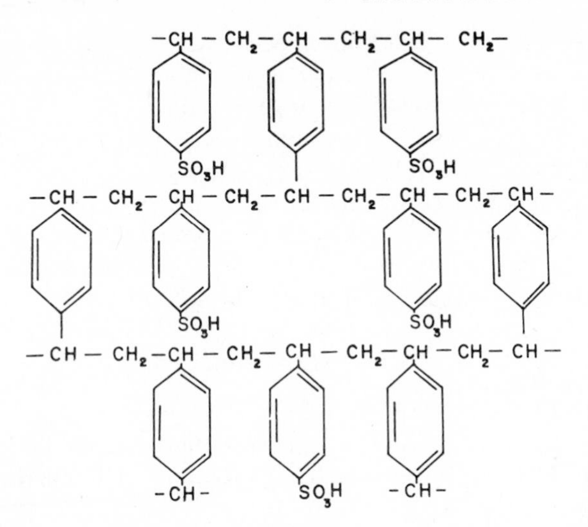

Fig. 4.2. Strong acid polystyrene type of cation-exchange resin.

Since the linear polystyrene chains can be linked together at different points, different structures may result, each structure having distinct physical characteristics. It is therefore convenient to define the degree of cross-linking. In the case of polystyrene ion exchangers, the degree of cross-linking is expressed as percentage of divinylbenzene contained in the polymer. The extent of cross-linking may vary from 1 to 16 %. Generally the higher the degree of cross-linking, the more rigid the structure. Resins of this type swell very little and are rather hard and brittle. Ion exchangers with a low degree of cross-linking have a more flexible structure, which enables them to take up considerable amounts of solvents; thus the structure is gelatinous.

SWELLING AND POROSITY

When a dry ion exchanger is placed in a solvent, the uptake of solvent molecules by the elastic three-dimensional polymer causes the substance to swell. This results in a certain porosity of the matrix which is negligible in the dry state. The amount of swelling is directly related to the type and number of functional groups affixed to the matrix, since different groups have different degrees of hydration. Swelling is a reversible process, and the equilibrium is determined by the two opposing forces: the elastic forces of the polymer matrix, and the tendency of the resin to undergo hydration, thereby increasing the osmotic pressure within the resin bead. It is also inversely proportional to the degree of cross-linkage and is greatest for exchangers with the fewest cross-linkages. With increasing degree of cross-linking, the resin structure becomes more rigid, the amount of swelling is small, and therefore the exchange process is slow. In addition, because of the small pores of the polymer network, only ions of certain sizes are free to penetrate the structure. Swelling also takes place when exchangers are brought in contact with electrolytes. The degree of swelling is found to decrease with increasing electrolyte concentrations and with increasing valency of the ions present in the electrolyte. Swelling characteristics of the resin are usually specified by the manufacturer and should be taken into account in designing experiments using a column setup, where allowance must be made for the change in volume accompanying the absorption of the solvent.

Chemical Properties of Ion-Exchange Resins

STOICHIOMETRY OF ION-EXCHANGE REACTIONS

Ion-exchange reactions involve ionic equilibria and are therefore subject to rules governing these reactions. They proceed stoichiometrically as (4.7) indicates.

$$RCOOH + Na^+ + OH^- \rightleftharpoons RCOO^-Na^+ + H_2O \qquad (4.7)$$

For every Na^+ ion removed from the external solution, a stoichiometrically equivalent number of H^+ ions diffuses out of the resin. Because of the requirement that electrical neutrality be maintained at all times, the exchange capacity of an ion-exchange resin is directly dependent on the electrical charge of the functional groups attached to the organic backbone rather than on the nature of the mobile ions.

BREAKTHROUGH AND TOTAL CAPACITY

When a solution of sodium chloride of a concentration C_0 is passed through a column containing a cation-exchanging resin in its acidic form, H^+ ions are exchanged for Na^+ ions. If the concentration of sodium ions in the effluent C_p is continuously monitored, a plot of C_p/C_0 versus the concentration of the sodium ions retained by the column gives a characteristic sigmoid curve called the "breakthrough curve" (Figure 4.3).

Point A of the curve is termed the "breakthrough point," and its value, expressed in milliequivalents of Na^+, is the "breakthrough capacity." This important parameter depends on several factors: the dimensions of the column, the flow conditions, the nature of the ion exchanger, the size of the beads, the temperature of the system, and the composition of the external electrolytes. At point B, C_p/C_0 is 0.5; this is the mean of the breakthrough and total capacity C. The total capacity ($C_p/C_0 = 1.0$) may be defined as the number of exchange sites equivalent per unit weight or volume of the material. When the total capacity is expressed in milliequivalents of exchange sites per gram of dry resin, it is called the "dry weight capacity." If the resin is in its swollen form, the term "wet volume capacity" is used, in reference to the

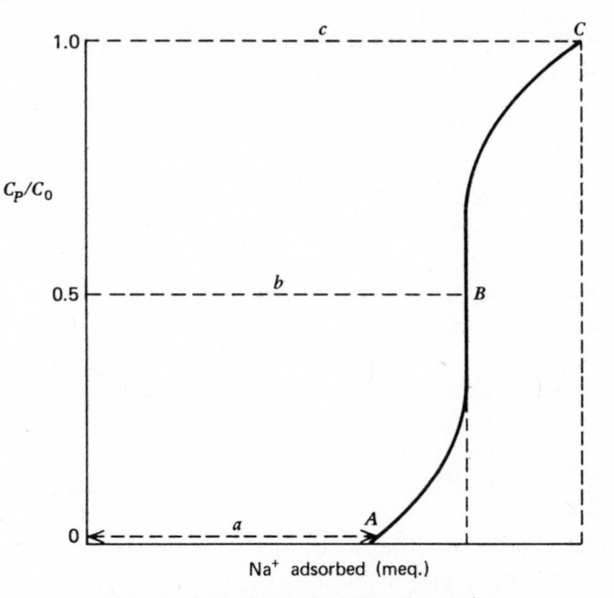

Fig. 4.3. Breakthrough and total capacity.

number of milliequivalents of exchange sites per unit volume of the swollen resin. Total capacity values can be determined by standard methods described in the literature [6]. They are also furnished by the resin manufacturers. Unless otherwise stated, the capacity values refer to the acidic form for cation exchangers and the chloride for anion exchangers.

AFFINITY OF THE RESIN FOR THE COUNTERION

Ion exchangers have different affinities for different counterions, and this is one of their most fundamental and useful properties. The equilibrium point for the following ion-exchange reactions

$$RSO_3^- Na^+ + Li^+ + Cl^- \rightleftharpoons RSO_3^- Li^+ + Na^+ + Cl^- \quad (4.8)$$

would depend not only on the relative concentration of the two competing ions but also on the preference exhibited by the resin for one ion over another. On the qualitative scale, ions can be arranged in a series called the selectivity sequence, which is very useful in designing an ion-exchange experiment. The order of affinities of some monovalent cations for a cation exchange resin follows the sequence

$$Li^+ < H^+ < Na^+ < NH_4^+ < K^+ < Rb^+ < Cs^+ < Ag^+ \quad (4.9)$$

Some divalent cations appear in the following order:

$$Cu^{2+} < Cd^{2+} < Ni^{2+} < Ca^{2+} < Si^{2+} < Pb^{2+} < Ba^{2+} \quad (4.10)$$

Selectivity series for anions are also available, and for some ions the sequence in increasing order of affinity is

$$F^- < \text{acetate} < \text{formate} < Cl^- < Br^- < I^- < NO_3^- < SO_4^{2-} < \text{citrate} \quad (4.11)$$

On a quantitative scale each ion can be assigned a numerical value depending on its position in the sequence. These values, expressed relative to a chosen ion, are known as selectivity coefficients. An example is given in Table 4.3, where the resin is initially in acidic form.

The selectivity coefficients apply to strong-cation and strong-anion types of resin. The strong cation exchangers contain sulfonic acid groups that are completely ionized, therefore have high affinity for cations. On the other hand, weak cation exchangers contain weakly dissociated carboxylic acid groups. Thus their affinity for cations is less than that of the stronger exchanger. The same trends are noted with strong, moderately strong, and weak anion exchangers.

The experimentally determined selectivity sequences can be correlated to measurable physical properties of ions [6–9]. The governing factor in ion-exchange reactions is the electrostatic force of attraction, which in turn depends mainly on the relative charge, the radius of the hydrated ions, and the degree of nonbonding interactions.

Table 4.3 Relative Selectivity Coefficients of Sulfonic Cation-Exchange Resins[a]

	Divinylbenzene (cross-link)		
Cation	4 %	8 %	10 %
Li	1.00	1.00	1.00
H	1.30	1.26	1.45
Na	1.49	1.88	2.23
NH_4	1.75	2.22	3.07
K	2.09	2.63	4.15
Rb	2.22	2.89	4.19
Cs	2.37	2.91	4.15
Ag	4.00	7.36	19.4
Tl	5.20	9.66	22.2

[a] Reproduced from *Amberlite Ion Exchange Resins—Laboratory Guide*, 1964, by permission of the Rohm & Haas Company.

Divalent ions have greater attraction for ions of the opposite charge than do monovalent ions. In addition, the affinity for the ions of the same valence is generally inversely proportional to their degree of hydration. However van der Waals interactions, complex formation, and the degree of polarization also play roles in determining the extent of sorption.

STABILITY OF ION-EXCHANGE RESINS

Both chemical and physical stability of the ion-exchange resins are necessary for the normal functioning of these materials. Resins should be chemically inert and not subject to moderate temperature changes. Most commercially available resins possess these properties, but slight variations may occur depending on the type of resin and the working conditions. More detailed data can be found in literature [10–12].

In spite of the relatively good stability of vinylbenzene cross-linked resins, strong oxidizing agents (boiling nitric acid or chromic-nitric acid mixtures) should be avoided, since they cause rapid decomposition of the resin. Temperature limitations imposed by the manufacturers should also be rigorously observed.

NOMENCLATURE

Ion-exchange resin notations may vary slightly from manufacturer to manufacturer, but they all contain essential physical and chemical parameters of the material. For example, the designation

$$\begin{array}{c}\text{Dowex 3-X4}\\ \text{50–100 mesh}\\ Cl^-\\ [RNH(CH_3)_2]^+Cl^-\end{array} \tag{4.12}$$

contains the trade name (Dowex 3), the percentage cross-linkage (4 %), the mesh size (50–100), the main ionic form (Cl^-), and the type of exchanger $[RNH(CH_3)_2]^+Cl^-$.

Mechanisms of Ion Exchange Reactions

Ion-Exchange Equilibria

The mechanism of ion-exchange reactions have been explained in several ways:

1. Ion-exchange reactions were treated as adsorption processes.
2. The law of mass action was applied.
3. The Donnan membrane theory was applied.

Let us treat these mechanisms in order.

1. The first treatment in which the Langmuir or Freundlich type isotherms has been invoked to explain the ion-exchange reactions gives a very incomplete interpretation of these phenomena. It is reported elsewhere in literature and is discussed in detail by Cassidy [13] and Morris and Morris [14]).

2. The equation representing an ion-exchange reaction may be written in general terms as follows:

$$RSO_3^-X^+ + Y^+ \rightleftharpoons RSO_3^-Y^+ + X^+ \tag{4.13}$$

where R represents the polymer matrix and X^+ and Y^+ are the exchangeable ions. Since an ion-exchange process is reversible, the law of mass action can be applied to the state of equilibrium. For the equilibrium constant to be a true thermodynamic value, activities should be used. Since the activities of the ions in the resin cannot be determined accurately, the equilibrium quotient K_q is expressed in concentration units:

$$K_q = \frac{[RSO_3^-Y^+][X^+]}{[RSO_3^-X^+][Y^+]} \tag{4.14}$$

In ion-exchange work it is customary to express concentrations in the resin phase as equivalents of counterion per unit weight of exchanger, while the concentrations of the ionic species in the aqueous phase are usually given in terms of normality. If the ions X^+ and Y^+ have the same charge, the concentration units cancel out in the equilibrium expression. In cases of unequal ionic charge, the equation representing the ion-exchange reaction contains coefficients that become exponents to which the concentration of the individual species must be raised:

$$3RSO_3^-X^+ + Y^{3+} \rightleftharpoons (RSO^-)_3Y^{3+} + 3X^+ \tag{4.15}$$

and

$$K_q = \frac{[(RSO_3^-)_3Y^{3+}][X^+]^3}{[RSO_3^-X^+]^3[Y^{3+}]} \tag{4.16}$$

In the equilibrium constant for an ion-exchange reaction involving unequal charges, the concentration units do not cancel out because of the exponential nature of the expression.

The values of K_q for a particular reaction vary with experimental conditions and the type of ions involved. They are constant only at low solute concentrations and within a narrow range of experimental conditions. The K_q values, which are termed "selectivity coefficients," are nevertheless very useful in expressing the affinity of a resin for a specific ion on both qualitative and quantitative scales. Since K_q values change with the initial state of the exchanger, for the sake of uniformity they are evaluated relative to a chosen reference standard. Therefore for the reaction

$$RSO_3^- Li^+ + X^+ \rightleftharpoons RSO_3^- X^+ + Li^+ \quad (4.17)$$

the selectivity coefficient is expressed as

$$K_{Li^+}^{X^+} \quad (4.18)$$

The Donnan membrane theory is derived from thermodynamic considerations and the requirements for electrical neutrality. An ion-exchange bead can be pictured as a concentrated electrolyte solution in which the ions of one sign are nondiffusible, whereas the counterions are free to move. The contact surface between the polymer bead and the external solution can be considered to be a semipermeable membrane. When equilibrium is attained between the external and internal solution, the concentration of free electrolyte within the bead will be less than that outside. This can be explained on the basis of the Donnan membrane theory. When a large nondiffusible ion is present on only one side of the membrane, the unequal distribution of diffusible electrolyte across a membrane that is permeable to both ions of the electrolyte is observed.

In case of a strong cation exchanger of the type $RSO_3^- Na^+$ in contact with a solution of sodium chloride, the polymer matrix with attached sulfonate groups constitutes a nondiffusible ion. The surface of the resin bead that acts as a membrane is permeable to both Na^+ and Cl^- ions. Figure 4.4 is a schematic representation of the surface of an ion-exchange bead. The dotted line represents the surface of the bead in contact with the external electrolyte. Since the electroneutrality principle has to be obeyed, the movable Na^+ ions within the bead counterbalance the negative charge of the sulfonate groups attached to the resin matrix. If a concentration gradient is present across the membrane surface, the diffusion of both Na^+ and Cl^- ions will take place (Fig. 4.4*a*). The ions from the external electrolyte move across the membrane as pairs because of the electroneutrality requirements. At equilibrium (Fig. 4.4*b*), the rate of diffusion is the same in both directions, and the ion-product concentration of the oppositely charged ions is the same on both sides of the membrane.

$$[Na^+]_R[Cl^-]_R = [Na^+]_S[Cl^-]_S \quad (4.19)$$

where subscripts R and S designate resin side and solution, respectively.

Charge balance requirements also have to be fulfilled:

$$[Na^+]_S = [Cl^-]_S \quad (4.20)$$

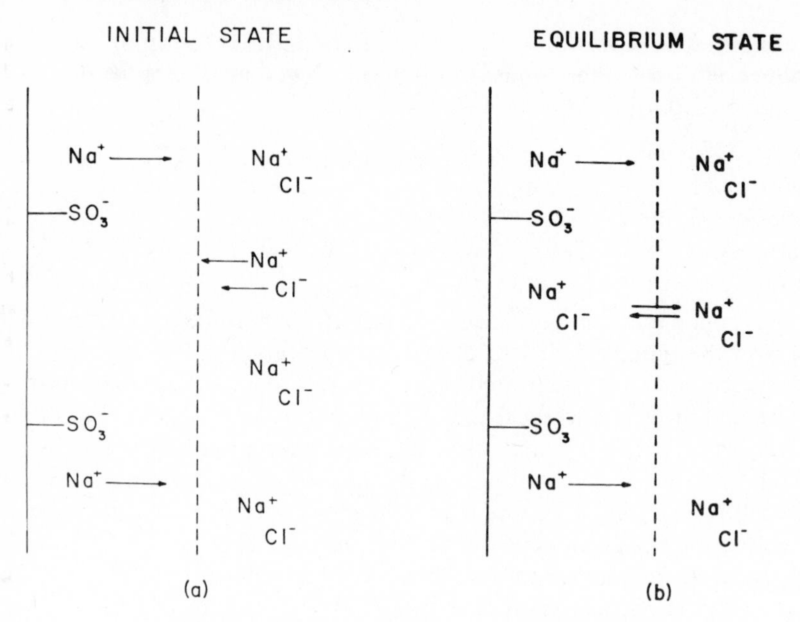

Fig. 4.4. Schematic representation of an ion-exchange bead.

and

$$[Na^+]_R = [Cl^-]_R + [RSO_3^-]_R \tag{4.21}$$

Combining (4.19) and (4.20), it follows that

$$[Cl^-]_S^2 = [Na^+]_R[Cl^-]_R \tag{4.22}$$

Substitution of (4.21) into (4.22) gives

$$[Cl^-]_S^2 = ([Cl^-]_R + [RSO^-]_R)[Cl^-]_R$$
$$= [Cl^-]_S^2 = [Cl^-]_R^2 + [RSO_3^-]_R[Cl^-]_R \tag{4.23}$$

which is equivalent to saying

$$[Cl^-]_S > [Cl^-]_R \tag{4.24}$$

or that the concentration of NaCl is greater in the external solution than within the resin bead. The Donnan membrane theory applies to anion exchangers as well as to cation exchangers, and it is also valid in ion-exchange reactions where more than one diffusible ion penetrates the membrane. In this case the concentration ratio of any given pair of ions is equal on both sides of the membrane.

For polyvalent ions the valence factor has to be considered in setting up the equilibrium expressions, if all other requirements are the same.

Ion-Exchange Kinetics

Although ion-exchange equilibria and thermodynamics have received extensive coverage, the kinetic studies of ion exchangers are still rudimentary. This is especially true in the quantitative treatment, which is restricted at present to idealized systems. However the general aspects of ion-exchange kinetics are well understood, and mathematical treatments are reported in literature [15–19]. Five possible steps may govern the kinetics of an ion-exchange reaction:

1. Diffusion of ions through the film adhering to the surface of the resin bead.
2. Diffusion of ions within the resin particle.
3. Exchange of counterions at the active site.
4. Diffusion of the exchange ions back to the surface of the resin bead.
5. Diffusion of the exchanged ions through the adherent surface film into the bulk of the solution.

In very dilute solutions ($<0.001\ M$) the rate-controlling step of the overall process is diffusion of the ions through the film. In more concentrated solutions the diffusion in the ion-exchanger particle itself controls the overall rate. Two other possible rate-controlling steps—passage through the particle-solution interface and exchange of ions at the fixed ionic groups—cannot be ruled out a priori. Although interfacial resistance to diffusion is unlikely for theoretical reasons, the actual exchange of counterions may be rate controlling in certain cases when counterions form sluggishly reacting complexes with fixed chelating groups.

Ion-Exchange Celluloses

Many different types of ion-exchange celluloses are now available. The degree of structure order may vary from amorphous to highly ordered "crystalline-type" fibers. Aggregates of glucosidic chains are held together by interchain bonding, which provides dimensional stability to the matrix. Various functional groups are introduced by chemical treatment of the matrix. The chemical groups react more readily in the localized amorphous areas than in the crystalline regions. Since the presence of functional groups disrupts the cellulose matrix, chemical treatments cannot be carried out to completion.

In addition to the fibrous cellulosic exchangers, microgranular cellulosic exchangers have found extensive use. They are usually prepared by subjecting cellulose to mild acid hydrolysis, which causes splitting of the chains and reduces or eliminates amorphous interfibrillar regions. This matrix can

Table 4.4 Commercially Available Cellulosic Ion Exchangers[a]

Ion Exchangers	*Ionizable Group*		*meq/g*
Anion exchanges			
AE–Cellulose	Aminoethyl-	$-O-CH_2-CH_2-NH_2$	0.3–1.0
DEAE–Cellulose	Diethylaminoethyl-	$-O-CH_2-CH_2-N(C_2H_5)_2$	0.1–1.1
TEAE–Cellulose	Triethylaminoethyl-	$-O-CH_2-CH_2-\overset{X}{N}(C_2H_5)_3$	0.5–1.0
GE–Cellulose	Guanidoethyl-	$-O-CH_2-CH_2-NH-\overset{NH}{\overset{\Vert}{C}}-NH_2$	0.2–0.5
PAB–Cellulose	*p*-Aminobenzyl-	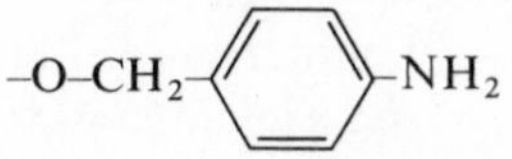 $-O-CH_2-C_6H_4-NH_2$	0.2–0.5
ECTEOLA–Cellulose	Triethanolamine coupled to cellulose through glyceryl and polyglyceryl chains; mixed groups		0.1–0.5
BD–Cellulose	Benzoylated DEAE–cellulose		0.8
BND–Cellulose	Benzoylated-naphthoylated DEAE–cellulose		0.8
PEI–Cellulose	Polyethyleneimine adsorbed to cellulose or weakly phosphorylated cellulose		0.1
Cation exchangers			
CM–Cellulose	Carboxymethyl-	$-O-CH_2-COOH$	0.5–1.0
P–Cellulose	Phosphate	$-O-\overset{O}{\overset{\Vert}{P}}(OH)-OH$	0.7–7.4
SE-Cellulose	Sulfoethyl-	$-O-CH_2-CH_2-\overset{O}{\overset{\Vert}{\underset{O}{\underset{\Vert}{S}}}}-OH$	0.2–0.3

[a] Reproduced from E. A. Peterson, *Cellulosic Ion Exchangers*, American Elsevier Publishing Company, Inc., New York, 1970, by permission of the author and the American Elsevier Publishing Company.

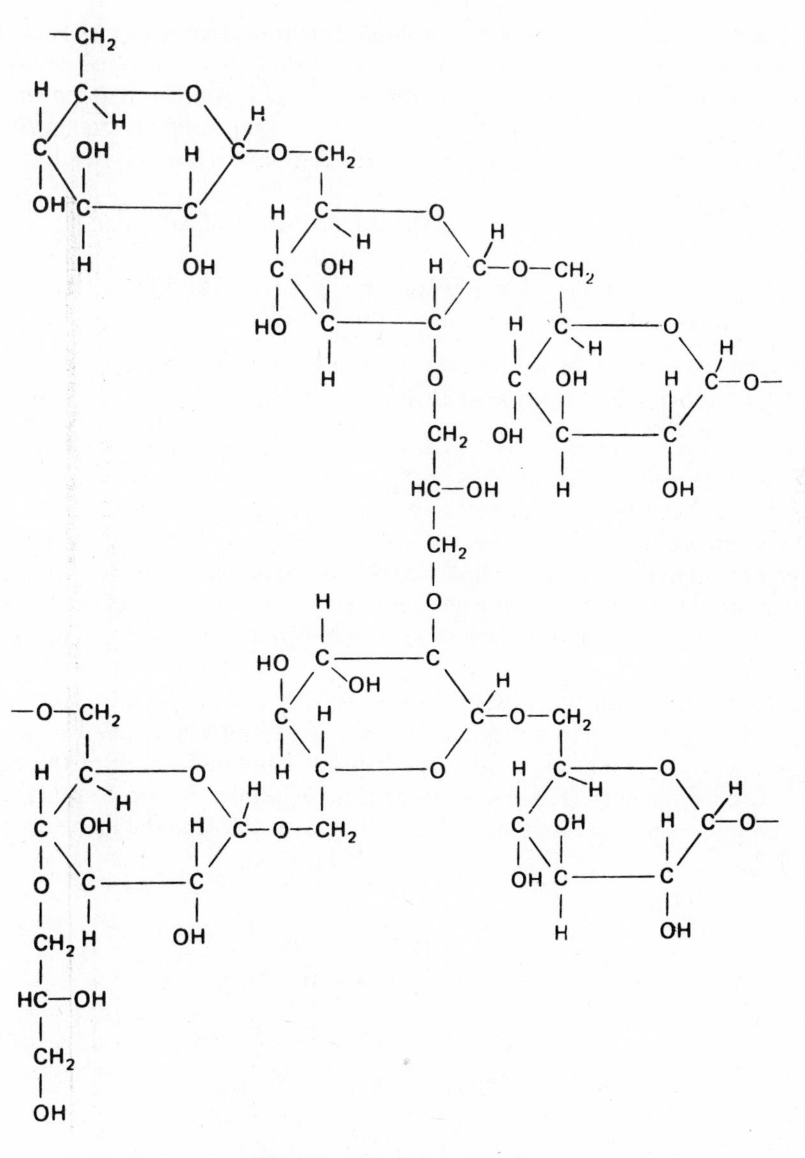

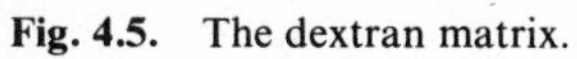

Fig. 4.5. The dextran matrix.

further be modified by cross-linking with bifunctional reagents before chemical attachment of the functional groups. Although the fibrous cellulosic exchangers are employed in separating large polyelectrolytes, the microgranular exchangers lend themselves very well to the study of small organic molecules. Some of the more common commercially available cellulosic ion exchangers are listed in Table 4.4.

Cellulosic exchangers find widespread use in purification of biological macromolecules such as enzymes, oligonucleotides and mononucleotides, nucleic acids, serum proteins, phospholipids, viruses, and hormones. More information can be found in the literature [20–27].

Ion-Exchange Dextrans and Polyacrylamide Gels

In contrast to the resinous exchangers whose matrix is hydrophobic, the dextran and polyacrylamide gels are hydrophilic. When immersed in water, they swell, forming insoluble gel-like structures.

Dextran exchangers are formed from linear polyglucose chains cross-linked through glycerin–ether linkages (Fig. 4.5). Polyacrylamide exchangers are prepared by polymerizing acrylamide in the presence of methylenebisacrylamide, which cross-links the polymerized acrylamide chains (Fig. 4.6).

If functional groups are not attached to the matrix, dextrans and polyacrylamide gels serve as molecular sieves in gel filtration. When used as ion exchangers, these materials have applications similar to those of the cellulosic ion exchangers. Table 4.5 gives some examples of the commercially available ion exchangers. More information is available in the literature [12, 28, 29].

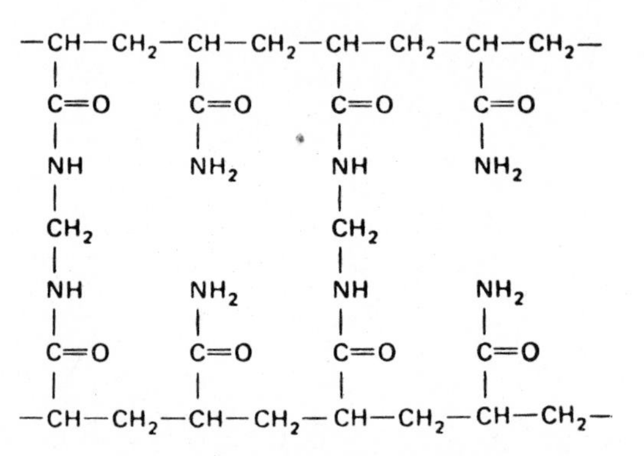

Fig. 4.6. Cross-linked polyacrylamide.

Table 4.5 Trade Names of the most Common Commercially Available Ion-Exchange Resins[a]

Resin Type	*Chemical Constitution*	*Usual Form as Purchased*	*Trade Names of Equivalent Ion Exchangers*[b]					
Strongly acidic cation exchanger	Sulfonic acid groups attached to a styrene and divinylbenzene copolymer	ϕ–$SO_3^-H^+$	Amberlite IR-120	Dowex 50W	Duolite C-20	Lewatit S-100	Ionac C-240 (or Permutit Q)	Zeocarb 225
Weakly acidic cation exchanger	Carboxylic acid groups attached to an acrylic and divinylbenzene copolymer	R–COO^-Na^+	Amberlite IRC-50	—	Duolite CC-3	Lewatit C	Ionac C-270 (or Permutit Q-210)	Zeocarb 226
Strongly basic anion exchanger	Quaternary ammonium groups attached to a styrene and divinylbenzene copolymer	$[\phi\text{–}CH_2N(CH_3)_3]^+Cl^-$	Amberlite IRA-400	Dowex 1	Duolite A-101D	Lewatit M-500	Ionac A-450 (or Permutit S-1)	Zeocarb FF (or De-Acidite FF)
Weakly basic anion exchanger	Polyalkylamine groups attached to a styrene and divinylbenzene copolymer	$[\phi\text{–}NH(R)_2]^+Cl^-$	Amberlite IR-45	Dowex 3	Duolite A-7	Lewatit MP-60	Ionac A-315 (or Permutit W)	Zeocarb G

[a] Reproduced from J. X. Khym, *Analytical Ion-Exchange Procedures in Chemistry and Biology—Theory, Equipment, Techniques*, Prentice-Hall, Englewood Cliffs, N.J., by permission of the author.

[b] Manufacturers: Amberlite, Rohm & Haas Company, Philadelphia; Dowex, Dow Chemical Company, Midland, Mich.; Duolite, Diamond Alkali Company, Redwood City, Calif.; Lewatit, Farbenfabriken Bayer, Leverkusen, Germany; Ionac (on Permutit of USA), Ionac Chemical Company, New York; Zeocarb, The Permutit Company, Ltd, London.

4 INORGANIC ION EXCHANGERS

A great variety of inorganic ion exchangers have been prepared in the past few years. The rebirth of the use of these substances was a result of the development of radiochemical engineering that required the use of ion-exchange materials resistant to radiation and heat.

Hydrous Oxide Exchangers

Hydrous oxides of certain tetravalent metals such as zirconium, thorium, tin, and tungsten can behave either as cation or anion exchangers, depending on the basicity of the central metal ion. A reaction scheme for making the cross-linked polymeric network either positively or negatively charged is

$$\gt\overset{+}{M} + OH^- \rightleftharpoons \gt M - OH \rightleftharpoons \gt M - \bar{O} + H^+ \tag{4.25}$$

$$\gt M - \overset{+}{O}H_2 \underset{}{\overset{H^+}{\rightleftharpoons}} \gt M - OH \overset{OH^-}{\rightleftharpoons} \gt M - \bar{O} + H_2O \tag{4.26}$$

At the isoelectric point these hydrous oxides behave as zwitterions and are capable of salt pickup. Depending on their form and type, these ion exchangers can be used to separate a variety of compounds, including oxyanions, halide ions, metals, and rare earth elements. More detailed study of these exchangers is available in the literature [30–35].

Acid Salt Exchangers

When some tetravalent metals such as zirconium, thorium, tin, and tungsten are reacted with the polyvalent ions like phosphate, tungstate, molybdate, or arsenate, salts of nonstoichiometric chemical composition are formed. Their ion-exchange properties are discussed in the literature [31].

Heteropoly Acid Salts

Heteropoly acids are compounds with the general formula $H_3[XY_{12}O_{40}] \cdot nH_2O$, where X can be phosphorus, arsenic, or silicon, and Y is molybdenum, tin, or vanadium. Some heteropoly acid salts like ammonium phosphomolybdate or phosphotungstate possess ion-exchange properties by the virtue of their crystallinity. Cations of these salts occupy large cavities and can be exchanged for other cations from the external solution if the latter is small enough to fit into the crystal network without causing distortions or bond breaking. These exchangers have been applied in selective extraction of some metals [12, 31, 36–38].

Metal Sulfides

For removal of trace amounts of transition metals and other heavy metals from aqueous solutions, certain sulfides (Ag_2S, CuS, CdS, PbS, As_2S_3, etc.) have proved to be very useful [39, 40]. The driving force in these "displacement-type reactions" is the formation of a sulfide that is less soluble than the exchanger sulfide.

Phosphate Gels

When combined with an appropriate filler of very small particles and large surface area, some inorganic phosphate gels such as hydroxylapatite can be used in column operation. Their major field of applications is in biochemical research.

5 MISCELLANEOUS

A new approach for preparing selective adsorbents has been developed in the Soviet Union [41]. Alumina or activated carbon is impregnated with selective precipitating agents such as dithizone, dimethylglyoxime, or phenylarsonic acid. Ion-exchanging foams of varying degree of homogeneity have also been used in oxidation-reduction reactions [42, 43]. It was found that certain preparations of adenosine tri- and monophosphates (ATP and AMP) with foreign ions incorporated into the structure appear to have ion-exchange properties [44, 45].

6 HPLC PACKING MATERIALS

Since the advent of HPLC, a great number of different commercial column packing materials have become available. Classical column chromatography utilized large diameter particles (>100 μm), but modern analytical liquid chromatography uses small diameter particles (down to 5 μm), narrow bore columns (as small as 1 mm i.d.), and high inlet pressures (up to 5000 psig).

Among the many different types of ion exchanger, the new bonded-phase materials deserve special attention. These packings contain particles of siliceous support to which the ion-exchange functional groups are attached by means of a covalent bond. An example is given in Fig. 4.7. Ion-exchange packing materials are available with several particle geometries. The microreticular resin (Fig. 4.8*a*) contains spherical particles of divinylbenzene cross-linked polystyrene. When immersed in a solvent, these particles have micro pores. When a resin also contains macro pores (several hundred angstroms order of magnitude) in addition to the micro pores, the resin is called macroreticular (Fig. 4.8*b*). Pellicular ion exchangers (Fig. 4.8*c*) consist of a thin

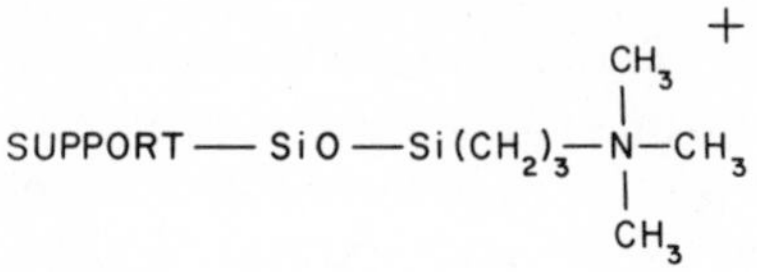

$$\text{SUPPORT}-\text{SiO}-\text{Si(CH}_2)_3-\overset{+}{\underset{\text{CH}_3}{\overset{\text{CH}_3}{\text{N}}}}-\text{CH}_3$$

Fig. 4.7. Structure of a bonded-phase ion exchanger.

film of styrenedivinylbenzene polymerized onto a spherical glass bead [46]. The particles are then chemically treated, to attach the functional groups to the matrix. The superficially porous ion-exchange resins (Fig. 4.8*d*) stem from Kirkland's work [47] and have an inert core with a porous surface to which the ion exchanger is bonded, either chemically or mechanically. In comparison with the totally porous materials, the thin coating of resin in the pellicular and superficially porous materials enables rapid solute mass transfer, higher efficiency, and faster equilibration; because of the very small

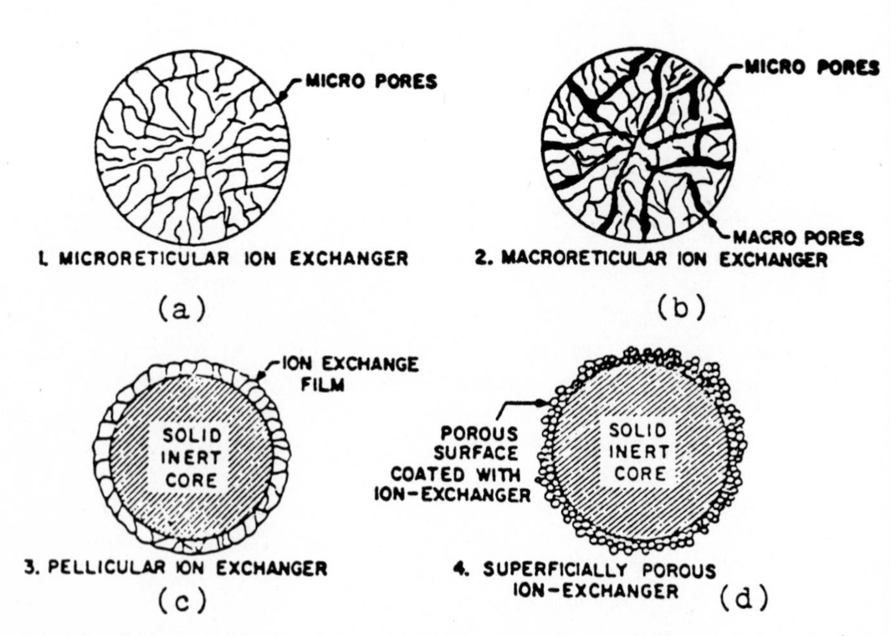

Fig. 4.8. Particle geometries of ion exchangers. From R. E. Leitch and J. J. de Stefano, *J. Chromatogr. Sci.*, **11**, 105 (1973). Reprinted by permission of the authors and the editor.

amount of available ion-exchange material. However, their capacities are lower than those of the totally porous packings. The characteristics of pellicular columns have been reviewed by Kirkland [47]. Tables 4.6 and 4.7 list, respectively, some porous ion-exchange resins and some superficially porous or pellicular ion-exchange resins for HPLC [48].

Liquid Chromatography Columns

Columns can be either purchased prepacked or made up in the laboratory, using the packing procedures described in literature [49–51].

The prepacked columns are constructed to fit almost any commercially available chromatograph. They come in a wide range of sizes and packing types. The most common column size is 25 cm long, 4 mm i.d. Larger column diameters are used in preparative work where the emphasis is on the capacity rather than on the efficiency.

Straight columns are preferred to coiled columns because the former are easier to pack and usually have higher efficiencies. The effect of the column shape on efficiency has been discussed in the literature [71, 72].

Footnotes for Table 4.6 (pages 220–221)

[a] Reproduced from R. E. Majors, *American Laboratory*, p. 13, October 1975, by permission of the author and the publisher.

[b]
1. Alltech Associates, Inc.
2. Altex Scientific, Inc.
3. Applied Research Laboratories
4. Applied Science Laboratories, Inc.
5. Beckman Instruments
6. Bio–Rad Laboratories
7. Durrum Chemical Corp.
8. E. I. DuPont de Nemours
9. E. Merck (Darmstadt, Germany)
10. Hamilton Co.
11. Hewlett-Packard
12. Macherey–Nagel & Co. (Germany)
13. Reeve Angel (Whatman)
14. The Separations Group
15. Touzart and Matignon (France)
16. Tracor, Inc. (Chromatec)
17. Varian Aerograph
18. Waters Associates

[c] DVB-PS, divinylbenzene-polystyrene.

[d] SB, strongly basic; WB, weakly basic; SA, strongly acidic, WA, weakly acidic.

Table 4.6 Porous Ion Exchangers for HPLC[a]

Type	*Name*	*Supplier*[b]	*Particle Size* (μm)	*Base Material*[c]	*Stre*
Anion	Aminex A-series	2, 6, 15, 17	A-14, 20 ± 3 A-25, 17.5 ± 2 A-27, 13.5 ± 1.5 A-28, 8 ± 2	A-14, 4% DVB-PS Others, 8% DVB-PS	S
	Durrum DA-X4F	7	11 ± 1	4% DVB-PS	S
	Durrum DA-X8F	7	8 ± 1	8% DVB-PS	S
	Hamilton 7800 series	3, 10, 15	20 ± 5	4, 6, or 8% DVB-PS	S
	Ionex SB	12	5–20	7% DVB-PS	S
	Partisil SAX	1, 13, 15	10	Silica gel	S
	Nucleosil-SB	12	5–10	Silica gel	S
	Vydac TP anion	2, 14, 15	10	Silica gel	S
	μ Bondapak-NH_2	18	10	Silica gel	W
	MicroPak-NH_2	17	10	Silica gel	W
	NucleosilMNH_2 or $-N(CH_3)_2$	12	5–10	Silica gel	W
Cation	Aminex A-series	2, 6, 15, 17	A-4, 20 ± 4 A-5, 13 ± 2 A-6, 17.5 ± 2 A-7, 7–11	8% DVB-PS	S
	Beckman AA-series	5	AA-20, 10.5 ± 1 AA-15, 11 ± 3	8% DVB-PS	S
	Beckman PA-series	5	PA-35, 11 ± 3 PA-28, 16 ± 4	7% DVB-PS	S
	Durrum-DC-1A	7	14 ± 2	8% DVB-PS	S
	Durrum-DC-4A	7	9 ± 0.5	8% DVB-PS	S
	Durrum-DC-6A	7	11 ± 1	8% DVB-PS	S
	Hamilton H-70	3, 10, 15	24 ± 6	8% DVB-PS	S
	Hamilton HP-series	3, 10, 15	AN 90, 13.5 ± 6.5 B 80, 8.5 ± 1.5	7% DVB-PS 7.75% DVB-PS	S S
	Inoex SA	12	10 ± 2 15 ± 2.5 20 ± 3	8% DVB-PS	S
	Chelex 100	6	25 ± 5	DVB-PS	W
	Nucleosil-SA	12	5, 7, 5, 10	Silica gel	S
	Partisil CSX	1, 13, 15	10	Silica gel	S
	Vydac TP Cation	2, 14, 15	10	Silica gel	S

Footnotes *a*, *b*, *c*, and *d* on page 219.

nctional Group	Ion-Exchange Capacity, Dry (meg./g)	Description
$-NR_3^+$	A-14, 3.4	A-28 can withstand pressures of 10,000 psi
$-NMe_3^+$	2	Also DA-X4, 20 $\pm$ 5 μ, for acidic biomolecules, peptides
$-NMe_3^+$	4	Also DA-X8, 11 $\pm$ 1 μ, for anions in urine
$-NR_3^+$	5	For carbohydrates, nucleotides
$-NR_3^+$	3	Moisture 55%
$-NR_3^+$		pH 1.5–10, T = 70°C
$-N(CH_3)_3^+$	1	pH 0–9; may be used with organic eluents
$-NR_3^+$		—
$-NH_2$		pH 2–9
$-NH_2$		pH 2–9
$-NH_2$		pH 2–9
$-N(CH_3)_2$		
$-SO_3^-$	5	A-7 can withstand pressure of 10,000 psi
$-SO_3^-$	5	Made for amino acid analysis
$-SO_3^-$	5	Made for amino acid analysis
$-SO_3$	5	For protein hydrolysates, amino acid analysis
$-SO_3^-$	5	For amino acid analysis, single column
$-SO_3^-$	5	For amino acid analysis, single column
$-SO_3^-$	5.2	For acidic and neutral amino acid analysis; hydrolysate analysis only
$-SO_3^-$	5.2	B-80 for basic amino acid analysis; AN-90 for acidic and amino acids
$-SO_3^-$	5	Moisture 55%
Immino-diacetate	2.9	Chelating resin, for ligand exchange; 68–76% water
$-SO_3^-$	1	pH 0–9: may be used with organic eluents
$-SO_3^-$		pH 1.5–10, T = 70°C
$-SO_3^-$		—

Table 4.7 Superficially Porous or Pellicular HPLC Ion-Exchange Resins[a]

Type	Name	Supplier[b]	Particle Size (μm)	Base[c]	Streng
Anion	AE-Pellionex-SAX	1, 13, 15, 16	44–53	PS-AE	SB
	AS-Pellionex-SAX	1, 15, 16, 17	44–53	PS-DVB	SB
	AL-Pellionex-WAX	1, 13, 15, 16	44–53	Al	WB
	Bondapak/AX/ Corasil	18	37–50	BP	SB
	Perisorb-AN	9, 11, 15	30–40	BP	SB
	Permaphase-AAX, –ABX	8	37–44	BP	SB
	Vydac SC anion	4, 12, 14, 15, 16	30–44	PS-DVB	SB
	Zipax-SAX	8	25–37		SB
	Zipax-WAX	8	25–37	PAM	WB
Cation	BondaPak CX/ Corasil	18	37–50	BP	SA
	HC- or HS- Pellionex-SCX	1, 13, 15, 16, 17	37–53	PS-DVB	SA
	Perisorb-KAT	9, 11, 15	30–40	BP	SA
	Zipax-SCX	8	FC	FC	SA
	Vydac SC cation	4, 11, 12, 14, 15, 16	30–44	PS-DVB	SA

[a] Reproduced from R. E. Majors, *American Laboratory*, p. 13, October 1977, by permission of the author and the publisher.

[b] 1. Alltech Associates
2. Altex Scientific
3. Applied Research Laboratories
4. Applied Science Laboratories
5. Beckman Instruments
6. BioRad Laboratories
7. Durrum Chemical Corp.
8. E. I. DuPont de Nemours
9. E. Merck (Darmstadt, Germany)
10. Hamilton Co.
11. Hewlett–Packard
12. Macherey–Nagel & Co. (Germany)
13. Reeve Angel (Whatman)
14. Separations Group

nctional Group	*Ion-Exchange Capacity, Dry (μ eq./g)*	*Description*
$-NR_3^+$		T = 70°C; pH 2–10; organic solvents acceptable
$-NR_3^+$	10	May be used to 85°C; pH 2–12; can be used with organic solvents
$-NH_2$	—	pH 2–7; T = 75°C
$-NR_3^+$	10	pH 2–7; organic solvents acceptable
$-NR_3^+$	30	pH 1–9
$-NR_3$	10	Stable to 75°C; pH 2–9; organic solvents acceptable; strong oxidizing agents; 1% stationary phase
$-NR_3^+$	100	Supplied in chloride form; organic solvents acceptable
$-NR_3^+$	12	Lauryl metharcylate polymer base; MW 500; pH 4–10
$-NH_2$	12	Polyamide base
$-SO_3^-$	30–40	pH 2–8; T = 60°
	HS, 8–10	pH 2–10; T = 40°C; organic solvents acceptable
$-SO_3^-$	HC, 60	
$-SO_3^-$	50	pH 1–9
$-SO_3^-$	5	Only small amounts of organic solvents; fluoropolymer base; MW 1000
$-SO_3^-$	100	Supplied in hydrogen form

5. Touzart and Matignon (France)
6. Tracor (Chromatec)
7. Varian Associates Aerograph
8. Water Associates

S-AE, polystyrene–aliphatic ester copolymer; PS-DVB, polystyrene-divinylbenzene copolymer; , aliphatic; BP, bond phase through silohexane bond; PAM, polyamide base; FC, orpolymer base copolymer.

B, strongly basic; WB, weakly basic; SA, strongly acidic.

The Unpacked Column

Some features of the unpacked column that can influence the final separation efficiency are

(*a*) ability to withstand pressure;
(*b*) smoothness of the internal walls;
and (*c*) length and diameter.

The tube itself should be made of material inert to the solvents used in a particular chromatographic separation. Among the materials commonly used are glass, steel, and plastic. Precision bore steel or glass columns have been the materials of choice in most applications. When chemisorption on steel is expected to occur, tantalum columns may be used as an alternative.

If the separations are to be performed at temperatures other than ambient, the columns can be jacketed to allow heating or cooling liquids to be pumped around the outer surface of the chromatographic tube.

Packing the Column

Since most ion-exchange materials swell considerably when brought in contact with water or other polar solvents, ion-exchange columns should not be packed dry.

The choice of the packing technique depends on the properties of the exchanger to be used. For particle sizes of 200 to 400 mesh, most commonly used in routine work, the "multistage batch-packing technique" is usually employed. It involves several steps:

1. The column with a porous support at the outlet is mounted in the upright position.
2. The exchanger is added as a slurry (1 volume of settled exchanger to 1 volume of water).
3. The exchanger is allowed to settle and the solvent is either drained out or aspirated from above, until a few centimeters of liquid remains above the exchanger bed.
4. Steps 1 through 3 are repeated until the column is filled to the desired height.

The "single-stage batch-packing method" is used for packing the column with particles of intermediate sizes. It is similar to the multistage batch-packing method except that a stirring motor is usually used in preparing the slurry to be fed into the column.

For some time it has been popular opinion that it is impossible to pack stationary phases of particle diameters below 20 μm. More recently, the slurry-packing procedure has been described in the literature and used by many researchers in the field of chromatography.

A balanced-density slurry is prepared using the packing material and a balanced-density solvent (or a mixture of solvents). The amount of solvent used in preparing the slurry is determined by trial and error. The direction of migration of the particles is observed and the amount of solvent added adjusted until the particles do not settle down. The dispersed slurry is then degassed by vibrating it for 15 min in an ultrasonic bath. Packing the column with the balanced-density slurry involves seven steps.

1. A frit of proper porosity is fitted at the end of the column, and the outlet is closed with a plug.
2. The column is filled with the slurring solvents, making sure that there are no trapped air bubbles.
3. The slurry is carefully fed into the reservoir connected to the top of the column.
4. The system is then pressurized to approximately 6000 psi.
5. The plug at the column outlet is removed and the slurry is pumped into the column until approximately 100 ml of the solvent has been collected.
6. The pump is shut off and the column carefully removed, without disturbing the bed by shaking or expansion.
7. The front fittings are installed, and the column is ready for chemical and thermal equilibration. This is done by passing through it the solvent to be used in a particular separation.

The chromatogram obtained on the first injection is unlikely to be the desired one. The k' values may be either too large (long retention times and barely discernible bands) or too small (poorly resolved peaks eluted near t_0). In most cases the k' values can be improved by changing the solvent strength.

If the efficiency of the system is too low (low column plate number), reducing the solvent flow rate will give better efficiency at the cost of increasing the separation time. Column length can also be increased if higher efficiency is needed. This leads to a proportional decrease in linear mobile phase velocity, and the separation time will increase with the square of the column length. Majors and MacDonald [73] have studied the practical implications of column performance in liquid chromatography. They showed that faster separations can be achieved with shorter columns at lower pressures, using smaller diameter particles. It is also possible to increase N without increasing the analysis time, simply by increasing the pressure.

For separations where the R_s value for the two peaks of interest is small (<0.5), and the k' factor is greater than 2, an increase in N may be impractical. In these difficult separations ($\alpha \approx 1$), necessary resolution can be achieved in reasonable time by changing the selectivity. This can be done by holding the solvent strength constant while changing its composition.

In the case of a general elution problem where the k' values of the components vary widely, the migration rates of the early emerging peaks must be slowed down and the migration of the late eluting peaks must be accelerated. This can be achieved in several ways: (*a*) solvent programming (stepwise or gradient elution), (*b*) coupled columns, (*c*) temperature programming, (*d*) repeated separation, and (*e*) flow programming.

In solvent programming, the composition of the solvent is changed over a period of time.

Coupled-column systems usually employ two columns with different stationary phases; the first one provides reasonable k' values for the late-emerging peaks; the front end of the chromatogram can then be well resolved on the second column.

Temperature programming, which involves increasing the temperature during the separation, often results in changes in solvent viscosity that might cause a loss in overall resolution. Therefore this approach is seldom used to solve the general elution problem.

In repeated separations, the conditions are selected for optimum resolution of the late-emerging peaks. The incompletely resolved initial peaks are collected as a fraction and rechromatographed under conditions optimized for their separation.

Finally, flow programming can also be used to solve a general elution problem. Starting with a low solvent flow to provide a better resolution of the initial peaks, the solvent flow is increased until the last peaks are eluted within a reasonable period of time.

7 ION-EXCHANGE TECHNIQUES AND APPLICATIONS

Choice of an Exchanger

In selecting an appropriate ion exchanger for a specific operation, several factors should be considered:

1. *The net charge of the solute.* Depending on whether the solute is anionic or cationic, an appropriate type of exchanger can be chosen. When a molecule is a zwitterion and its isoelectric point lies near neutrality, either a cation or anion exchanger can be used. All strongly acidic and strongly basic exchangers become dissociated in a pH range from about 1 to 13. Weakly acidic cation exchangers do not show any affinity for the cation of a neutral salt, and their capacity decreases rapidly below pH 4. Weakly basic anion exchangers are unstable at pH values below 7.

For the solutes whose pK_a or pK_b are known, the rule of thumb for a starting pH is to take the highest pK_a of all the solutes to be separated and add 1.5. For the bases, the lowest pK_b is taken and 1.5 is subtracted.

2. *The molecular weight of the solute.* The size of the solute molecule (its molecular weight) is an important factor in selecting a proper exchanger.

When synthetic resins of the vinyl cross-linked types are used, care should be taken to ensure that the pore size permits free movement of the solute molecules. For large organic molecules with molecular weight greater than 500, ion-exchange celluloses, dextrans, and similar resins are used.

3. *The size and the net charge of the solute.* Some large organic molecules, especially the ones of biological importance, can become irreversibly bound to the resinous exchangers because of their high net charge. Therefore instead of undergoing ion-exchange reactions, they are sorbed onto the resin. Thus cellulose exchangers, polyacrylamide, and dextran ion-exchange gels are used in this field because of their large pore size and low number of exchange sites.

4. *Physical and chemical properties of the medium from which the solute is to be extracted.* Certain reaction media demand the use of ion-exchange materials that are able to withstand temperatures above 100°C, high levels of radiation, and the presence of strong oxidizing or reducing agents. These requirements stimulated the development of new inorganic exchangers that fulfill these needs.

Ion-Exchange Processes

Three techniques are used for the practical application of ion-exchange procedures.

Batch Operation

In the batch operation, which is the simplest ion-exchange process, the contact between the exchanger and the electrolyte solution is made by stirring or shaking the mixture. After equilibrium has been established, the mixture is separated by filtration, settling, or centrifugation. The degree to which ion exchange takes place depends on the equilibrium constant of the exchange process. This technique can be successfully applied only to processes that have a favorable equilibrium constant (i.e., the ones that involve the formation of a weak electrolyte, an insoluble product, or a stable complex). Even then, the efficiency of the one-step exchange is usually quite low and the process must be repeated in what is known as a "cascade" or "multistage operation." This process is similar to discontinuous extraction and adsorption techniques.

A variation of this process is the RIP (resin-in-pulp) process, used for the industrial recovery of uranium. It involves an ion exchanger that is placed in the basket and moved up and down in the tanks filled with crude regenerating solution. This process seems to have potential for other applications [49].

Column Process

The column process is the most common and efficient ion-exchange method. It employs a fixed bed of resin which, in contrast to the batch method, does not form a homogeneous mixture with the electrolyte solution. The overall operation can be pictured as a multistage batch procedure in which multiple equilibria are established between the two phases. Thus a high-efficiency process results.

The routes by which the separation of ions can take place in column chromatography may differ. Accordingly, different elution techniques are defined on the basis of the procedures used for the column operation. The three fundamental types of column operation are as follows.

1. Frontal development, the simplest chromatographic procedure, consists of passing the sample solution continuously through the column. While the individual components migrate through the column, they are chromatographically separated in order of increasing selectivity. The appearance of the individual ions in the eluate is followed by any suitable detection method. Figure 4.9 is the chromatogram of the mixture of three components A, B, and C, where compound A has the least affinity and compound C has the greatest affinity for the resin. As the chromatogram indicates, only a portion of component A can be recovered from this mixture. Components that have the greatest affinity for the column are always contaminated by those with the least affinity. This limits the use of frontal development to qualitative analysis.

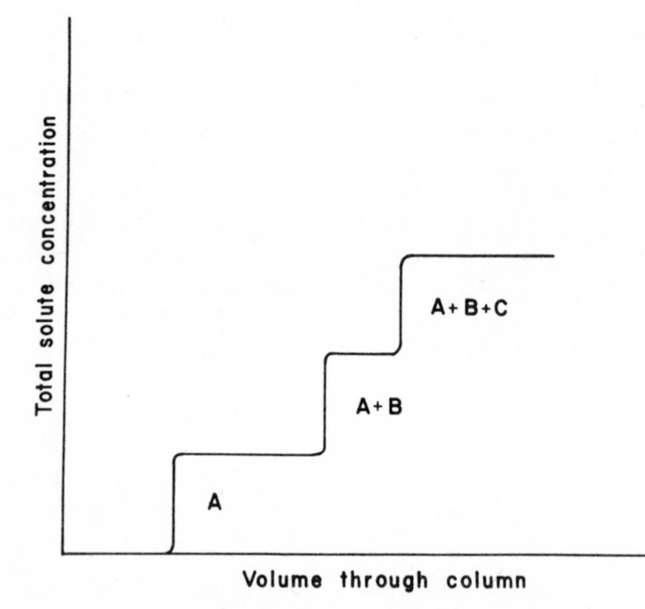

Fig. 4.9. Frontal development.

2. In the displacement technique the ions to be separated are charged on the column and exchanged with the resin counterions. To perform a chromatographic separation, this step is followed by washing with an electrolyte solution that contains ions having higher affinity for the resin than do the sample ions. These preferentially sorbed ions consequently displace all other ions. If, for example, sodium and potassium ions were to be separated on an acidic cation exchanger, a solution of calcium chloride could be used to perform the displacement, and all other ions could be displaced by calcium in accordance with the selectivity sequence $H^+ < Na^+ < K^+ < Cs^+ < Ca^{2+}$. Figure 4.10 illustrates this separation. As the chromatogram reveals, displacement technique furnishes zones of pure compounds and zones of mixed fractions. The mixed fractions can be subject to further separation by repeating the process. This method is mainly suitable for preparative separations.

3. In elution analysis the ions to be separated are charged on the column so that initially the sample occupies only a narrow band at the top of the column. The separation of the mixture is then performed by using an eluent of higher ionic strength. The ions with lower affinities move fastest through the column, and those with higher relative selectivity coefficients are retarded. The ions that are to be separated form zones or bands in the course of their migration through the column. The concentration gradient within each band

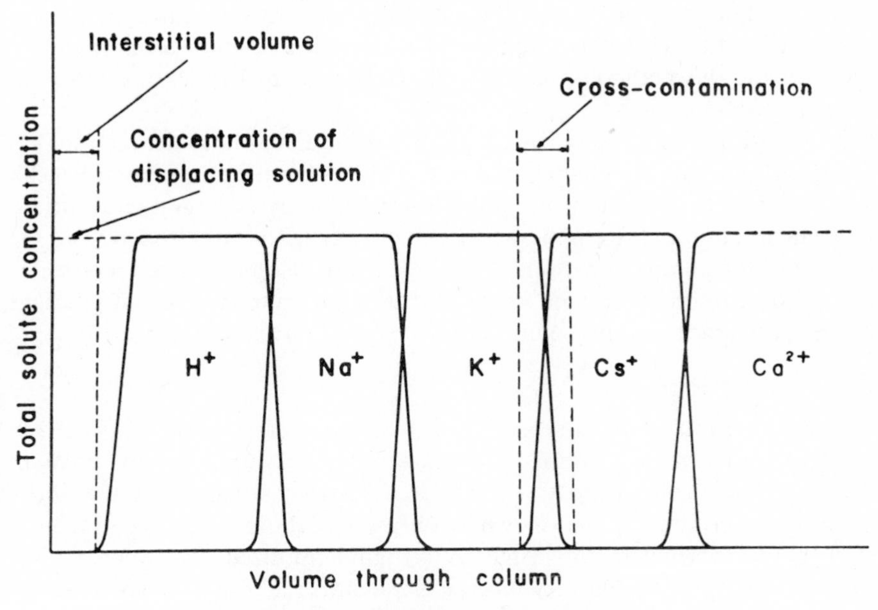

Fig. 4.10. Displacement development.

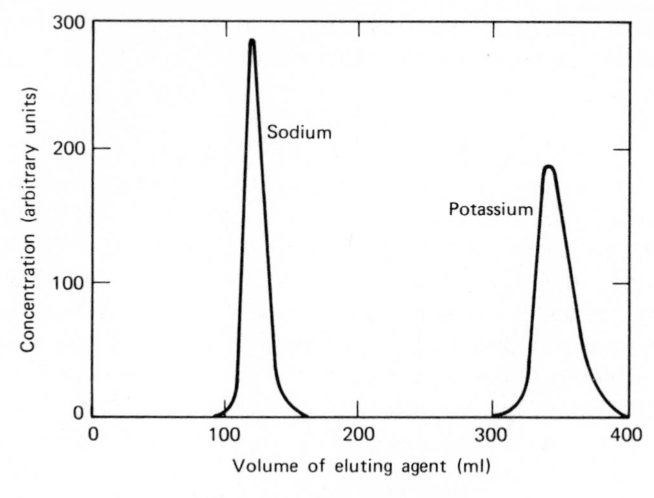

Fig. 4.11. Elution analysis.

is not sharp because diffusion occurs during migration. Therefore the elution curves are not square but bell-shaped (Gaussian curve of error). For example, the separation of sodium and potasium on a cation exchanger is illustrated in Fig. 4.11. The ion exchanger was in the acidic form and HCl was the eluent. The chromatogram shows that the elution mode furnishes pure fractions of the components. Thus this technique is a valuable analytical tool. The main drawback to the method is that only relatively small quantities of solutes can be analyzed.

A variation of elution chromatography is gradient elution, which makes use of a continuously changing eluent. This technique can minimize peak tailing, improve resolution, and provide a faster analysis. In gradient elution, concentration, ionic strength, or pH can change with time. In addition, the change of the anions or cations in the eluent can also be varied. Stepwise or continuous gradients can be used. With the continuous mode of gradient elution, the gradients can be linear, concave, or convex.

Continuous Process

In the continuous process type of ion-exchange operation, which is usually used in industrial operations, the exchanger and the solution move in countercurrent columns. This process was designed to eliminate the drawbacks of the column operation, where only one part of the bed is performing the separation, while the other part is virtually unutilized. In the continuous ion-exchange process, the consumed ion exchanger is continuously removed and fresh exchanger is supplied. The consumed ion exchanger is then regenerated and used again. A great advantage of this type of column operation is that a smaller amount of exchanger is needed to carry out a particular reaction.

Mixed-Bed Exchangers

In complex operations, instead of using an alternating sequence of cation and anion columns, mixed-bed exchangers composed of a cation and an anion exchanger are sometimes employed. In the mixed bed, cation and anion exchange are coupled into a single irreversible and complete reaction, and faster and more complete separations are achieved.

The regeneration of mixed-bed ion exchangers is a problem because hydraulic separation of the two exchangers is required, followed by consecutive treatment of the exchangers with the regenerating solutions. This method is rather troublesome, and in some instances the exchangers are discarded after one operation.

Since mixed-bed exchangers require little space and are relatively inexpensive, they are used in the preparation of ultrapure water with a conductivity of 0.05 μS/cm and a silicate content of 0.002 ppm.

Ion Exclusion

The ion-exclusion method is used for separating certain nonelectrolytes or weakly ionized solutes from electrolytes. This method is based on the following properties: organic nonionized components tend to distribute themselves in equal concentration between the resin interior and the external solution, but the concentration of the ionic substances becomes higher in the external solution than in the resin liquid, as predicted by the Donnan theory. Elution of such a mixture results in the appearance of the ionized solute, followed by the nonionized substance, since the latter has to displace both the dead space of the column and the resin liquid.

Ionophoresis

Ionophoresis is a method that employs a column of an ion exchanger to which a potential difference is applied. When the electrolyte is passed through the column, the ions are forced to move and are separated according to their mobilities.

Ion-Exchange Membranes

Ion exchangers can be produced in a membrane form that is permeable to water and/or other electrolytes. If the membrane allows the permeation of water only it is known as semipermeable; if it separates anions from cations, or vice versa, it is called permselective. There are three types of ion-exchange membrane: heterogeneous, interpolymer, and homogeneous.

Heterogeneous membranes are produced by mixing an ion-exchange granulate with an inert elastic binder and compressing the mixture into sheets. Interpolymeric membranes are made by casting a film from a homogeneous solution of two polymers, a polyelectrolyte and a water-soluble filmogenic material. The polyelectrolyte is firmly built into the chains of the matrix polymer so that immersion in water usually does not elute the polyelectrolyte. The homogeneous membranes are formed by copolymerization or polycondensation, followed by the introduction of functional groups into the matrix. These membranes have superior characteristics over the other two types.

All the membranes mentioned can be either cationic or anionic. It is also possible to construct composite membranes that contain both the cation- and anion-exchange groups. These membranes can be of two kinds: "mosaic" membranes, which are formed by parallel arrangements of the cationic and anionic regions, and the "sandwich membranes," featuring a serial arrangement of the two layers. The former have an unusually high salt permeability [53], whereas the latter behave like rectifiers [54, 55]. More information is available in the literature [56–60].

Oxidation-Reduction Polymers

Living organisms function by virtue of being both polymeric redox and ion-exchange systems: analogously, synthetic oxidation-reduction polymers (e.g., Fig. 4.12) have been produced [61]. Cassidy and co-workers [62–67] have investigated cross-linked copolymers of vinylhydroquinones and their derivatives and have prepared the polymers by protecting the hydroxyl groups of the monomer by esterification before free-radical polymerization. Lautsch et al. [68] incorporated porphyrin groups into large polymers. Other structures have been studied, involving either condensation polymers or conventional ion exchangers, which are used as the substrate for sorption of the redox-active ions or molecules. However redox resins have not yet reached a stage of commercial significance.

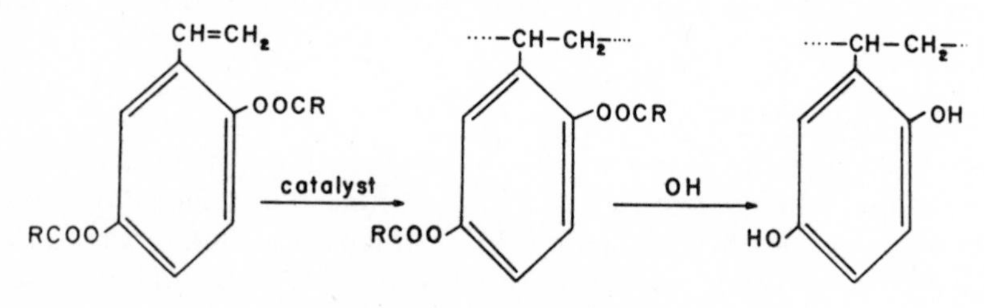

Fig. 4.12. Redox polymers.

Chelating and Ligand-Exchange Resins

Chelating ion exchangers are exchangers in which functional groups with properties of a specific reagent are introduced. Because of their specificity, these ion exchangers can sorb one ionic species to the exclusion of others. An example of this type of exchanger is the iminodiacetate ion exchanger (Dowex A-1, Fig. 4.13), in which the iminodiacetate groups are directly attached to the styrene matrix. Dowex A-1 can fix polyvalent ions with a high affinity by the formation of heterocyclic metal-chelate complexes. These exchangers have a variety of applications in trace metal analyses of natural waters, biological fluids, reagents, and so on.

Some cation exchangers containing Cu^{2+}, Ni^{2+}, or Ag^{+} as the counterion can function as unique chelating agents for ammonia, amines, amino acids, and other molecules that can act as a ligand. Furthermore these exchangers can act as ligand-exchange resins if there is a net gain in free energy going from one ligand to another. For example,

$$(R^-)_2[Cu(NH_3)_4^{2+}] + 4CH_3NH_2 \rightleftharpoons (R^-)_2[Cu(NH_2CH_3)_4^{2+} + 4NH_3 \quad (4.27)$$

Ion-Exchange as Catalysts

Both inorganic and organic ion exchangers have found use in catalysis. The catalytic activity of the inorganic exchangers depends on the presence of metals such as manganese, iron, chromium, and vanadium. These ion

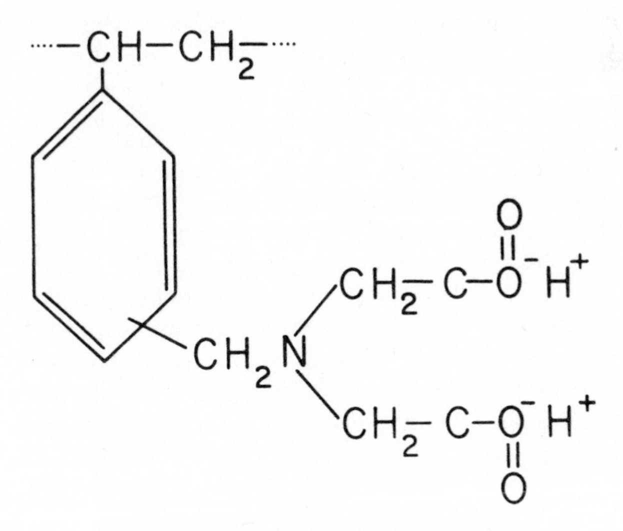

Fig. 4.13. Chelating resin, Dowex A-1.

exchangers have been used in the catalysis of oxidation reactions, hydrogenations, cracking processes, or alkylations.

Organic ion-exchange resins in the H^+-form and anion exchangers in the OH^--form can catalyze processes that are accelerated by acids and alkalies, respectively. Other exchangers of this kind also exhibit catalytic properties of their counterions.

Especially interesting are the macroreticular ion exchangers, which are particularly suited for heterogeneous systems because of their large surface area. The only limitation of the ion exchangers used in catalysis is their lower thermal and mechanical stability as compared with the traditional catalysts.

Ion Exchange in Nonaqueous Solvents

Certain ion exchangers can serve for purification of nonaqueous solvents or as catalysts in nonaqueous media. The behavior of the exchangers in nonaqueous systems depends mainly on the nature of the carbon chain of the resin as well as on its porosity and cross-linkage. Although the operation of ion exchangers in nonaqueous solvents is basically similar to that in water, nonaqueous solvents offer several disadvantages when compared with water. The swelling characteristics of the resins immersed in a nonaqueous solvent are lower; thus there may be chaneling or blocking of the column. In addition, the rate of mass transfer is slower than in water, and this will cause slower elution rates. In designing an ion-exchange process it should be borne in mind that the selectivities are different in nonaqueous media.

Ion-Exchange Chromatography

Ion-exchange chromatography, which has been growing in importance as an analytical method, involves the separation of ions due to the different affinity of the solute ions for the exchanger material. It is a liquid-solid technique, in which the ion exchanger represents the solid phase.

Principles of Chromatography

Liquid chromatography (LC) refers to any chromatographic process in which the *moving* or the *mobile phase* is a liquid, as contrasted to a gas in gas chromatography (GC). The nonmoving phase is called a *stationary phase.*

When a mixture of solutes is injected on the column, it occupies a very small volume. In the mobile phase the solutes move down the column. The solutes move at different rates because they undergo a series of sorption and desorption processes as a result of the interaction with both the stationary and the moving phase. At equilibrium, the distribution of any compound between the two phases is given by the distribution coefficient K.

$$K = \frac{[X]_s}{[X]_m} \qquad (4.28)$$

where $[X]_s$ and $[X]_m$ stand for the concentrations of the solute X in the stationary phase and the moving phase, respectively. The separation of a mixture is a result of the differences in the K values for each solute.

While traveling through the column, the solute molecules are dispersed. This spread in migration velocities results in a series of bell-shaped curves that form a chromatogram (Fig. 4.14). An elution band or a peak is characterized by two parameters: t_R value, which is the retention time of a particular compound, and t_W, the baseline width of the peak. These two parameters can be expressed in terms of volume units by using the flow rate of the mobile phase:

$$V_R = t_R f \tag{4.29}$$

where V_R = retention volume,
t_R = retention time,
f = flow rate, expressed in volume per unit time.

The capacity factor k' is related to the distribution coefficient K through the following expression:

$$k' = \frac{\text{total amount of X in stationary phase}}{\text{total amount of X in the moving phase}} \tag{4.30}$$

$$= \frac{V_s[X]_s}{V_m[X]_m} \tag{4.31}$$

$$= \left(\frac{V_s}{V_m}\right)K \tag{4.32}$$

where V_s and V_m refer to the volumes of the stationary phase and the moving phase, respectively. The retention time of the unretained component ($k' = 0$)

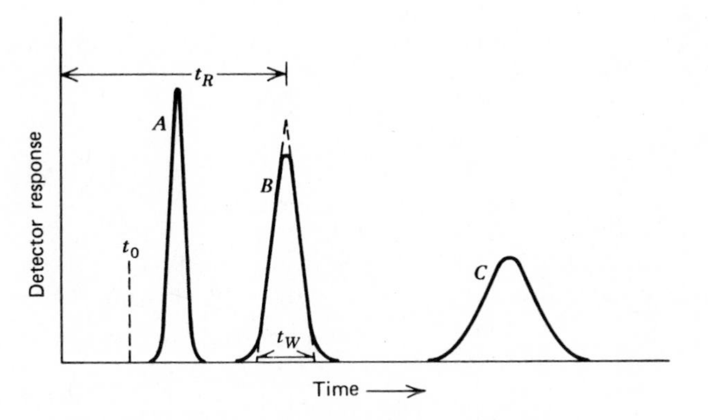

Fig. 4.14. Schematic representation of a chromatogram.

is a very important parameter; it is termed the *void volume* and designated by t_0.

It can be demonstrated that

$$t_R = t_0(1 + k') \tag{4.33}$$

or

$$V_R = V_m + V_s \cdot K \tag{4.34}$$

where V_m refers to the void volume. Equation 4.34 is a very important expression relating the retention volume to the distribution coefficient.

The other peak parameter t_W can be related to the number of theoretical plates N, which is an empirical measure of the column performance:

$$N = 16\left(\frac{t_R}{t_W}\right)^2 \tag{4.35}$$

or

$$N = 16\left(\frac{V_R}{W}\right)^2 \tag{4.36}$$

The N values are roughly constant for different solutes in a mixture and can be used for comparing column performance if all other chromatographic operating conditions are held constant.

The amount of band broadening t_W arises from several factors, including the column packing, column length and diameter, moving phase velocity, and other system parameters that are discussed in more detail in literature (see Supplementary Reading).

RESOLUTION IN LC

The objective in carrying out a chromatographic analysis is to achieve a good separation in a minimum amount of time. A measure of the effectiveness of separation of two adjacent peaks is the resolution parameter, given by the following expression:

$$R_s = \frac{t_{R_1} - t_{R_2}}{\sqrt{\frac{1}{2}(w_2 + w_1)}} \tag{4.37}$$

where the subscripts 1 and 2 refer to the two adjacent peaks. An R_s value of 1.5 stands for a complete baseline separation of the two adjacent peaks. A more fundamental expression can be obtained by relating the resolution to the conditions of separation:

$$R_s = \left(\frac{\sqrt{N}}{4}\right)\left(\frac{\alpha - 1}{\alpha}\right)\left(\frac{k'_2}{1 + k'_2}\right) \tag{4.38}$$

where N is a column efficiency factor, k' refers to the capacity factor, and α is the separation factor. The separation factor is given by

$$\alpha = \frac{k'_2}{k'_1} \tag{4.39}$$

where k'_2 is the capacity factor of band 2 and k'_1 is the value of band 1. The three factors affecting the resolution can be independently varied, to optimize the separation.

Column efficiency, which is one of the most important parameters in a chromatographic system, can be expressed in terms of the number of theoretical plates or the height equivalent to a theoretical plate (HETP). The terms "plate height" and "plate number" originate from the plate model of Martin and Synge [69] and are derived from the analogy between a chromatographic separation and a fractional distillation column. The "height equivalent to a theoretical plate" is given by the relation

$$\text{HETP} = \frac{\text{length of column}}{N} \tag{4.40}$$

The smaller the value of HETP, the better the given column.

The overall plate height can be expressed as a sum of incremental plate heights arising from four most important band-broadening phenomena: longitudinal diffusion (H_L), stationary phase mass transfer (H_S), mobile phase effects (H_M), and the mass transfer across the stagnant mobile phase (H_{SM}):

$$H = H_L + H_S + H_M + H_{SM} \tag{4.41}$$

Rather sophisticated approaches and quite different treatments of band-broadening phenomena are discussed in the literature [70–72].

In gas chromatography, (4.41) is reduced to the well-known van Deemter equation

$$H = A + \frac{B}{u} + C \cdot \bar{u} \tag{4.42}$$

where $\bar{u}$ is a linear velocity and A, B, and C are constants. In contrast to gas chromatography, a minimal H value is seldom encountered in liquid chromatography. The basic van Deemter equation relates H to particle size as well as to flow rates and is given by

$$H = 2\lambda dp + 2\psi\frac{D_G}{u} + \frac{8}{\pi^2}\frac{k'}{(k'+1)^2}\frac{df^2}{D_L}\bar{u} \tag{4.43}$$

where λ is the packing factor, dp the diameter of the particles in the bed, ψ the obstruction the bed presents to free molecular diffusion, D_G the diffusion coefficient for the solute in the stationary phase, $\bar{u}$ the average linear gas velocity, k' the partition ratio, df the average thickness of the stationary phase, and D_L the diffusion coefficient for the solute in the mobile phase [73].

Components of a Liquid Chromatograph

The essential elements of a high-pressure liquid chromatograph are given in Fig. 4.15.

INJECTION SYSTEM

The chromatographic column is the most vital part in achieving a desired separation. An integral part of the column is the injection system, located immediately before the column. Its main feature is a low dead volume, which prevents loss of resolution caused by band broadening. Several different injection systems can be used. There are three general methods for sample injection in liquid chromatography: (*a*) injection with solvent flow, (*b*) stop-flow injection, and (*c*) sample valve injection.

The first method uses a septum-disc injection port, and it is the simplest and most convenient at pressures below 1000 psi. In the stop-flow injection method the injector is equipped with a security cap that prevents leaking of the mobile phase through the septum. Since the diffusivity of solutes in liquids is very small, the stop-flow method does not cause any significant amount of band broadening.

The sample valve injection is most useful for analyses in which the sample size and column operating conditions can be kept constant. The valve is available with various sample volumes, ranging from 2 to 100 μm. These devices are extensively used in liquid chromatography, and they provide trouble-free operation without leakage at pressures up to 6000 psi.

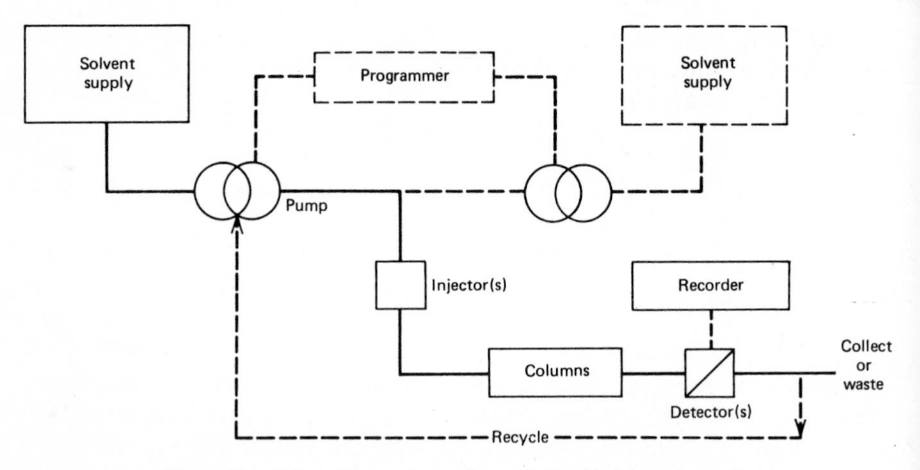

Fig. 4.15. Schematic representation of a liquid chromatograph

SOLVENT DELIVERY SYSTEM

During the past few years a great deal of effort has been directed toward obtaining high resolution and faster speeds. Small diameter columns and packing materials with particle sizes in the micrometer range necessitate high pressures at the column inlet. Also, since LC detectors are sensitive to variations in flow, considerable efforts have been made to produce pulseless and constant flow pumps.

Two different types of pump are available for liquid chromatographic operation: *constant pressure* (mobile phase is directly pressurized with an inert gas) and *constant flow*. The main advantage of the constant flow pumps is that a change in pressure has little effect on the separation, whereas the change in solvent flow affects retention times, resolution, and the baseline characteristics.

Constant flow pumps can be further subdivided into (*a*) reciprocating pumps, (*b*) diaphragm pumps, and (*c*) single-displacement pumps.

The piston in a reciprocating pump is in direct contact with the solvent. The liquid is withdrawn from the reservoir to fill the pump head during the intake stroke and pushed into the column during the discharge stroke. The check valves control the direction of flow. The main advantage of these pumps is that their delivery is not limited by a certain volume. They can operate continuously, and recycling the sample back through the pump is easily done. With a suitable damping system, or a flow feedback system, pulseless flow of unlimited volume can be achieved.

In a reciprocating diaphragm pump the working piston and the floating piston oppose each other in an oil chamber separated from the solvent chamber by a stainless steel membrane. The delivery of the pump is adjusted by limiting travel of the opposing piston to some fraction of the working piston.

Single displacement pumps operate with a screw-driven syringe and are capable of very high pressures and pulseless flow. The main disadvantage is a limited volume capacity.

Constant pressure pumps of the pneumatic type operate by the application of gas pressure to a hydraulic piston having a small surface area, thus providing a great pressure amplification. The apparatus is pulseless and capable of high pressures, but it must be refilled once the pumping cycle has been completed.

DETECTION DEVICES

The role of a detector is becoming more critical as separation work gains in sophistication. Most detection devices used in liquid chromatography continuously monitor the column effluent, giving a signal that is "translated" electronically into an elution pattern on a moving chart recorder. These flow-through cells must be designed to minimize turbulence and unswept areas;

the dead volume must be very low (microliter capacity), and there is a requirement for high sensitivity and low background noise.

Detection methods can employ measuring either optical (UV–visible, fluorescence), refractive index, electrical (electrical conductance, dielectric constant, electrochemical potential), or nuclear (radioactivity) properties of the solutes. The two most commonly used detectors are the photometric (single- and multiwavelength) detectors and refractometric detectors.

UV detectors detect only substances that absorb in the ultraviolet region. They have high sensitivity and are stable to small variations in temperature and flow rate.

Refractometric detectors can be used for detecting substances having a refractive index significantly different from that of the solvent. Their sensitivity is lower than that of the UV devices. Temperature changes cause significant deviations of the response. The main disadvantage is that this equipment cannot be used with gradient elution.

Applications: Organic Substances

AMINO ACIDS, PEPTIDES, AND PROTEINS

Ion-exchange chromatography holds a special position among the various methods used for the separation and identification of amino acids and related compounds. Since the original work of Moore and Stein [74], who had succeeded in separating a mixture of amino acids on Dowex 50-8X (Na^+ form) using 0.2 *M* sodium citrate buffers of progressively increasing pH (from 3.4 to 11.0), many modifications of their chromatographic system have been tried.

The method was later automated and a special instrument, an amino acid analyzer, was developed by Hamilton [75]. This instrument is equipped with a fully automatic sample injector and can be operated by a one- or two-column system. It can be used in the analysis of protein hydrolysates, physiological fluids, culture media, and so on. The operating parameters most commonly varied to carry out a separation of amino acids are the pH and the ionic strength of the mobile phase. Figure 4.16 illustrates the sensitivity of these autoanalyzers in determining the amino acid content of a single wet thumb print.

Two types of ion-exchange material have been used to achieve a separation of peptides and proteins: resinous exchangers [76, 77] and modified cellulose ion exchangers [79–83]. A complete separation of a peptide mixture is quite complex and usually cannot be achieved by a single chromatographic procedure. Both cation and the anion exchangers can be used for the fractionation of the peptide or protein mixtures because of zwitterionicity. Resinous materials have proved to be unsuitable for the separation of high molecular

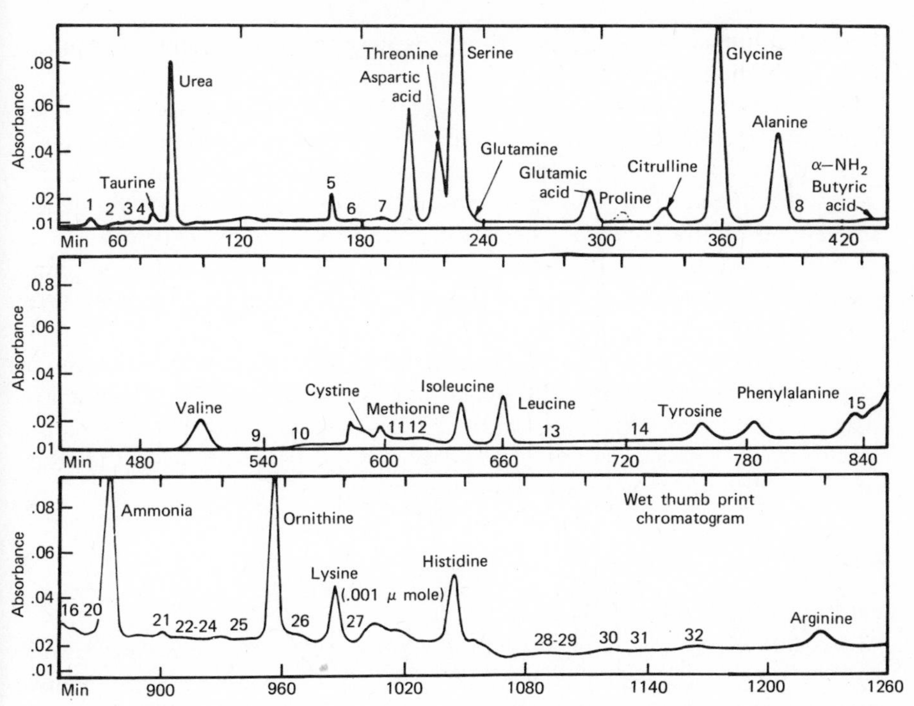

Fig. 4.16. Amino acids found by ion-exchange chromatography in a single thumb print made on a wet glass surface. For visual comparison of peaks, the amount of lysine (0.011 μmole) is indicated in the bottom diagram. On the abscissa, 1 min is equivalent to 0.5 ml of column efficient volume. Conditions: column, 0.64 × 125 cm of Dowex 50-X8, 17.5-μ beads; elution with sodium citrate buffers at 30 ml/hr. Reproduced from P. B. Hamilton, *Nature*, **205**, 284 (1965); by permission of the authors and the editor.

weight peptide oligomers and proteins because they have high charge density and tend to bind irreversibly to the resin. However chemically modified cellulose gives very satisfactory results in peptide and protein analyses.

CARBOHYDRATES AND CARBOHYDRATE DERIVATIVES

Because carbohydrates and their derivatives form negatively charged complexes with borate, these complexes can be separated by ion exchange, using an anion-exchange resin [84–86]. These analyses were first semiautomated and are now fully automated [87–90]. Figure 4.17 illustrates the separation of some simple saccharides.

ORGANIC ACIDS

Ion-exchange chromatography has been applied in analysis of various aliphatic and aromatic acids [91–96]. Until recently, these analyses, as well

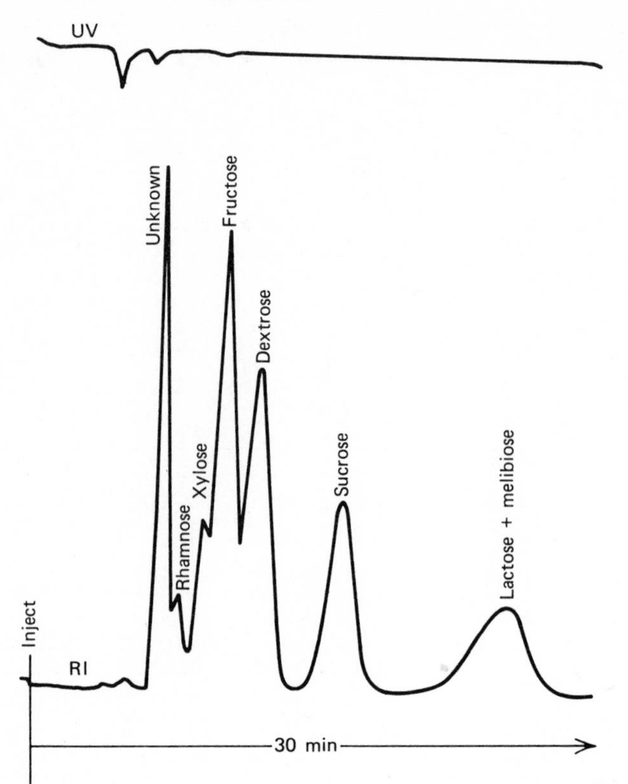

Fig. 4.17. Separation of some simple saccharides. Conditions: Bondapak AX/Corasil (2 mm i.d. × 122 cm); solvent, water–ethyl acetate–propanol 2; detector, UV at 254 nm and RI. Reproduced from *Technical Data*, Waters Associates Inc., Milford, Massachusetts, with the permission of the company.

as the subsequent automated analyses, were hampered by a lack of adequately sensitive and convenient detection methods. Present-day instrumentation is able to provide a meaningful resolution of more than 50 lower molecular weight constituents in urine. Higher molecular weight components are usually removed and analyzed by other methods. Figure 4.18 is a chromatogram of some organic acids present in human urine.

AMINES

Aliphatic, aromatic, and heterocyclic amines can be separated on weakly acidic cation cellulosic exchangers and on weak and strong acid resins [97–100]. A separation of some phenethylamines appears in Fig. 4.19. Separation methods can also be automated using an analyzer originally designed for amino acid work.

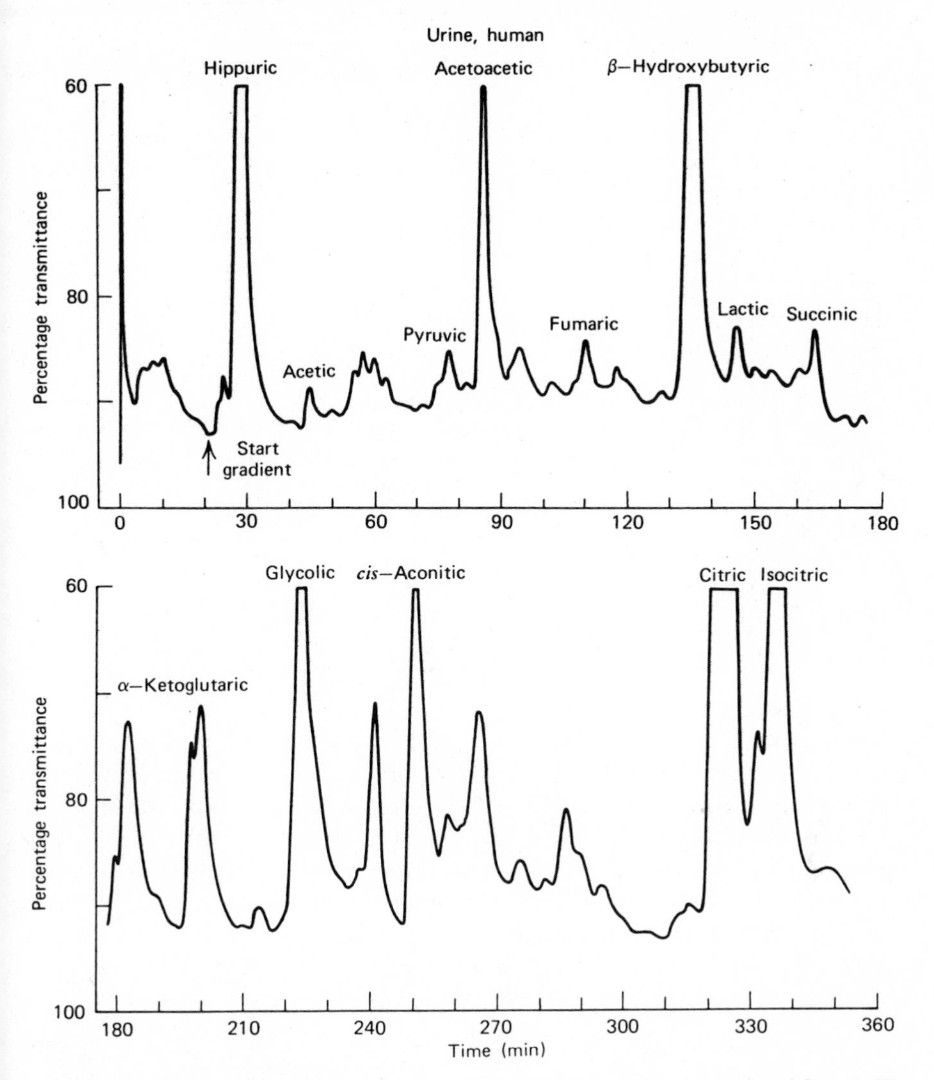

Fig. 4.18. Chromatogram of organic acids in 0.1 ml of human urine. Reproduced from J. W. Rosevear et al., *Clin. Chem.*, **17**, 721 (1971), by permission of the authors and the editor.

NUCLEIC ACID COMPONENTS

Since Cohn [100] first separated the nucleoside bases in 1949, ion-exchange chromatography has been an important tool in the separation and quantitation of nucleic acid fragments. For the separation of nucleotides, anion-exchange chromatography has been the most widely used method [101, 102],

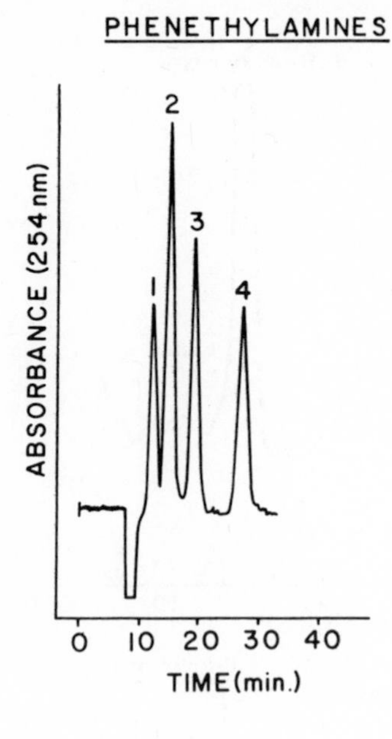

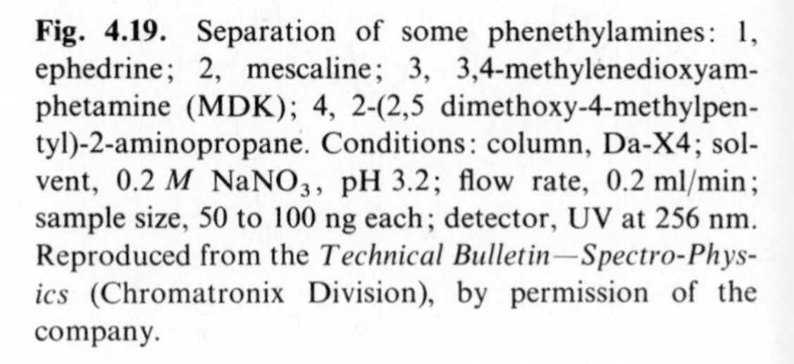

Fig. 4.19. Separation of some phenethylamines: 1, ephedrine; 2, mescaline; 3, 3,4-methylenedioxyamphetamine (MDK); 4, 2-(2,5 dimethoxy-4-methylpentyl)-2-aminopropane. Conditions: column, Da-X4; solvent, 0.2 *M* $NaNO_3$, pH 3.2; flow rate, 0.2 ml/min; sample size, 50 to 100 ng each; detector, UV at 256 nm. Reproduced from the *Technical Bulletin—Spectro-Physics* (Chromatronix Division), by permission of the company.

but cation exchangers have also been employed [101, 103, 106]. Anion-exchange chromatography has been used for the separation of the nucleosides and their bases [107, 108], but cation-exchange chromatography has proved to be more efficient [109–113]. Many of the separations were time-consuming. Using HPLC, however, rapid, efficient, and reproducible analyses of nucleotides [107–117], nucleosides, and bases [118–126] were achieved. A major breakthrough in obtaining better separations was the development of pellicular resins by Horvath and Lipsky [46], followed by the introduction of superficially porous, chemically bonded resins [47]. Recently microparticle, chemically bonded, totally porous resins have given excellent results and have expanded the potential of the technique for the analyses of nucleotides [116]. However it was found that the reverse phase partition mode of HPLC gave better separations if the nucleosides and bases were present together in a cell extract, since these compounds do not have the highly ionized phosphate group that makes possible the excellent separation of nucleotides by ion exchange.

The separation of purine and pyrimidine bases on a pellicular anion-exchange resin, and the separation of the mono-, di-, and triphosphate nucleotides obtained on a pellicular anion-exchange resin and on a microparticle chemically bonded anion exchanger, are illustrated in Figs. 4.20 and 4.21, respectively.

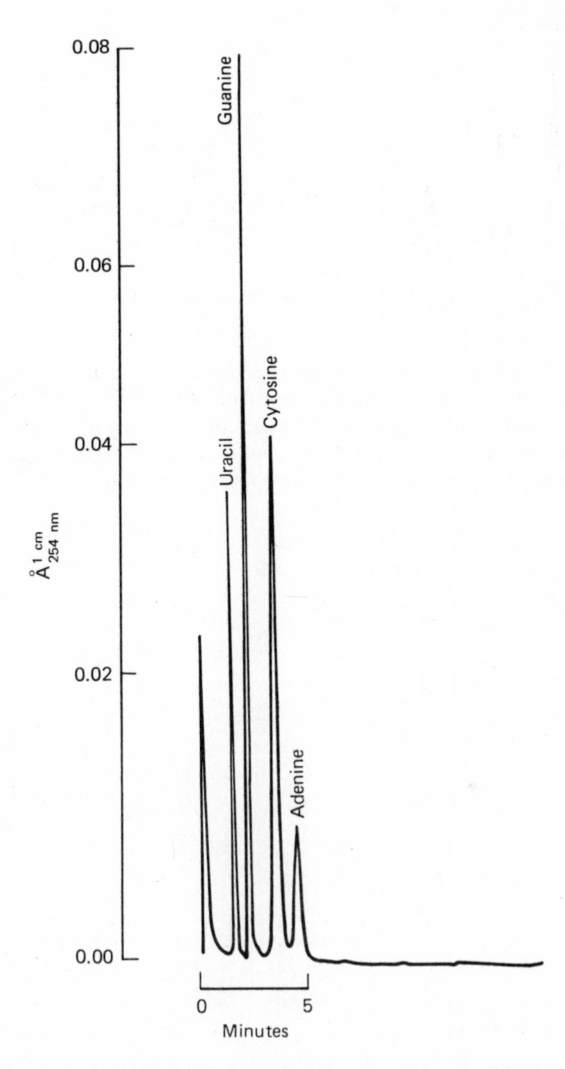

Fig. 4.20. Purine and pyrimidine bases. Conditions: column, pellicular cation-exchange resin; eluent, 0.025 *M* $NH_4H_2PO_4$, pH, 4.4; flow rate, 36 ml/hr. Reproduced from C. A. Burtis and D. R. Geere, *Nucleic Acid Constituents by Liquid Chromatography*, by permission of Varian Aerograph.

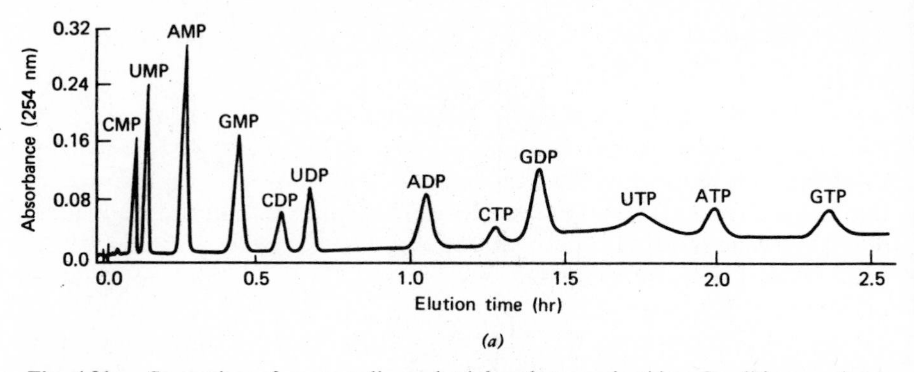

(a)

Fig. 4.21*a*. Separation of mono-, di-, and triphosphate nucleotides. Conditions: column, pellicular anion-exchange resin; starting eluent, 0.01 *M* KH_2PO_4, pH 3.25; gradient eluent, 1.0 *M* KH_2PO_4, pH 4.2; gradient delay, 7.5 min; column flow rate, 24 ml/hr; gradient flow rate, 12 ml/hr. Reproduced from C. A. Burtis and D. R. Geere, *Nucleic Acid Constituents by Liquid Chromatography*, by permission of Varian Aerograph.

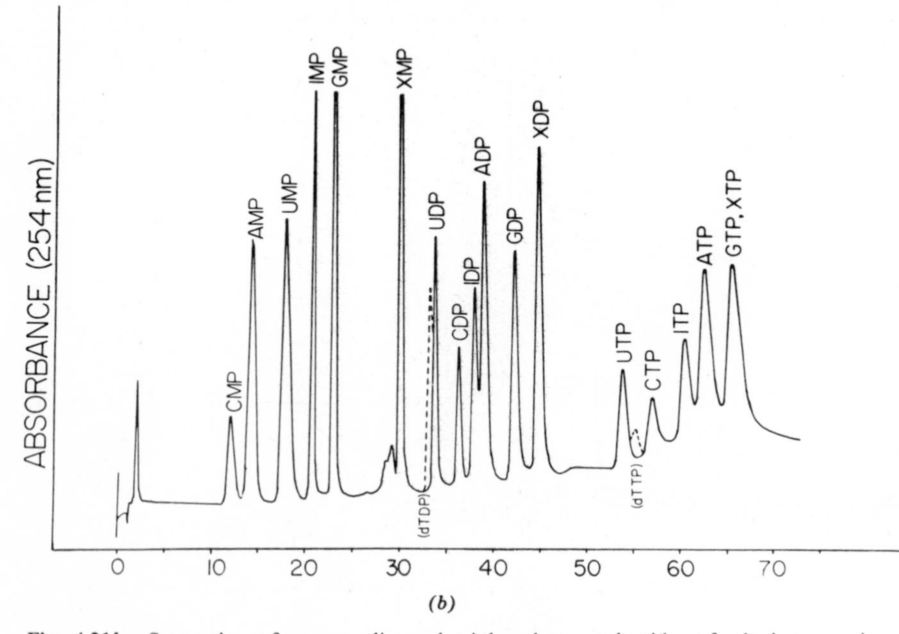

(b)

Fig. 4.21*b*. Separation of mono-, di-, and triphosphate nucleotides of adenine, guanine, hypoxanthine, xanthine, cytosine, uracil, and thymine. Conditions: column, Partisil-10SAX; temperature, ambient; eluents (low) 0.007 KH_2PO_4, pH 4.0; (high) 0.25 *M* KH_2PO_4, 0.50 *M* KCl, pH 4.5; gradient, linear, 0 to 100% of high-concentration eluent in 45 min; flow rate, 1.5 ml/min; detector, UV at 254 mm. Dashed lines indicate elution positions of *d*TDP and *d*TTP. Contributed by R. A. Hartwick and P. R. Brown (personal communication).

Inorganic Substances

CATIONS

Alkali and Alkaline Earth Metals. Alkali metals have been analyzed in the complexed or uncomplexed forms by different chromatographic procedures, using either resinous materials or inorganic exchangers. Complete separations have been done on a Dowex 50W-X16 (highly cross-linked, sulfonated divinylbenzene resin) [128], on Duolite C-3 (a sulfonated phenolformaldehyde resin) [129], on inorganic exchangers [130, 131], and on cellulose phosphate [132].

Figure 4.22 presents an HPLC analysis of a mixture of radioactive alkali metals using a conventional cation exchanger and a radioactive detector.

Alkaline earth metals have been separated using cation exchange resins [128, 133] and anion exchangers when complexed with the citrate ion [134].

Transition Metals. When transition metals are complexed with the tartrate ion [135], they can be separated by anion exchange. A typical separation appears in Fig. 4.23.

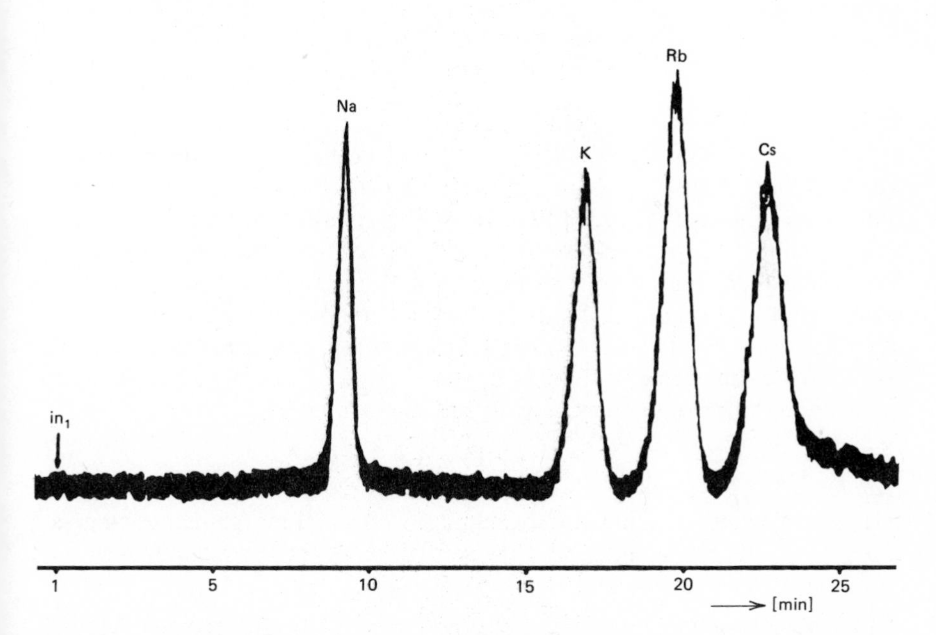

Fig. 4.22. Separation of alkali metals by cation-exchange chromatography. Separation of a mixture of ^{24}NaCl, ^{42}KCl, ^{86}RbCl, and ^{137}CsCl. Conditions: column, Aminex Q-1505; eluent, 1.61 *M* HCl; flow rate, 128 cm^3/hr; sample size, 20 μl. Reproduced from J. F. Huber and A. M. Van Urk-Schoen, *Anal. Chim. Acta*, **58**, 395 (1972), by permission of the authors and the editor.

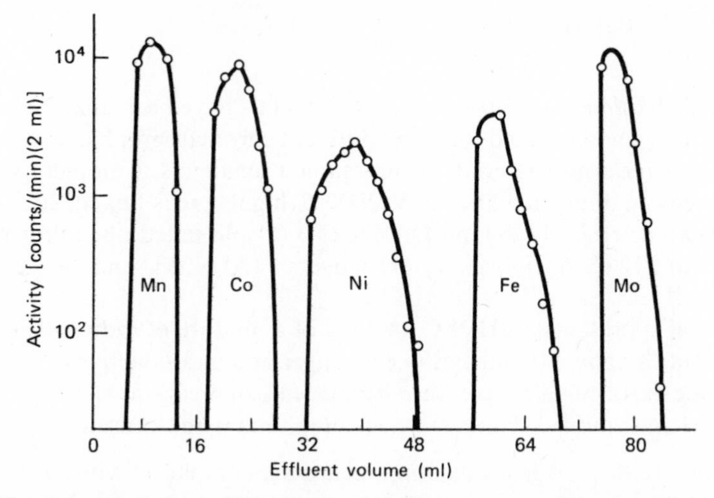

Fig. 4.23. Separation of transition metals as their tartrate complexes by anion-exchange chromatography. Conditions: column, Dowex 2-X8, tartrate form; elution with 8.5×10^{-2} tartrate solution at pH 4.0 for manganese; cobalt; and nickel, 0.1 *M* HCl for iron, and 3 *M* NaOH for molybdenum at 3 drops/min. Reproduced from G. P. Morie and T. R. Sweet, *J. Chromatogr.*, **16**, 201 (1964), by permission of the authors and the publisher.

Rare Earth Elements. The importance of the role of ion exchangers in the separation of the rare earth elements cannot be too strongly emphasized. The first successful separations of these elements using ion-exchange methods were done in 1947 [136]. It is interesting to note that the element with atomic number 61 (promethium) was first isolated and chemically identified by ion exchange. Since then, a variety of ion-exchange procedures have been proposed [137, 138] and tried, using different complexing agents and working conditions. Figure 4.24 presents chromatograms of the separation of some rare earth elements obtained on a series of anion-exchange resin materials of different degrees of cross-linkage.

ANIONS

Halides. The separation of chloride, bromide, and iodide ions has been successfully achieved using anion-exchange resins and inorganic exchangers [140, 141]. Figure 4.25 illustrates the separation of the three halide ions on an anion-exchanger column.

Phosphorus Oxyanions. Anion-exchange chromatography has proved to be useful in the analysis of complex polyphosphate mixtures [142, 143] as well as for mixtures containing lower phosphorus anions, thiophosphates, imidophosphates, and so on [144, 145]. These methods offer many advantages over the tedious and sometimes ineffective classical separations. A separation of the three lower phosphorus anions is given in Fig. 4.26.

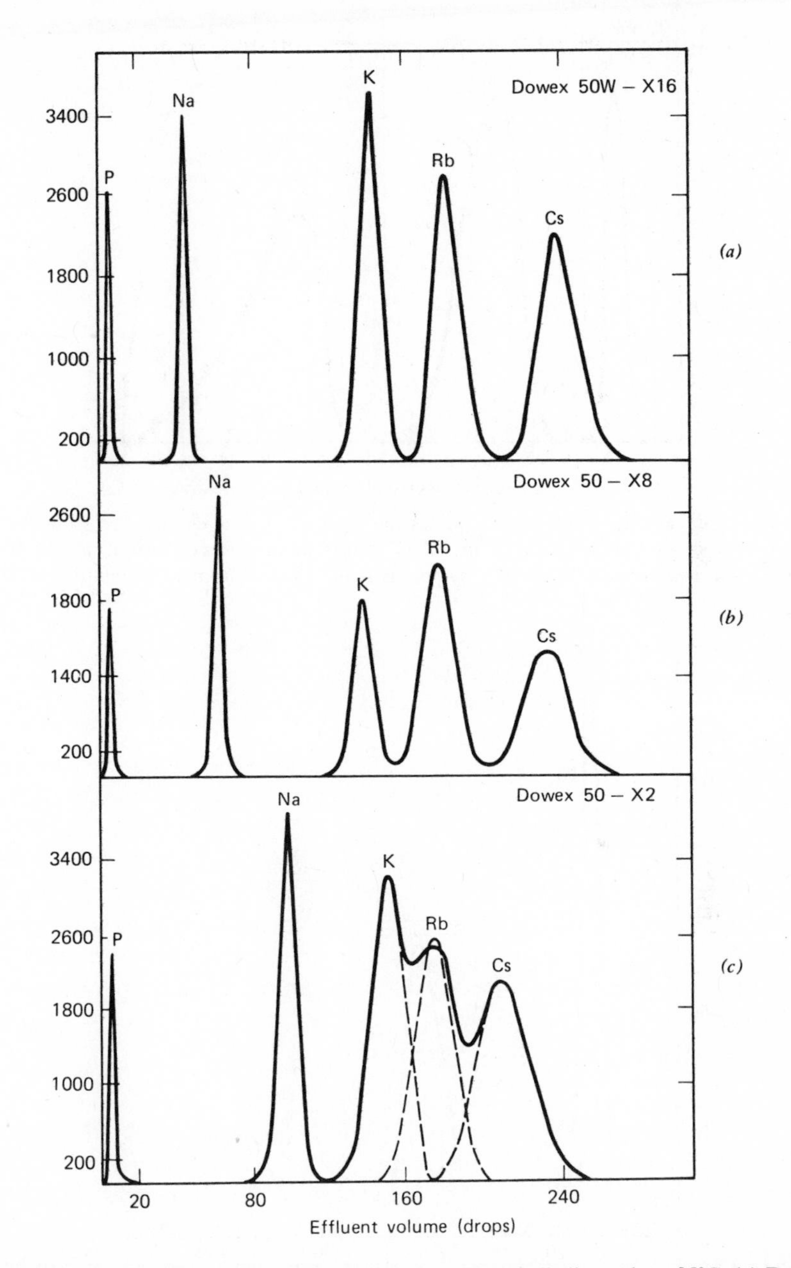

Fig. 4.24. Effect of resin cross-linking on the separation of alkali metals at 25°C. (*a*) Dowex 50 W-X16 (H^+): eluent, 0.0679 *N* HCl; flow rate, 0.72 cm/min: (*b*) Dowex 50 W-X8 (H^+): eluent, 0.261 *N* HCl; flow rate, 0.72 cm/min. (*c*) Dowex 50 W-X2 (H^+): eluent, 0.636 *N* HCl; flow rate, 0.71 cm/min. Reproduced from R. Dybczynski, *J. Chromatogr.*, **71**, 507 (1972), by permission of the author and the publisher.

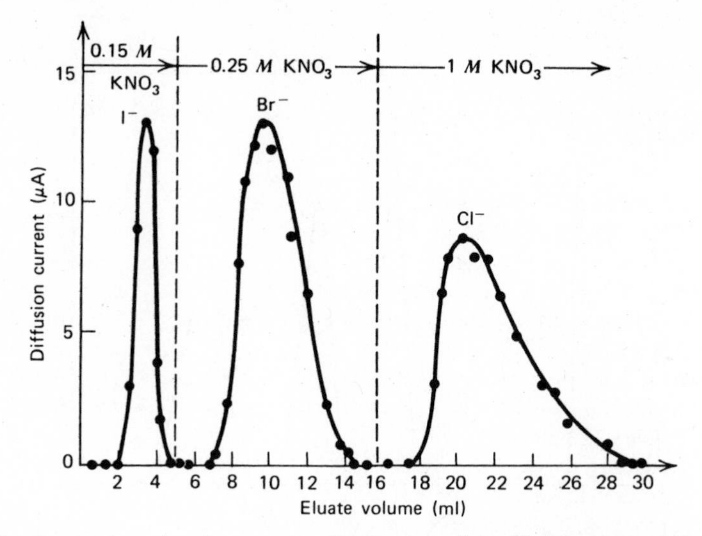

Fig. 4.25. Comparison of exchanger types for the separation of halide mixtures. Conditions: column, hydrous zirconium oxide; eluent; KNO_3 at 0.14 ml/min. Reproduced from S. Tustercowski, *J. Chromatogr.*, **31**, 268 (1967), by permission of the author and the publisher.

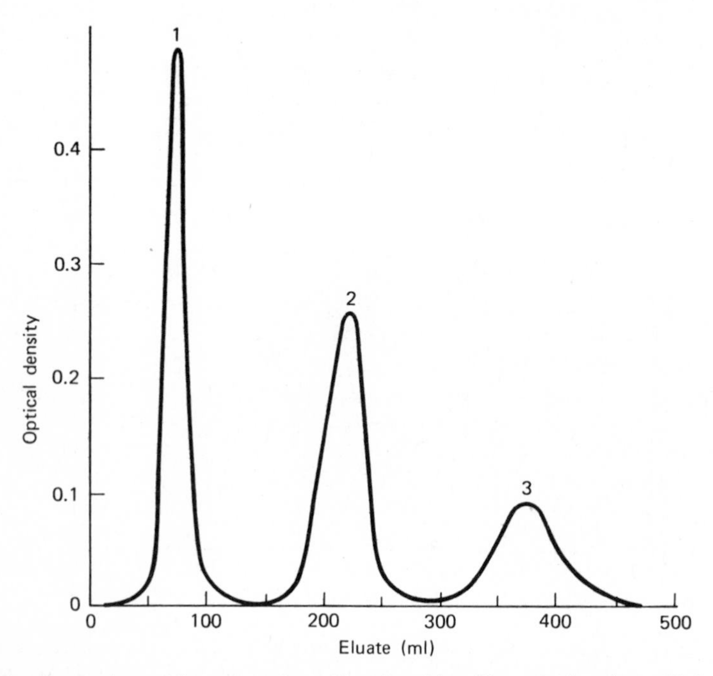

Fig. 4.26. Separation of hypophosphate (1), phosphite (2), and phosphate (3) by anion-exchange chromatography, at pH 11.4 at 2°C. Conditions: column, Dowex 1-X8, Cl^- form; elution with a gradient of KCl solution at 25 ml/hr. Reproduced from F. H. Pollard et al., *J. Chromatogr.*, **9**, 227 (1962), by permission of the authors and the publisher.

References

1. H. S. Thompson, *J. Roy. Agric. Soc.*, **11**, 68 (1850).
2. J. T. Way, *J. Roy. Agric. Soc.*, **11**, 313 (1850).
3. R. Gans, *Chem. Ind.* (London), **32**, 197 (1909).
4. R. Gans, *Chem.-Ztg., Chem. App.*, **31**, 355 (1907).
5. B. A. Adams and E. L. Holmes, *J. Soc. Chem. Ind.*, **54**, 1 (1935); British Patents 450308, 450309 (1936).
6. F. Helfferich, *Ion Exchange*, McGraw-Hill, New York, 1962, Chapter 4.
7. O. Samuelson, *Ion-Exchange Separations in Analytical Chemistry*, Wiley, New York, 1963, Chapters 2–4.
8. R. Kunin, *Ion-Exchange Resins*, 2nd ed., Wiley, New York, 1958, Chapters 2, 3, 5.
9. H. F. Walton, in *Chromatography*, 2nd ed., Reinhold, New York, 1967, Chapter 12.
10. O. Samuelson, *Ion-Exchange Separations in Analytical Chemistry*, Wiley, New York, 1963, Chapters 2, 8.
11. F. Helfferich, *Ion Exchange*, McGraw-Hill, New York, 1963, Chapter 2.
12. BioRad Laboratories, "Materials for Ion-Exchange-Gel Filtration-Adsorption," Richmond, Calif., June 1971.
13. H. G. Cassidy, *Fundamentals of Chromatography*, Wiley-Interscience, New York, 1957, Chapter 9.
14. C. J. O. R. Morris and P. Morris, *Separation Methods in Biochemistry*, Wiley-Interscience, New York, 1963, Chapter 8.
15. E. W. Berg, *Physical and Chemical Methods of Separation*, McGraw-Hill, New York, 1963, Chapters 10, 11.
16. F. Helfferich, *Ion Exchange*, McGraw-Hill, New York, 1962, Chapters 1, 4–6.
17. F. Helfferich, in *Ion Exchange*, Vol. 1, J. A. Marinsky, Ed., Dekker, New York, 1966, Chapter 2.
18. G. E. Boyd and B. Soldano, *J. Am. Chem. Soc.*, **75**, 6091 (1953).
19. G. E. Boyd, A. W. Adamson, and L. S. Myers, *J. Am. Chem. Soc.*, **69**, 2836 (1947).
20. E. A. Peterson, in *Cellulose Ion Exchangers*, T. S. Work and M. Work, Eds., American Elsevier, New York, 1970.
21. Reeve Angel Scientifica Division, "Whatman Advanced Ion-Exchange Celluloses" (Data Manual and Catalog-2000), Clifton, N.J.
22. H. A. Sober and E. A. Peterson, *J. Am. Chem. Soc.*, **76**, 1711 (1954).
23. H. A. Sober and E. A. Peterson, *J. Am. Chem. Soc.*, **78**, 751 (1956).
24. J. Porath, *Ark. Kern.*, **11**, 97 (1957).
25. G. Semenza, *Helv. Chim. Acta*, **43**, 1057 (1960).
26. C. J. Epstein, C. B. Anfinsen, and M. Sela, *J. Biol. Chem.*, **23**, 1825 (1962).
27. C. J. Epstein, C. B. Anfinsen, and M. Sela, *J. Biol. Chem.*, **237**, 3459 (1962).
28. H. Determann, *Gel Chromatography*, Springer-Verlag, New York, 1968.
29. Pharmacia Fine Chemicals, Inc., "Sephadex Ion Exchangers," Piscataway, N.J., March 1971.
30. M. J. Fuller, *Chromatogr. Rev.*, **14**, 45 (1971).
31. C. B. Amphlett, *Inorganic Ion Exchangers*, American Elsevier, New York, 1964, Chapters 4, 5.
32. A. Brandone, S. Meloni, S. Girardi, and F. Sabbioni, *Analysis*, **2**, 300 (1973).

33. Y. D. Dolmatov, Z. N. Bulavina, and M. Y. Dolmatova, *Radiokhimiya*, **14**, 526 (1972).
34. E. Hallaba, N. Z. Misak, and H. N. Salama, *Indiana J. Chem.*, **11**, 580 (1973).
35. A. L. Ruvarac and M. I. Trtanj, *J. Inorg. Nucl. Chem.*, **34**, 3893 (1972).
36. J. E. Salmon, *Progress in Nuclear Energy*, Vol. 2, Series IX, Analytical Chemistry, Pergamon Press, Oxford, 1961.
37. J. Smit, R. Van, W. Robb, and J. J. Jacobs, *J. Inorg. Nucl. Chem.*, **12**, 95 (1959).
38. J. Smit, R. Van, W. Robb, and J. J. Jacobs, *J. Inorg. Nucl. Chem.*, **26**, 509 (1964).
39. H. O. Phillips and K. A. Kraus, *J. Chromatogr.*, **17**, 549 (1965).
40. H. O. Phillips and K. A. Kraus, *J. Am. Chem. Soc.*, **85**, 486 (1963).
41. A. M. Gurvich and T. B. Gapon, *Zh. Anal. Khim.*, **27**, 933 (1972).
42. T. Braun, O. Bekeffy, I. Haklits, K. Kadar, and G. Majors, *Anal. Chim. Acta*, **64**, 45 (1973).
43. T. Braun, A. B. Farag, and A. Klimes-Sznik, *Anal. Chim. Acta*, **64**, 11 (1973).
44. H. Buchwald and W. P. Thistlewaite, *J. Inorg. Nucl. Chem.*, **5**, 341 (1958).
45. J. Dritil and V. Dourim, *J. Inorg. Nucl. Chem.*, **12**, 367 (1959).
46. C. G. Horvath, B. A. Preiss, and S. R. Lipsky, *Anal. Chem.*, **39**, 1422 (1967).
47. J. J. Kirkland, *J. Chromatogr. Sci.*, **7**, 361 (1969).
48. R. E. Majors, *Am. Lab.*, May 27, 1972.
49. R. E. Majors, *Anal. Chem.*, **44**, 1722 (1972).
50. J. J. Kirkland, *J. Chromatogr. Sci.*, **9**, 206 (1971).
51. J. J. Kirkland, *J. Chromatogr. Sci.*, **10**, 593 (1972).
52. R. F. Hollis and C. K. McArthur, *Min. Eng.*, **9**, 442 (1957).
53. F. de Körösy and J. Shorr, *Nature*, **197**, 685 (1963).
54. G. Kolf, *Ber. Bunsenges. Phys. Chem.*, **71**, 877 (1967).
55. G. Bähr, *Ber. Bunsenges. Phys. Chem.*, **71**, 883 (1967).
56. R. E. Kesting, *Synthetic Polymeric Membranes*, McGraw-Hill, New York, 1971.
57. W. F. Blatt, A. Dravid, A. S. Michaels, and L. Nelsen, in *Membrane Science and Technology*, Plenum Press, New York, 1970.
58. S. B. Tuwiner, *Diffusion and Membrane Technology* (American Chemical Society Monograph 156), Reinhold, New York, 1962.
59. U. Eisner and H. B. Mark, *Talanta*, **16**, 27 (1969).
60. J. Inczedy and L. Erdey, "Some Analytical Applications of Ion-Exchange Membranes," *Symp. Balatonszelplak*, **1963**, 207 (1965).
61. H. G. Cassidy and K. A. Kun, *Oxidation-Reduction Polymers*, Wiley-Interscience, New York, 1965.
62. H. G. Cassidy, *J. Am. Chem. Soc.*, **71**, 402 (1949).
63. M. Ezrin and H. G. Cassidy, *Ann. N.Y. Acad. Sci.*, **57**, 79 (1953).
64. I. H. Updegraff and H. Cassidy, *J. Am. Chem. Soc.*, **71**, 407 (1949).
65. K. A. Kun and H. G. Cassidy, *J. Org. Chem.*, **27**, 841 (1963).
66. K. A. Kun and H. G. Cassidy, *J. Polym. Sci.*, **56**, 83 (1962).
67. H. Kamogawa and H. G. Cassidy, *J. Polym. Sci.*, **1**, 1971 (1963).
68. W. Lautsch, W. Broser, W. Biedermann, and H. Gnichtel, *J. Polym. Sci.*, **17**, 479 (1955).
69. A. J. P. Martin and R. L. M. Synge, *Biochem. J.*, **35**, 1358 (1941).
70. E. Grushka, L. R. Snyder, and J. M. Knox, *J. Chromatogr. Sci.*, **13**, 25 (1975).

71. R. P. W. Scott, D. W. J. Blackburn, and T. Wilkins, *J. Gas Chromatogr.*, 183 (1967).
72. H. Barth, E. Dallmeir, and B. L. Darger, *Anal. Chem.*, in press.
73. J. J. Van Deemjer, F. S. Zuiderweg and A. Klingenberg, *Chem. Eng. Sci.* **5**, 271 (1956).
74. S. Moore and W. H. Stein, *J. Biol. Chem.*, **192**, 663 (1951).
75. P. B. Hamilton, in *Automation in Analytical Chemistry*, Vol. 1, Technicon Symposia, 1967, Midiad, Inc., New York, 1968.
76. W. A. Schroeder, in *Methods in Enzymology*, Vol. 25, Part 13, C. H. W. Hirs and S. N. Timasheff, Eds., Academic Press, New York, 1972.
77. W. A. Schroeder, R. T. Jones, J. Corsuick, and K. McCalla, *Anal. Chem.*, **34**, 1570 (1962).
78. W. A. Schroeder and B. Robberson, *Anal. Chem.*, **37**, 1583 (1965).
79. A. E. Peterson and H. A. Sober, *J. Am. Chem. Soc.*, **78**. 756 (1956).
80. J. B. Clegg, M. A. Naughton, and D. J. Weatherall, *J. Mol. Biol.*, **19**, 91 (1966).
81. E. A. Peterson, *Cellulosic Ion Exchangers*, American Elsevier, New York, 1970.
82. A. M. Smith and M. A. Stahman, *J. Chromatogr.*, **41**, 228 (1969).
83. J. Hurwitz, M. Gold, and M. Anders, *J. Biol. Chem.*, **239**, 3462 (1964).
84. J. X. Khym, L. P. Zill, and W. E. Cohn, in *Ion Exchangers in Organic and Biochemistry*, C. Calmon and T. R. E. Kressman, Eds., Wiley-Interscience, New York, 1957, Chapter 20.
85. L. P. Zill, J. X. Khym, and G. M. Cheniae, *J. Am. Chem. Soc.*, **75**, 1339 (1953).
86. J. X. Khym, R. L. Jolley, and C. D. Scott, *Cereal Sci. Today*, **15**, 44 (1970).
87. G. J. Green, *Nat. Cancer Inst. Monogr.*, **21**, 447 (1966).
88. C. D. Scott, R. L. Jolley, W. W. Pitt, and W. F. Johnson, *Am. J. Clin. Pathol.*, **53**, 701 (1970).
89. R. B. Kesler, *Anal. Chem.*, **39**, 1416 (1967).
90. A. Floridi, *J. Chromatogr.*, **59**, 61 (1971).
91. H. Bush, R. B. Hurlbert, and V. R. Potter, *J. Biol. Chem.*, **196**, 717 (1952).
92. J. X. Khym and D. G. Doherty, *J. Am. Chem. Soc.*, **74**, 3199 (1952).
93. K. S. Lee and O. Samuelson, *Anal. Chim. Acta*, **37**, 359 (1967).
94. B. Carlsson and O. Samuelson, *Anal. Chim. Acta*, **49**, 247 (1970).
95. E. Martinsson and O. Samuelson, *Chromatographia*, **3**, 405 (1970).
96. S. Katz, W. W. Pitt, Jr., and G. Jones, Jr., *Clin. Chem.*, **19**, 817 (1973).
97. H. Hatano, K. Sumizu, S. Rokushika, and F. Murakami, *Anal. Biochem.*, **35**, 377 (1970).
98. R. A. Wall, *J. Chromatogr.*, **60**, 195 (1971).
99. L. Lepri, P. G. Desideri, V. Coas, and D. Cozzi, *J. Chromatogr.*, **49**, 239 (1970).
100. W. E. Cohn, *Science*, **109**, 377 (1949).
101. E. Volkin, J. X. Khym, and W. E. Cohn, *J. Am. Chem. Soc.*, **73**, 1533 (1951).
102. I. C. Caldwell, *J. Chromatogr.*, **44**, 331 (1969).
103. S. Katz and D. P. Comb, *J. Biol. Chem.*, **328**, 3065 (1963).
104. F. Manley and G. J. Manley, *J. Biol. Chem.*, **235**, 235 (1960).
105. F. R. Blattner and H. P. Erickson, *Anal. Biochem.*, **18**, 220 (1967).
106. E. Junowics and J. H. Spencer, *J. Chromatogr.*, **37**, 518 (1968).
107. J. C. Green, C. E. Nunley, and N. G. Anderson, *Nat. Cancer Inst. Monogr.*, **21**, 431 (1966).

108. N. G. Anderson, J. G. Green, M. L. Barber, and F. C. Ladd, Jr., *Anal. Biochem.*, **6**, 153 (1963).
109. M. Uziel, C. K. Koh, and W. E. Cohn, *Anal. Biochem.*, **25**, 77 (1968).
110. A. C. Burtis, M. N. Munk, and F. R. MacDonald, *Clin. Chem.*, **16**, 667 (1970).
111. C. F. Crampton, F. R. Frankel, A. M. Benson, and A. Wade, *Anal. Biochem.*, **1**, 249 (1960).
112. E. W. Busch, *J. Chromatogr.*, **37**, 518 (1968).
113. C. Horvath and S. R. Lipsky, *Anal. Chem.*, **41**, 127 (1969).
114. R. A. Hartwick and P. R. Brown, *J. Chromatogr.*, **112**, 651 (1975).
115. P. R. Brown, *J. Chromatogr.*, **52**, 257 (1970).
116. P. R. Brown, J. Herod, and R. E. Parks, Jr., *Clin. Chem.*, **19**, 919 (1973).
117. R. E. Parks, Jr., and P. R. Brown, *Biochemistry*, **12**, 3294 (1973).
118. J. J. Kirkland, *J. Chromatogr. Sci.*, **8**, 72 (1970).
119. H. Breter and R. Zahn, *Anal. Biochem.*, **54**, 346 (1973).
120. J. Mrochek, W. Butts, W. Reiney, Jr., and C. Burtis, *Clin. Chem.*, **17**, 72 (1971).
121. F. Blattner and H. Erickson, *Anal. Biochem.*, **18**, 220 (1967).
122. R. P. Singhal, *Biochim. Biophys. Acta*, **319**, 11 (1973).
123. J. X. Khym, *J. Chromatogr.*, **97**, 277 (1974).
124. R. Singhal, *Eur. J. Biochem.*, **43**, 245 (1973).
125. R. Singhal and W. Cohn, *Biochim. Biophys. Acta*, **262**, 565 (1972).
126. P. R. Brown, *J. Chromatogr.*, **99**, 587 (1974).
127. C. G. Horvath and S. R. Lipsky, *Anal. Chem.*, **39**, 1922 (1967).
128. R. Dybczynski, *J. Chromatogr.*, **71**, 507 (1972).
129. F. Nelson, D. C. Mickelson, H. O. Phillips, and K. A. Kraus, *J. Chromatogr.*, **20**, 107 (1965).
130. C. J. Coetzel and E. F. C. H. Rhower, *Anal. Chim. Acta*, **44**, 293 (1969).
131. J. D. Donaldson and M. J. Fuller, *J. Inorg. Nucl. Chem.*, **32**, 1703 (1970).
132. J. G. van Raaphorst and H. H. Karemaker, *Talanta*, **17**, 345 (1970).
133. F. H. Pollard, G. Nickless, and D. Spincer, *J. Chromatogr.*, **13**, 224 (1964).
134. F. Nelson and K. A. Kraus, *J. Am. Chem. Soc.*, **77**, 801 (1955).
135. G. P. Morie and T. R. Sweet, *J. Chromatogr.*, **16**, 201 (1964).
136. E. R. Tompkins, J. X. Khym, and W. E. Cohn, *J. Am. Chem. Soc.*, **69**, 2767 (1947).
137. L. Wodkiewicz and R. Dybczynski, *J. Chromatogr.*, **68**, 131 (1972).
138. J. Minczewski and R. Dybczynski, *J. Chromatogr.*, **7**, 98 (1962).
139. L. Wodkiewicz and R. Dybczynski, *J. Chromatogr.*, **32**, 394 (1968).
140. R. C. Degeiso, W. Rieman, III, and S. Lindenbaum, *Anal. Chem.*, **26**, 1840 (1954).
141. S. Tustanowski, *J. Chromatogr.*, **31**, 268 (1967).
142. J. A. Grande and J. Baukenkamp, *Anal. Chem.*, **28**, 1497 (1956).
143. G. Kura and S. Ohaski, *J. Chromatogr.*, **36**, 111 (1971).
144. F. H. Pollard, G. Nickless, D. E. Rogers, and M. T. Rothwell, *J. Chromatogr.*, **17**, 157 (1965).
145. F. H. Pollard, D. E. Rogers, M. T. Rothwell, and G. Nickless, *J. Chromatogr.*, **9**, 227 (1962).
146. J. F. K. Huber and A. M. Van Urk-Schoen, *Anal. Chim. Acta*, **58**, 395 (1972).

147. J. W. Rosevear, K. T. Pfaff, and E. A. Moffit, *Clin. Chem.*, **17**, 721 (1971).
148. C. D. Scott, R. L. Folley, A. B. W. Wilson Pitt, and W. F. Johnson, *Am. J. Clin. Pathol.*, **53**, 701 (1970).

SUPPLEMENTARY READING: ION-EXCHANGE LITERATURE

J. J. Kirkland, Ed., *Modern Practice of Liquid Chromatography*, Wiley-Interscience, New York, 1971.

P. R. Brown, *High Pressure Liquid Chromatography–Biochemical and Biomedical Applications*, Academic Press, New York, 1973.

H. F. Walton, Ed., "Ion-Exchange Chromatography," in *Benchmark Papers in Analytical Chemistry*, Vol. I, Dowden, Hutchinson & Ross, Stroudsburg, Pa., 1967.

R. P. W. Scott, *Contemporary Liquid Chromatography*, in Techniques of Chemistry, Vol. 11, Wiley-Interscience, 1967.

"High Pressure Liquid Chromatography in Clinical Chemistry," in *Proceedings of a Symposium Held at Kings College Hospital Medical School, December 15–16, 1975*, P. F. Dixon, C. H. Gray, C. K. Lim, and M. S. Stoll, Eds., Academic Press, New York, 1976.

K. Dorfner, in *Ion Exchangers: Properties and Applications*, A. F. Coers, Ed., Ann Arbor Science Publishers, Ann Arbor, Mich., 1972.

B. L. Karger, L. R. Snyder, and C. Horvath, *An Introduction to Separation Science*, Wiley-Interscience, New York, 1973.

J. X. Khym, *Analytical Ion-Exchange Procedures in Chemistry and Biology—Theory, Equipment, Techniques*, Prentice-Hall, Englewood Cliffs, N.J., 1974.

W. Rieman, III, and H. F. Walton, *Ion Exchangers in Analytical Chemistry*, Pergamon Press, New York, 1970.

J. C. Giddings, *Dynamics of Chromatography*, Dekker, New York, 1965.

Chapter V

AFFINITY CHROMATOGRAPHY

Sheldon W. May

1 Introduction 257

2 The Support 259

Polysaccharide Supports 260
Other Supports 269

3 The Arm 273

4 The Ligand 275

5 Specialized Procedures, Techniques, and Approaches 284

6 Commercially Available Materials 287

7 Reviews and Technical Information 287

1 INTRODUCTION

The terms "affinity chromatography," "biospecific adsorption," and "ligand-specific chromatography" currently denote separation techniques based on the specific binding properties of biological molecules. The use of such techniques for solving the intricate and difficult separation problems often encountered in biochemical research has undergone a veritable explosion in popularity in less than a decade, and applications extending beyond the traditional bounds of biological research are now beginning to appear. Since a number of general concepts and techniques have already emerged from work in this area, and we have every reason to expect that this trend will continue, it is important that researchers in other areas of chemistry be made aware of the "state of the art" in affinity chromatography.

Although the *widespread* use of affinity chromatography can rightfully be called a recent advance in biochemistry, the basic idea of exploiting the specific binding properties of biomolecules in heterogeneous systems has been around for some time. In 1910 Sarkstein [1] adsorbed the starch-splitting enzyme amylase onto insoluble starch, but several decades then passed with only sporadic reports of similar procedures. Lerman's report in 1953 [2] on the adsorption of tyrosinase onto a resorcinol-cellulose conjugate was especially significant because it represented an example of the intentional synthesis of a biospecific conjugate by covalent attachment of an inhibitor to an insoluble support, and also because the enzyme was successfully eluted from the column at alkaline pH. Other workers applied affinity chromatography to proteins, polynucleotides, antibodies, and antigens in the late 1950s and early 1960s, but it was the introduction of the CNBr activation method for coupling amines to polysaccharide supports by Axen et al. in 1967 [3] that opened up the field to the biochemical community at large. This activation method obviated the necessity for the complex chemical synthesis and activation of affinity supports, and it is still by far the most commonly used reaction for producing affinity conjugates.

Conceptually the methodology of affinity chromatography is simple. First, a specific ligand is attached to an insoluble support to form a conjugate, and this is contacted with a "feed" either in a column or a batch configuration. (Alternatively, a water-*soluble* macromolecular matrix may be employed; the conjugate then is used in conjunction with an ultrafiltration membrane of the appropriate porosity.) In a totally idealized situation, contact of the conjugate and the feed would result in only the component that interacts specifically with the ligand becoming attached to the "affinity adsorbent," everything else simply passing through the column in the void volume. Finally, the desired component is eluted from the conjugate by imposing conditions that dissociate it from the immobilized ligand, and the column can be immediately recycled.

In practice, a number of complications of this simple idealized picture are almost always present. In the first place, "nonspecific" adsorption of components of the feed to matrix or the ligand or both commonly occurs, and indeed, even the desired material may become adsorbed by virtue of both specific and nonspecific (e.g., ionic, hydrophobic) interactions. Depending on the particular system, such interference effects may be minimized by judicious choice of operating conditions (e.g., pH, temperature, or ionic strength). As we discuss later, however, such "interference" effects are not always detrimental, and they can even be exploited to enhance the usefulness of a given conjugate. In any event, either nonspecifically adsorbed materials must be gently washed from the conjugate before elution of the desired component, or else elution conditions that leave such materials behind on the conjugate must be chosen. In the latter case, some treatment of the conjugate usually is required before it can be recycled.

Another type of complicating situation is often encountered when several different components of the feed are "specifically" adsorbed by virtue of a common affinity for the ligand. For example, when a "group specific" adsorbent such as immobilized AMP is used (see p. 351), several different dehydrogenases and kinases all interact with the nucleotide and bind to the conjugate. In such cases it is necessary to use *specific* elution procedures, such as pulses or gradients of soluble nucleotides, to effect the desired separation. Another very interesting situation occurs when some feed component A interacts with the immobilized ligand and another soluble molecule B to form a ternary ligand–A–B complex. This allows one to control the adsorption and elution of A onto the conjugate by simply adjusting the concentration of B in the column eluent, a highly useful circumstance. Thus, for example, A is eluted by omitting B from the feed without changing environmental conditions such as pH or ionic strength, whereas other components with either specific or nonspecific affinity for the ligand itself are left behind. A number of such sophisticated procedures based on multiple-component complex formation with the ligand have been developed and used successfully in affinity chromatography. It should be mentioned also that although the foregoing discussion deals with equilibrium associations (reversible complex formation) between the conjugate and components of the feed, it is also possible to use ligands that actually form covalent bonds with the desired component, but both elution and recycling of the conjugate then become significant problems (see p. 350).

2 THE SUPPORT

The characteristics desirable in a good support for affinity chromatography have been summarized elsewhere by May [4] as follows.

1. The support should be homogeneous in form—preferably relatively rigid porous beads— to allow for good flow properties in column operations. Porosity allows selective entry of components and provides a large effective surface area for attachment of the ligand and for interaction of the immobilized ligand with the desired component of the feed.
2. The support should exhibit good mechanical, chemical, and physical stability under operating conditions, and in the widest possible range of pH, temperature, and solvents. This allows flexibility in the design of experiments and in optimization of adsorption and elution conditions.
3. The support should be highly insoluble in the operating medium and resistant to breakdown by processes such as microbial attack.
4. Neutrality is a highly desirable characteristic because it eliminates interference effects due to nonspecific ion-exchange adsorption.
5. An important concern is the degree of hydrophobicity or hydrophilicity of the support material, and considerable attention has been directed toward this matter. In biochemical operations the activity of the material

being isolated can easily be destroyed by the drastic alterations in three-dimensional structure attributable to such factors. The isolation of soluble macromolecules is probably best carried out using a highly "wettable" hydrophilic support.

6. It must be possible, of course, chemically to couple ligands to the support, and the necessary chemical conversions should require mild conditions and should not destroy the structural integrity of the support. Also, it is very useful to be able to control readily the coupling process, to be able to achieve a preselected degree of modification. An area of some concern is the stability of the chemical linkages formed between the support and ligand, and it is likely that severe shortcomings are encountered in this regard with some of the more widely used supports.

7. Availability at reasonable cost is clearly an important consideration.

An "ideal" support with all these characteristics obviously does not exist, yet affinity chromatographic procedures are being successfully carried out in many laboratories throughout the world. Among the support materials that have been used are polystyrene, cellulose, cross-linked dextrans (e.g., Sephadex, a product of Pharmacia Fine Chemicals, Inc., Piscataway, New Jersey 08854), agarose, polyacrylamide, derivatized controlled pore glass, and more recently, Ultrogel, a product of LKB Instruments, Inc. (Rockville, Maryland 20852). This section considers the advantages and disadvantages of several of the more commonly used support materials, but a detailed discussion of characteristics and chemistry of activation and ligand coupling is confined to the most commonly used polysaccharide support materials. A number of review articles and books contain detailed discussions of the properties and chemistry of other support materials, and the reader is referred to Section 7 for appropriate references. Excellent technical literature is also available from a number of manufacturers (see p. 366).

Polysaccharide Supports

The chemical and physical properties of cellulose, cross-linked dextrans (Sephadex), and agarose, and the relative advantages of each of these as a support for affinity chromatography have been considered in detail elsewhere [4–7]. Beaded agarose gels with various exclusion limits are by far the most commonly used support materials, and Sepharose 4B or its equivalent, Biogel A-15M (BioRad Laboratories, Richmond, California 98404), with 4 % agarose and an exclusion limit of about 1.5×10^9 daltons, have been particularly popular. The major advantages of agarose are that it is a hydrophilic material that is readily activated and chemically functionalized, it exhibits good flow properties and high porosity, and it is relatively free of nonspecific adsorption effects. Among the disadvantages of this support

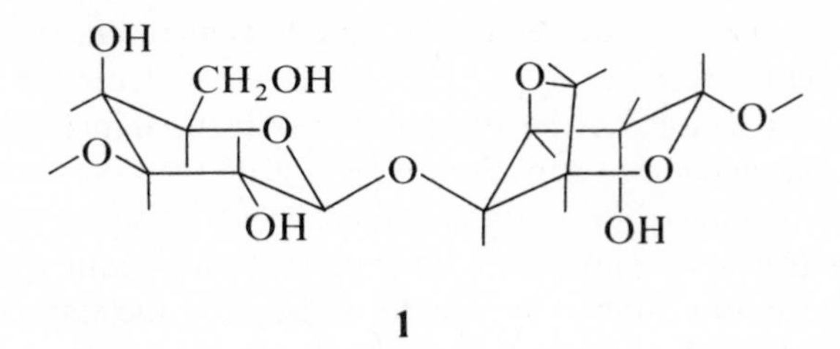

1

material are its sensitivity to extremes of pH, temperature, and solvents, the small amount of negative charge it possesses, its susceptibility to microbial attack, and its instability with respect to mechanical disruption and "ligand leakage" on prolonged washing. For further discussion of these and related points, refer to the literature references cited earlier and to the excellent technical literature provided by Pharmacia. Recently Pharmacia has made available Sepharose, which has been cross-linked with 2,3-dibromopropanol,

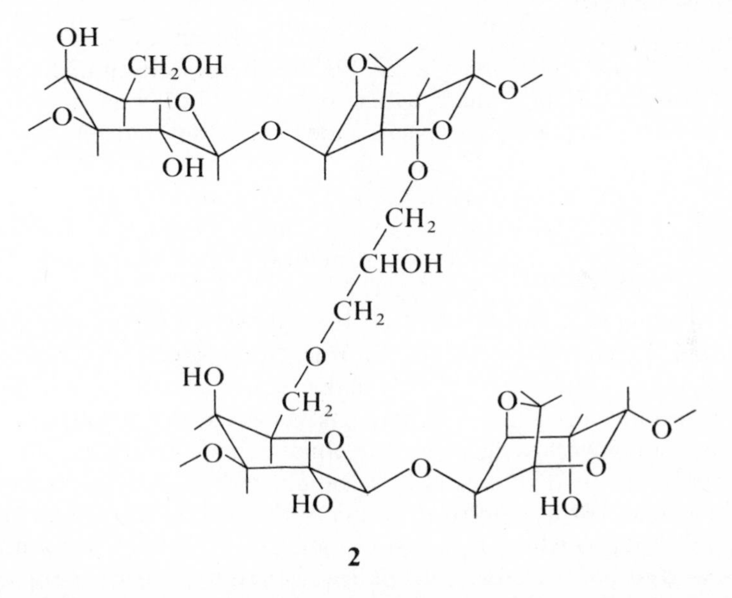

2

2 (Sepharose CL), a procedure that was developed by Porath's group [8, 9]. Cross-linking has a negligible effect on the permeability of the agarose but substantially increases its stability toward extremes of pH and temperature, solvents, and mechanical leakage. As suggested by Porath, commercial Sepharose CL is desulfated at 120°C under reducing alkaline conditions after cross-linking, thus providing a more neutral matrix. Undoubtedly cross-linking increases substantially the attractiveness of agarose as a support for affinity chromatography. Normally agarose gels are stored refrigerated in aqueous suspension in the presence of a bacteriocide (e.g., 0.02 % sodium azide).

As mentioned previously, the most commonly used procedure for the activation of polysaccharide supports is the CNBr method first introduced by Axen et al. [3]. This reaction is conveniently carried out in alkaline solution, and a number of variations, modifications, and "improvements" on the original procedure have been put forth by various workers. A major concern is that the CNBr continuously hydrolyzes under the alkaline activation conditions, thus consuming base equivalents. Also, the activation reaction itself generates protons (see below). This is generally compensated for either by continuous addition of base (manually or with an automatic titrator) to maintain the pH constant, or by the use of a high concentration of buffer. Other considerations such as the effect of variations in pH on activation, and the extent of matrix cross-linking that occurs with excess CNBr have been discussed elsewhere [6, 7, 10–12]. A representative activation procedure is as follows:

First, the percentage of dry agarose in the gel being used is noted and designated ρ (e.g., 2 and 4 for Sepharose 2B and 4B, respectively). The agarose is washed with 2 *M* phosphate buffer, pH about 12.1, and filtered on a Büchner funnel. Ten grams of gel is suspended in 2.5 ρ ml cold 5 *M* phosphate buffer, pH about 12.1, and diluted with water to a volume of 20 ml. Over a 2-min period, ρ ml of a 100 mg/ml CNBr solution is added and the reaction is allowed to proceed for 10 min at 5 to 10°C with gentle stirring. The product is washed by filtration with cold water to neutrality and coupled to the ligand as soon as possible.

The foregoing procedure (from Ref. 10) is typical of procedures that use high buffer concentrations for pH control. An important modification by Cuatrecasas and co-workers [11, 12] makes use of a solution (2 mg/ml) of CNBr in acetonitrile, an inert solvent that is miscible with water. They recommend suspending washed agarose in an equal volume of water, adding this 1 : 1 slurry to an equal volume of 2 *M* sodium carbonate, then adding 0.05 volume of the CNBr-acetonitrile solution to the stirring suspension at 5°C. Other published procedures make use of solid CNBr, preferably ground with a mortar and pestle just before use, but these have the disadvantage of requiring the investigator to weigh out and transfer small quantities of a volatile, highly toxic solid. Detailed descriptions have also been provided for activation procedures that employ direct base addition from a burette or autotitrator with the aid of a pH meter for pH control [6, 11].

A modicum of control can be exerted on the CNBr activation reaction by careful manipulation of experimental conditions such as the relative amounts of CNBr and gel. With high relative concentrations of the former, cross-linking of the matrix can occur, which may lead to loss of adsorption capacity. Also, high loading of the ligand will result and, particularly with macromolecular ligands, multiple bonds between the ligand and conjugate will form ("multipoint attachment"). Although this may stabilize the conjugate toward ligand leakage, it can also result in conformational alterations that adversely affect the biological activity of the ligand. With an eye toward this possibility, macromolecular ligands are often coupled in the presence of inhibitors that bind at the active site, thereby, it is hoped, protecting its integrity and conformation. May has suggested that a similar situation occurs in the immobilization of the nonheme iron protein rubredoxin, where the integrities of the occupied iron-binding sites are protected during immobilization [13]. It should be noted that a low degree of matrix activation is needed for some applications related to affinity chromatography (e.g., determinations of whether individual subunits of multimeric enzymes exhibit catalytic activity). Also, low loading of either affinity chromatography ligands or immobilized enzymes normally gives higher efficiency of utilization of the bound material in either adsorption or catalysis.

On the basis of chemical and spectroscopic evidence, as well as studies with model compounds, it is generally assumed that reactive imidocarbonate structures are formed upon activation of polysaccharide supports with CNBr [3, 14–20]. Presumably these arise from cyanates formed by initial attack of matrix hydroxyl groups on the CNBr, and cyclization to the active imidocarbonate functionality must compete with hydrolysis to the inert carbamate (Fig. 5.1). In the case of dextrans (α,D-glucose units), the cyclic imidocarbonate structures presumably include the 2,3 and the 3,4, and the 4,6, as well as the noncyclic intermolecularly cross-linked species, but only the structures analogous to the latter two are possible for agarose (see Refs. 4 and 5).

Usually, immediately following activation the gel is washed by filtration using cold water and buffer, then coupled to the ligand of choice. However CNBr-activated agarose that has been freeze-dried in the presence of dextran and lactose to preserve the bead form of the gel is available from Pharmacia. The material is supplied in 15-g airtight packs, and the manufacturer claims a shelf life of about 18 months for this material at 8°C. In our laboratory unopened commercial material has been used successfully as much as a year after the expiration date. The commercial material must be reswollen and washed extensively (Pharmacia suggests washing with 1 mm HCl) before use. Axen and Ernback [16] reported that activated agarose suitable for storage can be prepared by replacing the water around the gel with acetone and storing the acetone-swollen material at room temperature (10 % activity loss in 2 months). Also, they reported that activated cellulose or Sephadex can be stored by washing with acetone-water mixtures of increasing acetone concentration and finally with pure acetone, which is then evaporated *in vacuo*, after which the product is stored in a sealed vessel in the refrigerator.

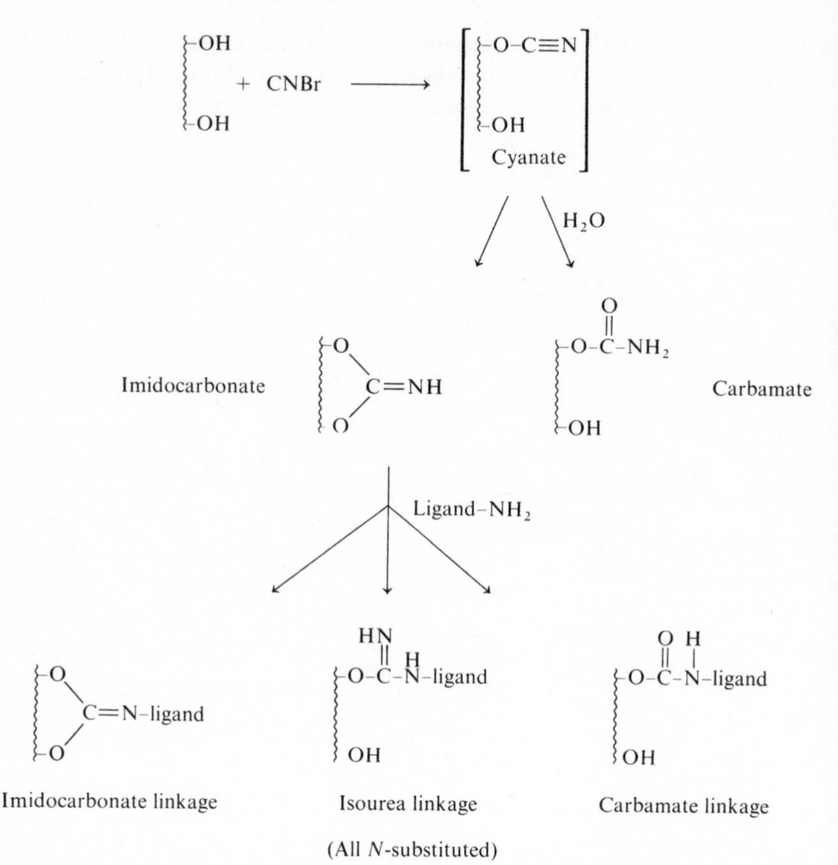

Fig. 5.1. Activation of polysaccharides with CNBr and coupling of amino ligands [4].

Coupling of amino-containing ligands to CNBr-activated polysaccharide matrices is normally carried out in a basic buffer solution (e.g., 0.25 *M* $NaHCO_3$, pH 9.0) by rotating the suspension end over end in a cylindrical vessel in the presence of excess ligand for a couple of hours at room temperature or overnight in the cold room. Work by Cuatrecasas [21], confirmed by many other investigators, established that coupling efficiency drops markedly at either low (below ca. 8) or high (above ca. 10) pH. Since the reactivity of the amino ligand depends on its pK_a, primary aliphatic amines are best coupled near pH 10, whereas more acidic amines are readily coupled with high efficiency at pH 8 to 9. The decreased coupling efficiency observed above pH 10 most likely reflects instability of the activated matrix in this region.

Thus the ability to vary the coupling pH gives the investigator a measure of control of the extent of ligand loading in the conjugate. The relative concentration of ligand present during coupling, of course, also affects loading in the conjugate. Especially with macromolecular ligands, certain other factors, discussed below, must be taken into account.

Additional considerations regarding the coupling reaction are the following:

1. The stability of the ligand itself may be poor at the alkaline optimal coupling pH, resulting in the attachment of large amounts of inactive material. In such cases coupling efficiency should be sacrificed and the coupling reaction carried out in a less basic medium.
2. Particularly with macromolecular ligands, multipoint attachment and cross-linking may occur. Although this may be detrimental to the activity or binding properties of the immobilized ligand, one would expect multipoint attachment of a ligand to solidify its connection to the matrix and reduce "ligand leakage," often a highly desirable situation. In any case multipoint attachment can be minimized by coupling at a less favorable pH or by using a higher relative concentration of free ligand in the coupling reaction.
3. Coupling reactions do not generally proceed to 100% completion, and it is good practice to block any active groups that remain on the matrix after attachment of the ligand. Usually this is effected by shaking the newly synthesized conjugate in the presence of glycine, ethanolamine, or some similar molecule for a couple of hours at room temperature.
4. It is now recognized (but was not explicitly considered in much of the early work) that the imidocarbonate and/or isourea linkages formed upon ligand coupling introduce an often nonnegligible degree of positive charge into the conjugate. Section 3 discusses alternate chemistry that has been developed for the synthesis of more neutral conjugates.

Questions have been raised regarding the inherent long-term stability of ligand-Sepharose conjugates produced from CNBr-activated agarose. This problem was discussed by Cuatrecasas and Parikh in 1972 in connection with their report on the use of *N*-hydroxysuccinimide ester derivatives of agarose for coupling of amino-containing ligands [22]. The relevant chemistry is summarized in Fig. 5.2. In examining the stability of alanine-Sepharose prepared by both this method and by direct CNBr coupling, these investigators found limited stability for both conjugates. They concluded that the liability resides in bonds resulting from the CNBr linkage step (imidocarbonate and isourea), not from the more distal amide linkage in the former conjugate. In a model study quoted by March et al. [23], alanine was coupled to CNBr-activated agarose, and after extensive washing to remove noncovalently adsorbed material, the conjugate was stored at 4°C for 30 days; while in

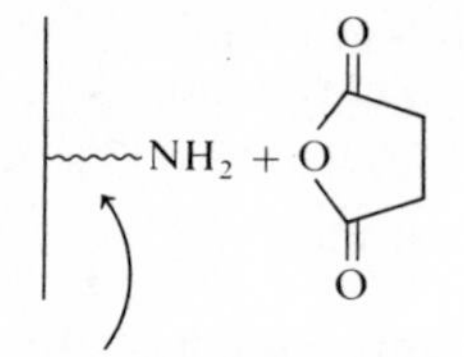

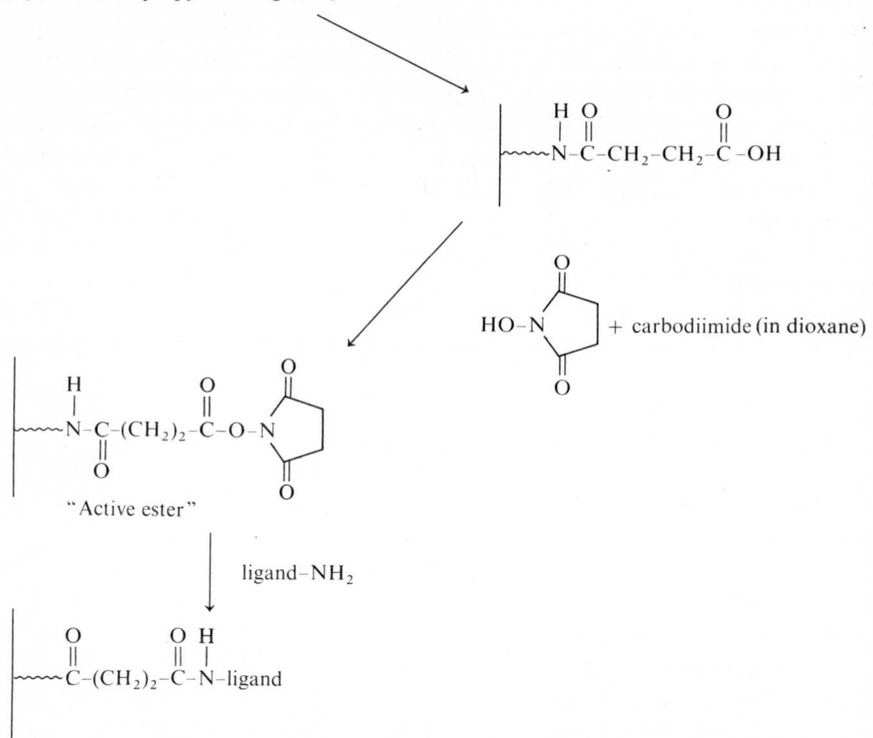

Fig. 5.2. Chemistry of *N*-hydroxysuccinimide ester of agarose.

storage, it lost about 0.1 % of the coupled alanine per day. Similar observations were reported by Tesser et al. [24], who observed leakage of picomolar quantities of a cyclic AMP analog after coupling to CNBr-activated Sepharose. Many other authors have commented on the general problem of ligand leakage, some reporting that it is undetectable under their particular experimental conditions (see Section 7 and Refs. 25–28).

A detailed analysis of the ligand leakage problem has been presented by Parikh et al. [11]. Among the points brought out by these authors are the following:

1. From an operational point of view, the presence of a small amount of "leaked" ligand will not seriously interfere with the affinity chromatography procedure if the K_d for the complex is greater (i.e., less favorable) than about 10 μM and if the bound and free ligands interact similarly with the component being isolated. In such cases, they claim, concentrations of free ligand as high as 1 to 5 % of the bound species will not significantly compete with adsorption. With very high affinity ligands (e.g., K_d of 1 nM), however, and in cases where free ligand is more effective in complex formation than bound ligand, the net result is essentially complete complex formation in solution. Thus adsorption to the affinity column may be completely prevented. In addition, assay procedures for detection of the desired material may not respond to the *complexed* form actually present in solution, leading to the erroneous conclusion that the material remains bound to the gel, whereas in actuality no binding at all took place. This kind of situation is most likely to occur in hormone-receptor systems that exhibit high affinity. Thus the authors suggest that for such systems, the chemistry of ligand attachment be so designed that the least stable bond be that formed by the CNBr activation. This would increase the probability that "leaked" and "bound" ligands will be similar in both structure and affinity for the protein, thereby minimizing complications.

2. To minimize "leakage" effects, adsorbents used for "high affinity" systems should contain the least amount of ligand compatible with effective adsorption. It may thus be necessary to dilute the conjugate with unsubstituted agarose before actually carrying out the chromatography.

3. Experimental conditions may be adjusted to minimize leakage. For example, decreasing the temperature or using increased flow rates over shorter periods of time may be advantageous.

4. What appears to be persistent, continuous leakage of covalently bound ligand may actually be release of ligand molecules that had been noncovalently adsorbed onto the matrix only. Depending on the particular compound in question, exhaustive washing procedures extending over days (or even weeks), and utilizing various solvents or specially formulated solutions, may be required.

5. The authors carried out a standard series of quantitative leakage tests using a simple ligand (alanine), a protein ligand (albumin), and a polypeptide ligand (insulin) and concluded as a rule of thumb that simple monoamines are lost at an initial rate of 0.1 % per day, whereas proteins leak about one-fifth as rapidly. Also, they assert that the rates of ligand leakage may decrease by about sixfold after the first 30 days.

The basis for an alternative approach to the control of ligand leakage is provided by the investigations of Wilchek and co-workers [30, 31]. These authors described the preparation of various polymer-agarose conjugates, such as polylysyl-Sepharose, which exhibit high stability because of multipoint attachment of the polymer to the support. Aside from their minimal hydrophobicity—thus minimizing nonspecific adsorption effects and other complications discussed in Section 3—these macromolecular spacers provide greater separations of the ligand from the support than do conventional alkyl-chain "arms." Various ligands can then be coupled to the hydrophilic polymeric spacers by standard chemical procedures to give specific conjugates that should exhibit a much reduced degree of "ligand leakage." Other polymers containing acyl hydrazide groups (e.g., polyglutamic acid hydrazide) can be used to provide more "neutral" conjugates after reaction with CNBr-activated agarose (p. 332). Another interesting variation involves reacting D,L-alanine-*N*-carboxyanhydride with polylysyl-Sepharose to give short (5 to 10 residues), straight extensions out from the stable, multipoint-attached conjugate **3**. The ligand of interest can then be attached to the ends of these

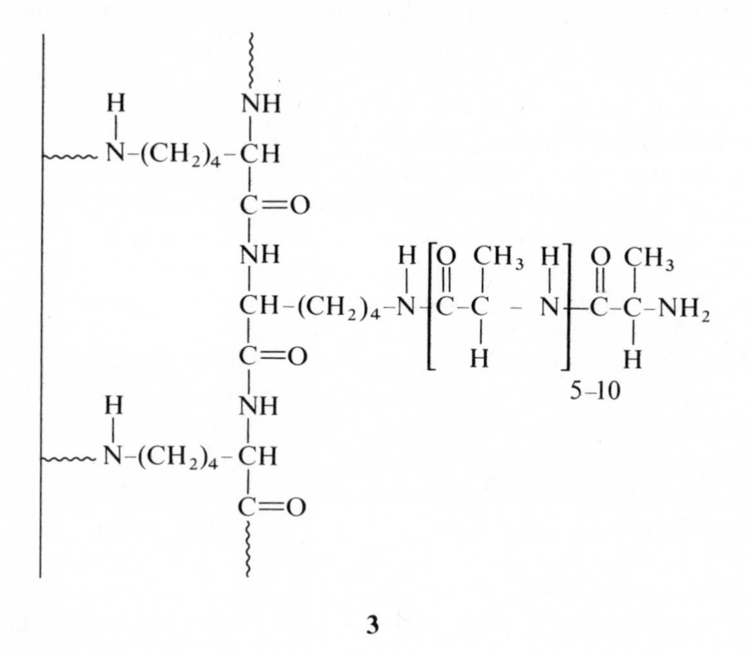

3

extensions. Several polymer-agarose conjugates are now available commercially from Miles Research Products, Elkhart, Indiana 46514.

Table 5.1 presents information about some of the alternate activation chemistry that has been developed for agarose and other polysaccharide matrices. These methods provide the means of producing derivatives without

5.1 Alternate Activation of Polysaccharide Supports

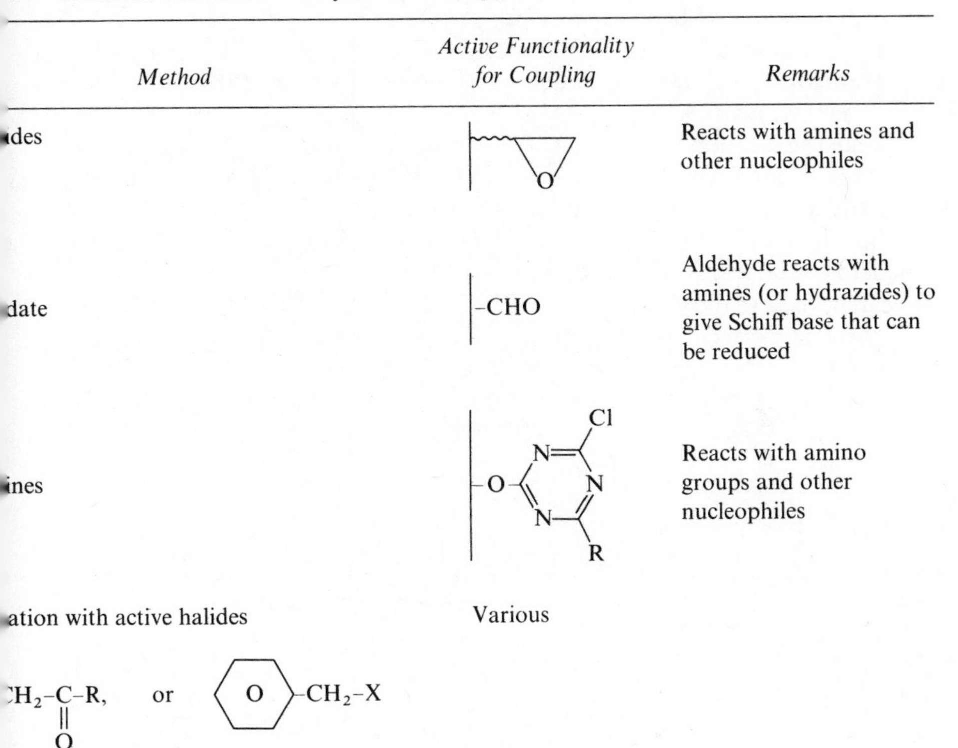

Method	*Active Functionality for Coupling*	*Remarks*
ides		Reacts with amines and other nucleophiles
date	–CHO	Aldehyde reacts with amines (or hydrazides) to give Schiff base that can be reduced
ines	–O–(triazine: N, N, N; Cl; R)	Reacts with amino groups and other nucleophiles
ation with active halides $CH_2–C(=O)–R$, or (O)–CH_2–X	Various	

the complications inherent in CNBr-activation, but each method has its own set of advantages and disadvantages. Figure 5.3 summarizes chemistry that can be used to build up a large variety of reactive functionalities on the matrix after initial attachment of an "arm" that terminates with, say, an amino or carboxyl group, thus facilitating subsequent coupling of the desired ligand. It should be kept in mind that the relative advantages and disadvantages of different types of spacer arm have been considered extensively in the affinity chromatography literature.

Other Supports

A number of other support materials have been used successfully in affinity chromatography procedures. The physical properties and activation chemistry for these materials have been considered elsewhere in detail and the reader

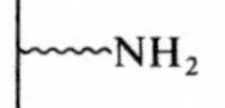

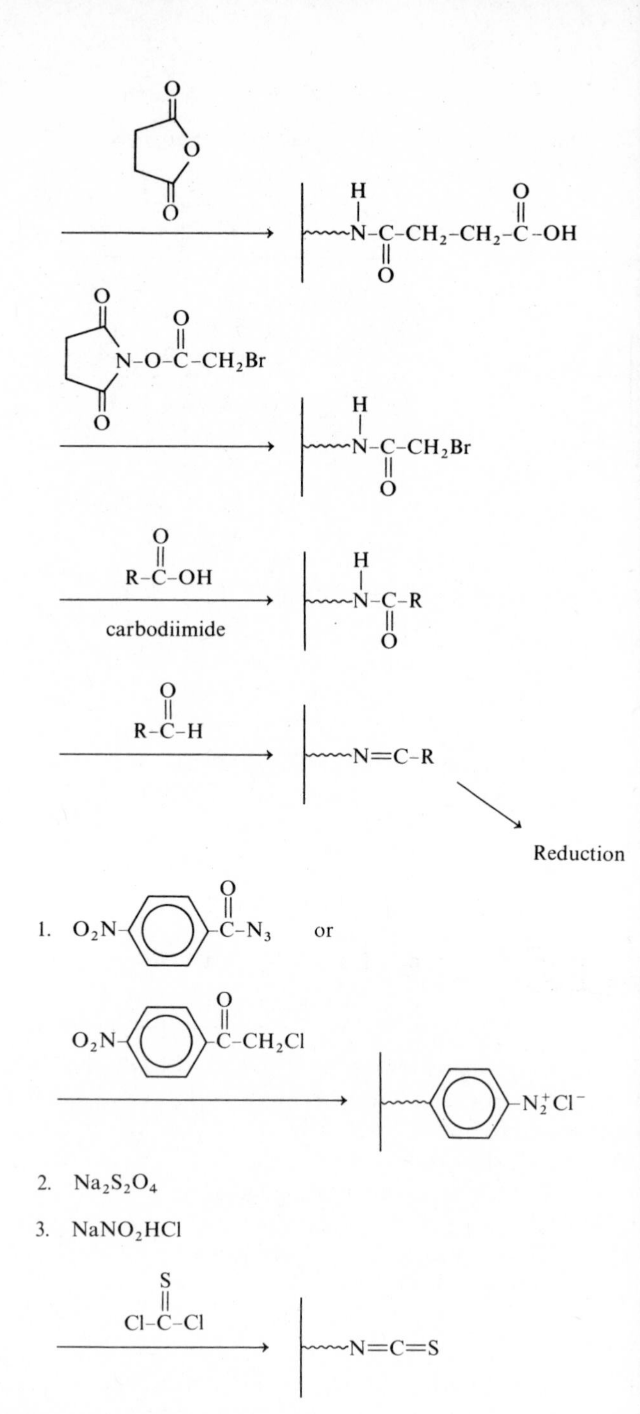

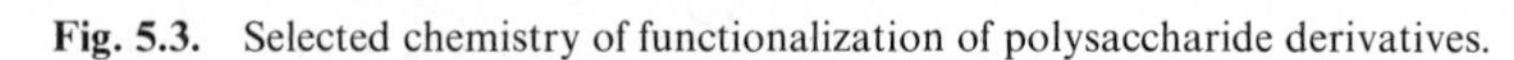

Fig. 5.3. Selected chemistry of functionalization of polysaccharide derivatives.

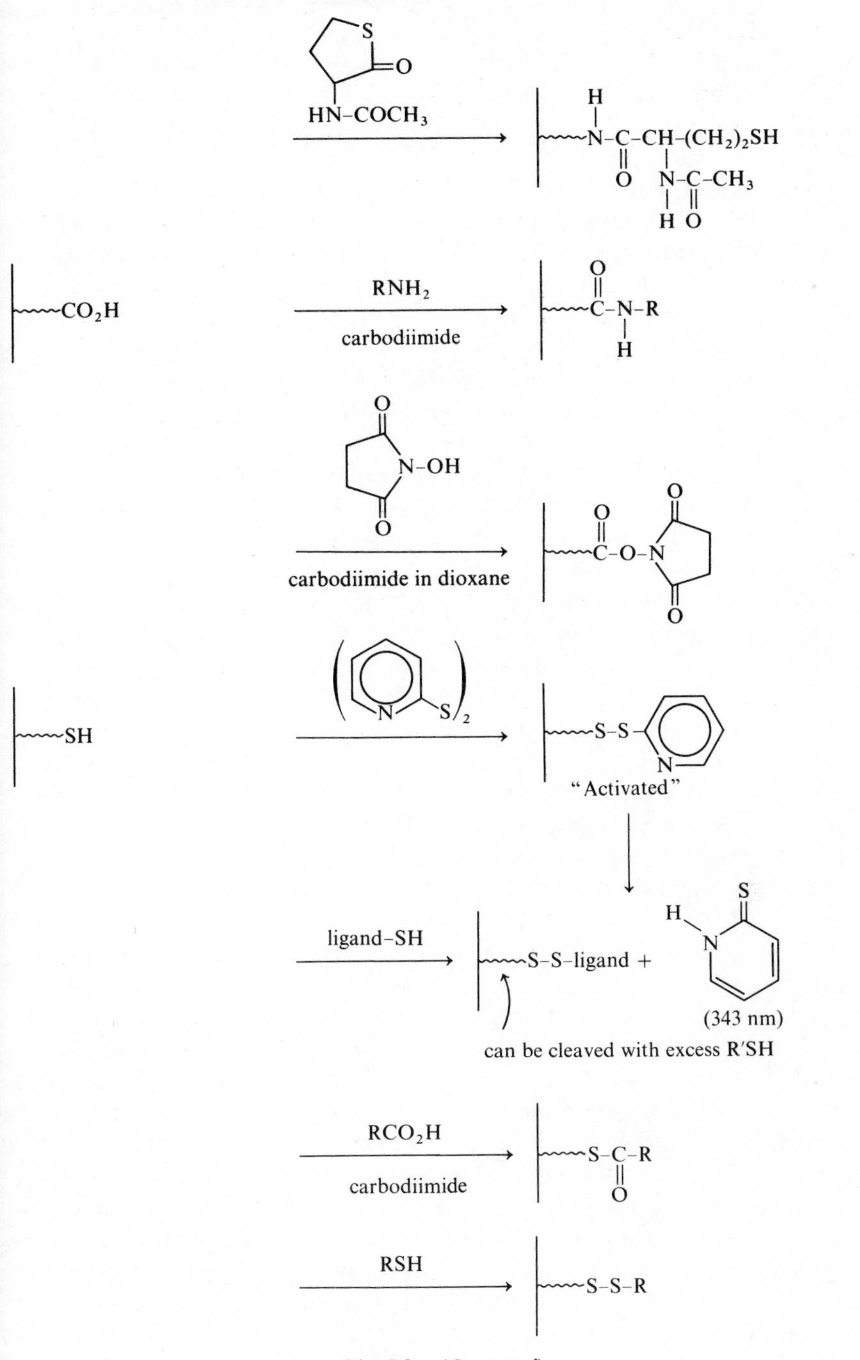

Fig. 5.3. (*Continued*)

is directed to the list of review articles in Section 7. A brief consideration of the major advantages and disadvantages of the more popular of these materials follows.

1. *Polyacrylamide Gels*. These are copolymers of acrylamide and *N,N*-methylenebisacrylamide and are available as spherical beads in various porosities from BioRad Laboratories (Biogel P). Among the advantages of this material are uniformity, chemical stability, high concentration of activatable carboxamide groups, and low tendency toward nonspecific adsorption of proteins.

2. *Glass*. A large number of controlled pure glass derivatives are manufactured by Corning Glass Works and distributed by Pierce Chemical Company (Rockford, Illinois 61105). Many of these materials are already prepared specifically for affinity chromatographic procedures, and activation chemistry for attaching virtually any ligand to glass derivatives is now available. Among the advantages of glass are rigidity, stability toward biological and chemical degradation, and uniformity in size and variety in porosities of the beads that are commercially available. Disadvantages of this support include the reported propensity toward nonspecific adsorption of proteins and "solubility," which can lead to ligand leakage, especially at high temperature in basic solution. Also, problems with flow rates of glass-bead columns and mechanical trapping of particulate material from large samples have been reported [32]. Procedures have been suggested for minimizing the nonspecific adsorption problem [33], and zirconium-clad beads are available commercially for use in the alkaline pH range.

3. *Ultrogel*. This new product of LKB, Inc. (United States office: Rockville, Maryland 20852) is composed of a three-dimensional polyacrylamide lattice with an interstitial agarose gel and is sold as preswollen rigid beads with diameters within the narrow size range of 60 to 140 μm. This small, uniform size, plus the low compressibility of the beads, allows for good flow rates and high resolution in gel filtration applications. Ultrogel is supplied in four types, designated ACA (for acrylamide-agarose) 22, 34, 44, and 54. The two numbers indicate the concentrations of acrylamide and agarose, respectively; thus the 22 gel (2 % acrylamide, 2 % agarose) is the most porous, and the 54 gel is the least porous. Fractionation ranges vary from 60,000 to 1,000,000 for the 22 gel to 6000 to 70,000 for the 54 gel. Ultrogel possesses certain advantages that make it potentially more desirable as an affinity chromatography support than either agarose or polyacrylamide alone. Its physical properties and flow characteristics are better than those of agarose, and the low porosity that appears to be associated with functionalized polyacrylamide is circumvented in this combination gel. Also, Ultrogel contains both agarose hydroxyl and polyacrylamide carboxamide functionalities that can be indepen-

dently activated. Thus combination conjugates can be prepared by combining two different ligands through independent series of reactions. In view of these considerations, Ultrogel promises to become an increasingly popular support material for affinity chromatography.

3 THE ARM

The literature of affinity chromatography is permeated with the notion that interposing a spacer or "arm" between the ligand and the support is beneficial. Presumably the arm increases the ligand's freedom of motion by extending it away from the relatively restricted environment of the matrix, thus minimizing steric restrictions that could otherwise result in the affinity of the immobilized ligand being drastically less than that observed in free solution (higher "operational" value of the dissociation constant K_d). Work by Cuatrecasas and co-workers [21, 34] provided clear examples of the dramatic effects of arms, particularly in making possible the successful use of "weak" ligands ($K_d = 10^{-3}$ M) in affinity chromatography, and similar observations were reported soon thereafter by many other investigators (Section 7). Despite some isolated reports on negligible or even detrimental effects of long arms (e.g., Ref. 35), the use of at least moderately long spacers quickly gained virtually universal acceptance.

During the past 5 years or so, there has been a complete reexamination of the use of arms—particularly those composed of hydrophobic and/or charged functionalities—in affinity chromatography. May has reviewed the developments in this area in detail elsewhere [4]. For the purposes of this chapter, only a summary of the major conclusions and operational considerations is presented. References 36 to 47 provide an entrance into the pertinent recent literature.

1. It is clear that the hydrocarbon chains commonly used as alkyl "arms" are themselves capable of contributing significantly to the binding of macromolecules such as proteins to affinity columns. Hydrophobic interactions are certainly involved in such binding. However since arms are generally designed to terminate with reactive functionalities such as amino or carboxyl groups, to allow subsequent ligand attachment, and since such attachment procedures do not proceed with 100% conversion, the conjugate may contain terminally charged "dangling arms," which can cause ionic adsorption to the column. Even more significantly, the imidocarbonate and/or isourea functionalities formed upon coupling of any amine to CNBr-activated matrices are positively charged (pK_a = 10.5). Thus it appears that a combination of ionic and hydrophobic interactions with the functionalized support—essentially those of an "insolubilized detergent"—are operative, and studies on

the effects of various eluents on adsorption to such "hydrophobic" columns are consistent with this notion. It is very important to note that by coupling acyl hydrazides, rather than amines, to CNBr-activated agarose, "neutral" derivatives are produced, since the pK of the resulting linkages is about 4. Thus a major source of interference effects can be eliminated.

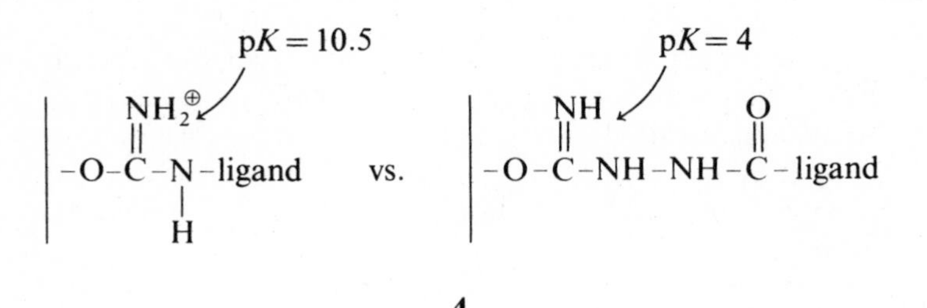

4

2. Although the adsorptive properties of alkyl arms obfuscates the interpretation of a number of reported affinity chromatography procedures, many investigators quickly recognized the potential of simple alkyl-Sepharose conjugates for the isolation of specific macromolecules. Thus the concept of "hydrophobic chromatography" evolved, and this technique has been assimilated very rapidly into the arsenal of standard biochemical purification techniques. Note, however, that possible denaturation effects and other complications are possible with alkyl ligands capable of acting as insoluble detergents, rendering "neutral" derivatives prepared from acyl hydrazides quite attractive.

3. Hydrophilic arms—such as those in which every other atom is part of an amide, amino, carbinol, or other hydrophilic grouping—are attractive because conceptually, interference effects due to "hydrophobic" and "detergent" chromatography are eliminated, leaving "ligand-specific" interactions as the sole adsorptive mechanism operative. Evidence has been presented that short hydrophilic arms are as effective as longer ones; this is not surprising, since the desirability of long arms was established with alkyl chains that are now known to present complications. Also, with hydrophilic arms, concern regarding "dangling arms" is lessened considerably because the major source of interference (hydrophobicity) is eliminated, and minor charge effects associated with the arm of the matrix can be overcome by appropriate adjustment of ionic strength.

4. It has been postulated that "compound affinity" effects, by which a weak specific interaction between an enzyme and a ligand is synergistically strengthened by nonspecific hydrophobic interactions, occur in a number of

cases. Thus the usefulness of a "weak" ligand attached by way of a hydrophobic arm may be enhanced by adjusting the salt concentration in the buffer just above the level at which nonspecific adsorption to the arm begins to cause noticeable interference.

Developments such as those just summarized have led to the introduction of a number of new commercial products useful in affinity chromatography applications. A number of alkyl- and alkylamino-agarose derivatives are available from Miles Research Products, PL Biochemicals, Inc. (Milwaukee, Wisconsin 53205), ICN Pharmaceuticals, Inc. (Cleveland, Ohio 44128), BioRad Laboratories, and Pharmacia. A variety of "neutral" agarose derivatives based on acyl hydrazides are available from Miles, which also sells polymer-agarose derivatives (multipoint ligand attachment). Several of the companies named also sell hydrophilic ligand-substituted polysaccharide supports of various descriptions.

4 THE LIGAND

Much has been written about the characteristics of an "ideal" ligand for affinity chromatography, but in practice, of course, virtually all ligands have some drawbacks. Among the major considerations to be kept in mind in evaluating potential ligands are the following.

1. *Specificity of Binding*. Specific binding minimizes adsorption of impurities with structures or recognition sites similar to those of the desired component. However relatively nonspecific ligands can be used successfully when coupled with highly specific adsorption and/or elution procedures, with the result that undesired material either is bound less readily or is left behind on the column. Techniques of this type have been developed for use with immobilized cofactors and other "group-specific" conjugates possessing the distinct advantage that a single conjugate can be used for isolation of a number of different macromolecules by judicious choice of adsorption and elution conditions (see below).

2. *Strength of Binding*. For reversible ligands the quantity of interest is the operational value of K_d, the *dissociation* constant for binding of the component to the ligand *in the immobilized state*. This value can differ considerably from the dissociation constant measured in free solution, and the extent of this difference certainly varies considerably from system to system and is strongly dependent on the configuration of ligand attachment in the conjugate in question. Thus statements in the literature to the effect that ligands with a K_d of 1 mM or less are suitable for affinity chromatography should be viewed with caution, since they usually refer to the K_d measured

in free solution. Among the techniques suggested to increase the effectiveness of weak ligands are the use of either highly loaded conjugates or very long arms. The former technique has the disadvantage that a substantial number of ligand molecules may become inaccessible to the feed, whereas in the latter case effects such as "compound affinity" are undoubtedly operative. O'Carra et al. [36] have suggested the following: optimal retardation on affinity columns, expressed in column-volume units, should approximately equal the concentration of the immobilized ligand divided by K_d (as measured in free solution). Thus for a ligand with $K_d = 5 \times 10^{-3}\ M$, immobilized at a concentration of 0.6 mM, a retardation of only 0.12 column volume is expected, and the protein should emerge immediately after the void volume. Steric hinderance by the matrix would decrease the extent of retardation below the expected value, whereas nonspecific adsorption effects would cause increased retardation on the column.

3. *Reversibility of Binding.* Although affinity chromatography is normally based on the formation of a reversible complex, there are now a number of examples of ligands that form covalent bonds with the molecule being isolated being used successfully [48–51]. Elution requires cleavage of this bond with a chemically reactive molecule (e.g., a nucleophile); thus *direct* recycling of the conjugate is not normally possible.

4. *Interference Effects.* A factor that commonly contributes to nonspecific adsorption is the presence of substantial charge on the ligand (or the arm), which leads to the ion-exchange type of adsorption onto the conjugate. In such cases elution procedures must be employed that distinguish between material bound ionically and material bound through specific affinity for the ligand. Examples of such procedures are provided by the work of Robert-Gero and Waller with a methionine-Sepharose conjugate [52] and of Silverstein with a lysine-Sepharose conjugate [53]. An interesting example of the use of a negatively charged ligand to purposely *exclude* a desired component while adsorbing undesired material, followed by further chromatography of the excluded fraction, is presented by the work of Yon on the purification of aspartate transcarbamylase [54].

In contrast to these concerns about maximizing the specificity of the ligand chosen for inclusion in an affinity conjugate, the concept of using "general" or "group-specific" ligands is now becoming increasingly popular. The strategy here is to take advantage of the fact that many enzymes operate only in the presence of certain organic molecules, coenzymes, which participate directly in the catalytic mechanism as actual cosubstrates. An example of this type of enzyme is lactate dehydrogenase (LDH), which catalyzes the following reaction:

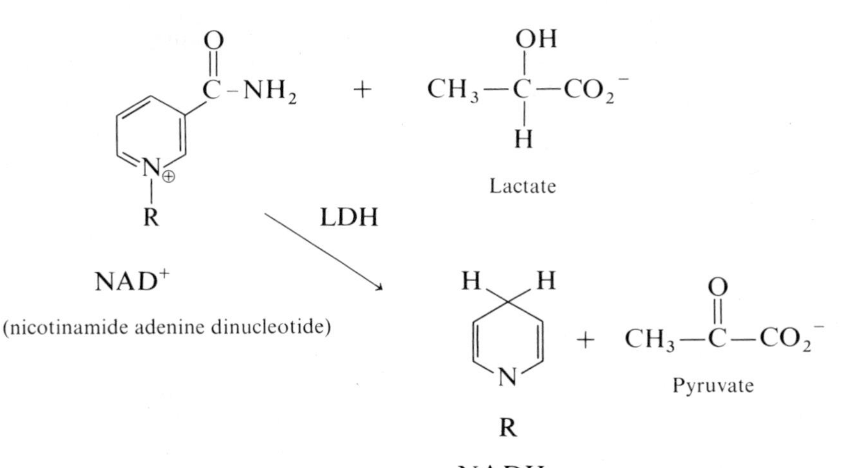

where the coenzyme structure has been abbreviated to show only the site of hydrogen (hydride) transfer. Both the "substrate" (lactate or pyruvate) and the "coenzyme" (NAD$^+$ or NADH) are bound by specific, discrete sites at the active site of LDH where the redox process is catalyzed, and this binding occurs in a compulsory ordered fashion—the pyridine-nucleotide coenzyme binding first. Thus we anticipate that a properly constructed affinity conjugate with a pyridine nucleotide ligand will be capable of binding LDH or, for that matter, any other dehydrogenase that binds this type of coenzyme in a compulsory ordered fashion. On the other hand, binding of LDH to a conjugate with a pyruvate or lactate analog as a ligand should occur only in the presence of coenzyme, in which case ternary complex formation would be responsible for adsorption.

Although there are a number of coenzymes of vastly differing structure, several commonly occurring ones are related structurally to the "adenosine monophosphate half" of the NAD$^+$ molecules. Thus AMP, ADP, ATP, and cyclic AMP differ only in the number and position of the phosphate groups attached to this half of the NAD$^+$ molecule, and NADP$^+$ is simply NAD$^+$ phosphorylated at the position indicated in **5**. The strategy in affinity chromatography experiments with enzymes that utilize such cofactors thus becomes clear. If cleverly constructed, a small number of nucleotide-matrix conjugates could bind to a very large number of *functionally unrelated* enzymes, since these enzymes all utilize structurally related cofactors. This obviates the necessity of synthesizing a separate ligand-matrix conjugate for each enzyme—clearly an operational advantage and an impetus for commercial suppliers to produce standard coenzyme conjugates that can be utilized

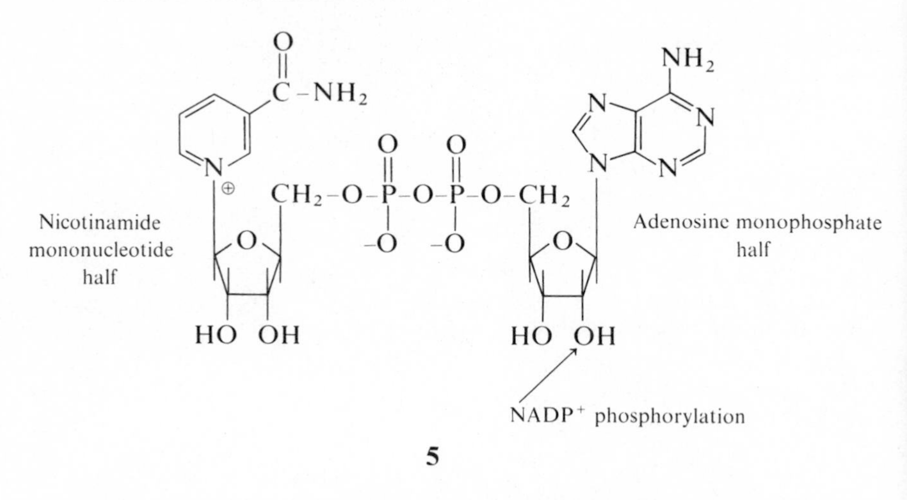

5

by a large number of investigators. On the other hand, highly specific adsorption and/or elution procedures must be developed when such "general" ligands are used, to allow isolation of a particular enzyme from a crude mixture containing many other macromolecules capable of binding to the coenzyme conjugate.

Mosbach and co-workers [55–58], Lowe and Dean [59, 60], as well as others, have demonstrated that it is possible to overcome these obstacles and successfully utilize columns of immobilized NAD^+ and AMP analogs in the purification of a variety of enzymes. Although the exact structures of the immobilized nucleotides used in the early work of these investigators were somewhat uncertain, and a number of complicating problems were encountered, the initial studies clearly established the potential of this approach. Among the elution procedures developed by these workers are the use of pulses of oxidized or reduced cofactors at appropriate concentrations, and the use of linear gradients of NADH. In both cases, since the adsorbed enzymes differ in their affinities toward NADH or NAD^+, they are selectively eluted at different concentrations of these cofactors.

Another powerful strategy in the design of selective adsorption and elution procedures takes advantage of the ability of many enzymes to form ternary complexes with coenzyme and substrate (or substratelike inhibitor), with this complex formation process proceeding in a compulsory ordered fashion. Thus if two enzymes are bound to an immobilized nucleotide column, they might be separately eluted by first applying an eluent containing cofactor plus a substrate of one enzyme, and subsequently applying an eluent containing the same cofactor plus a substrate of the other enzyme. Each enzyme will elute only when *both* the appropriate substrate and cofactor are present in solution, since this allows an equilibrium shift toward the soluble ternary complex, which can form only in the compulsory ordered (coenzyme-first) fashion. Soluble substrate alone would not effect elution. A number of variations on this theme are possible for both the adsorption and the elution steps, and the effects of such manipulations on elution can often provide valuable mechanistic information in addition to effective elution.

Having established the basic principle, Mosbach and co-workers [61] synthesized and fully characterized the compounds **6** and **7**, N^6-carboxymethyl-NAD^+ and N^6-[N-(6-aminohexyl)-acetamide]-NAD^+, respectively, to facilitate the use of immobilized coenzymes in affinity chromatography on a routine basis. Both NAD^+ analogs exhibited substantial coenzymatic

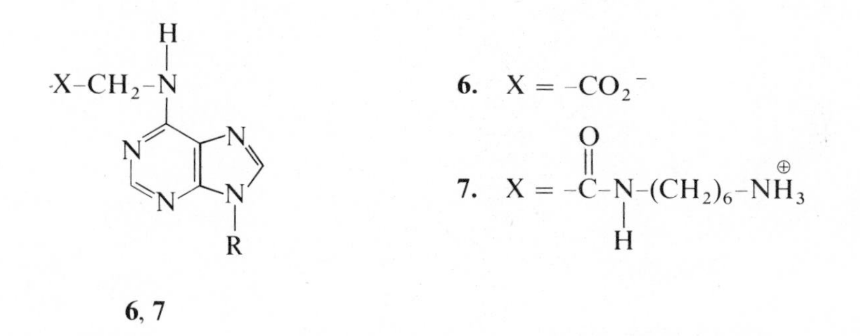

activity when tested with several dehydrogenases. Derivative **7** was coupled to CNBr-activated Sepharose and was shown to retain coenzymatic activity in the immobilized state, and a column of this conjugate was used to resolve a mixture of albumin, alcohol dehydrogenase, and lactate dehydrogenase. Alcohol dehydrogenase was eluted with a 1 mM solution of soluble NAD^+, while lactate dehydrogenase remained bound to the column and was subsequently eluted with 0.5 mM NADH. This result illustrates the application of a pulsed elution technique, which owes its success in this instance to the difference of about one hundredfold between the dissociation constants of lactate dehydrogenase and the oxidized and reduced cofactors, respectively. A slightly different NAD^+ derivative, N^6-(6-aminohexyl)-NAD^+, was prepared by Craven et al. [62] [from N^6-(6-aminohexyl)-5′-AMP] and coupled to CNBr-activated Sepharose. The affinities of NAD^+ dependent-dehydrogenase, but not of $NADP^+$-dependent dehydrogenases for this conjugate were found to be greater than those which had been observed with either the earlier, less well-defined NAD^+ conjugates or with an immobilized AMP derivative.

In related work, Mosbach's group [63] also synthesized and fully characterized the N^6-(6-aminohexyl) derivatives of adenosine-2′,5′-biphosphate

(**9**) and adenosine-3′,5′-biphosphate (**10**) for comparison with the previously prepared N^6-(6-aminohexyl)-adenosine-5′-monophosphate (**8**). In solution studies, NAD^+-dependent enzymes (e.g., alcohol dehydrogenase) were found to be inhibited only by **8**, whereas several $NADP^+$-dependent enzymes (e.g., glucose-6-phosphate dehydrogenase) were inhibited only by **9**, which carries the same adenosine moiety as $NADP^+$. Derivatives **8** and **9** were coupled

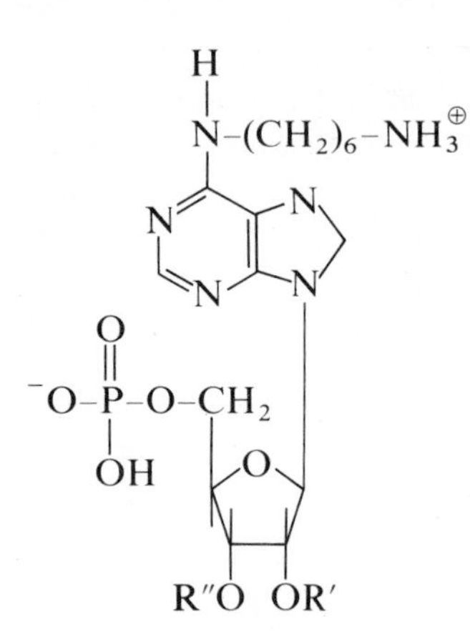

8, 9, 10

	R′	R″
8.	H	H
9.	PO_3H	H
10.	H	PO_3H

separately to CNBr-activated Sepharose, and mixtures containing lactate dehydrogenase (which is NAD^+ dependent), glucose-6-phosphate dehydrogenase, and 6-phosphogluconate dehydrogenase (both $NADP^+$ dependent) were applied to each of the conjugates. Both the $NADP^+$-dependent enzymes were retained on the Sepharose-**9** conjugate, and lactate dehydrogenase emerged in the void volume. Pulsed addition of $NADP^+$ eluted the bound enzymes, whereas a pulse of NAD^+ was without effect, as expected. On the other hand, lactate dehydrogenase but not 6-phosphogluconate dehydrogenase, was retained on the Sepharose-**8** conjugate, and a pulse of NADH (but not of NADPH) was required for elution. Interestingly, $NADP^+$-dependent glucose-6-phosphate dehydrogenase also was retained on the latter conjugate and subsequently was eluted with NADPH, although its binding to this adsorbent was considerably weaker than that to the Sepharose-**9** conjugate. The authors ascribe this finding to a weak inhibitory binding to this enzyme by adenosine-5′-monophosphate derivatives. Parallel experiments with other enzyme mixtures, and with the Sepharose-**10** conjugate, revealed a good correlation between inhibition potency and retention on the various immobilized derivatives. As a demonstration of the utility of such derivatives, a crude extract from *Candida utilis* was applied to the Sepharose-**9** conjugate, and elution with a linear gradient of $NADP^+$ allowed the separation and purification of a number of $NADP^+$-dependent enzymes in a single chromatographic step.

Lowe and Dean and co-workers [64–70] also carried out a series of studies on the use of immobilized nucleotides in affinity chromatography. N^6-(6-Aminohexyl)-5′-AMP (**8**) was conveniently prepared from the 6-mercapto derivative of 5′-AMP in high yield. The advantage of this procedure lies in the commercial availability of the starting 6-mercapto derivative, which is also a by-product of *Brevibacterium ammoniagenes* metabolism. The Sepharose-**8** conjugate prepared by the CNBr procedure was characterized by titration data, phosphate analysis, and alkaline phosphatase degradation, and it exhibited toward various dehydrogenases and kinases binding characteristics that were similar to those of the Sepharose-**8** conjugate prepared by Mosbach, although some differences attributable to differing ligand loadings in these two adsorbents were observed. Among the interesting results reported by these investigators in a series of reports on the use of their immobilized nucleotide were the following.

1. A comparison of the conjugates **11** and **12** revealed differences in their affinities for various enzymes, with yeast alcohol dehydrogenase and glycerokinase binding only to **11** and myokinase and glyceraldehyde-3-phosphate dehydrogenase binding only to **12**. Presumably this reflects differences in the roles of the various parts of the nucleotide structure in productive interactions with these enzymes.

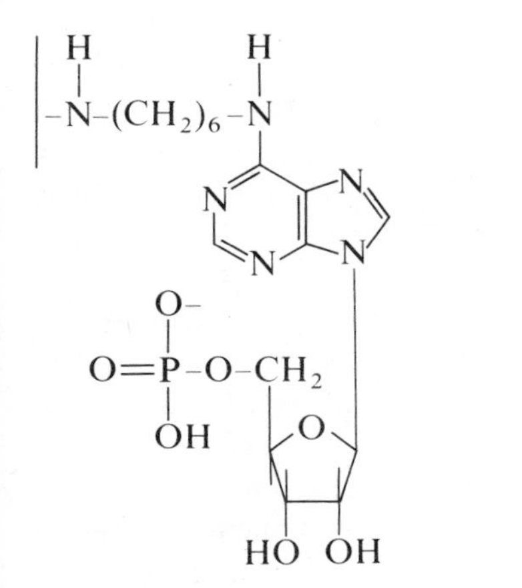

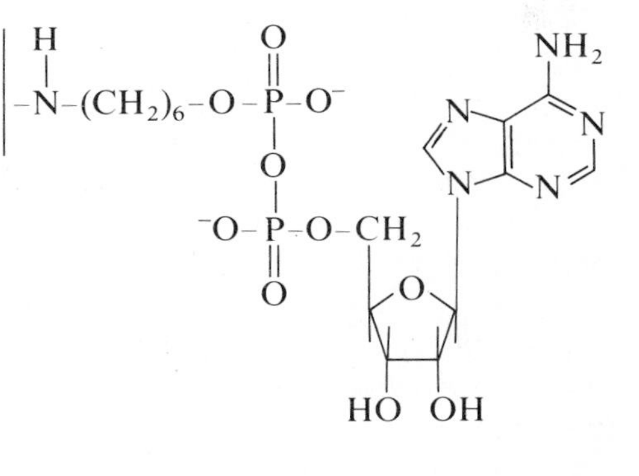

11 **12**

2. Examination of the effect of nucleotide loading on the binding capacities of adsorbents **11** and **12** revealed a general trend of increased binding effectiveness (expressed as units bound per micromole of nucleotide) for columns with higher ligand concentrations. Thus, for example, lactate dehydrogenase was readily eluted by a salt gradient from a column of **11** with a loading of 0.125 μmole of ligand per milliliter of Separose, but came off only upon application of a 5 m*M* NADH pulse from a 1.5 μmole/ml column of the same adsorbent. Interestingly, the geometry of the column was found to exert a significant effect on its binding capacity, but this effect appears to decrease in significance as the ligand concentration is increased. Thus it is important to make careful note of ligand concentration, total ligand content, and geometry of an affinity column if reproducible results are to be obtained. Batchwise procedures were found to be sensitive to factors such as enzyme concentration and equilibration time, and the authors note that column procedures can often function more effectively at lower flow rates.

3. The effectiveness of alterations in pH and temperature as tools for eluting enzymes bound to nucleotide columns was examined in great detail. In principle, use of such simple techniques would obviate the necessity of separating the eluted protein from the soluble NADH or soluble inhibitor that would be present after the use of NADH pulses or ternary complex elution. However it should be pointed out that effective elution conditions would have to be worked out for each particular application, and it is by no means clear that the *best* pH or temperature range for elution from an immobilized ligand could be predicted in advance, even if the behavior of the desired enzyme in free solution has been well studied. In addition, care would have to be exercised to avoid denaturation of the desired material, and this concern is especially crucial with the many enzymes that are relatively unstable.

4. As an example of the use of **11** for the isolation of enzymes that had proved difficult to resolve by conventional techniques, alcohol dehydrogenase and phosphofructokinase were successfully obtained in high purity from a partially purified extract of *Bacillus stearothermophilus* by differential elution from the immobilized nucleotide. A purification scheme was developed involving DEAE-cellulose and hydroxylapatite columns followed by adsorption onto **11**. Elution with NADH released homogeneous alcohol dehydrogenase, and subsequent elution with ATP-Mg^{2+} gave partially purified phosphofructokinase, which was reapplied to the column and eluted with either a salt or pH gradient to give homogeneous protein. In addition, glyceraldehyde-3-phosphate dehydrogenase was isolated from the material that had not adsorbed on to **11** by applying this material to a Sepharose-NAD^+ column and eluting it with soluble NAD^+. This example points out that a variety of elution techniques (soluble cofactors, pH or salt changes, etc.) can

be used in mutually complementary fashion to effectively isolate a number of different types of enzyme from these general "group-specific adsorbents" in a single procedure.

There are numerous other examples of the use of immobilized nucleotides for the isolation of specific macromolecules. An example of alternate coupling chemistry is provided by the work of Zappelli et al. [71], who prepared the 8-(2-carboxyethylthio) derivative of NAD^+ and coupled this molecule to both soluble and insoluble macromolecular supports using carbodiimides. Barry and O'Carra [72] have summarized a number of pertinent considerations to be kept in mind in interpreting results with immobilized nucleotides, and they also demonstrate how the effectiveness of "general" nucleotide ligands can be drastically affected by the chemistry of the coupling process. An interesting example of the reversal of the usual situation is provided by the work of Das et al. [73], who successfully used a column of immobilized alcohol dehydrogenase for the isolation of NAD^+ from nucleotide mixtures or from a preultrafiltered crude yeast homogenate. The Sepharose–alcohol dehydrogenase conjugate retained 60 % of its initial effectiveness for NAD^+ separation from nucleotide mixtures after 34 reuses over a period of weeks.

A number of immobilized coenzyme conjugates are now available commercially from Pharmacia, PL Biochemicals, and other vendors.

Increasing attention is now being directed toward the potential of immobilized or entrapped enzymes for practical applications, and work in this area is often referred to as "enzyme technology." Among the interesting areas of concern has been the problem of retaining and regenerating expensive, dissociable coenzymes in reaction mixtures involving immobilized or membrane-entrapped enzymes, and approaches such as the prior attachment of coenzymes to soluble macromolecular carriers or the development of membranes with more selective permeability have been pursued [74, 75]. Viewed from this perspective, the recent work of Mosbach's group in Sweden [76] provides an elegant example of the bridge between affinity chromatography and enzyme immobilization—two parts of the general field of "multiphase biochemistry." When these investigators coupled preincubated mixtures of alcohol dehydrogenase plus the reduced form of the NAD^+ analog **7** to CNBr-activated Sepharose, they obtained conjugates with ratios of bound coenzyme molecules to bound enzyme subunits ranging from 0.5 to 140. In the *absence* of added soluble NAD^+, these conjugates exhibited enzymatic activity—internal activity—which is presumed to arise from enzyme-nucleotide complexes being immobilized onto the matrix, with the coenzyme fixed in position at the active site of the dehydrogenase. A number of controls and supplemental experiments were carried out to support this explanation and to demonstrate that proper localization of the nucleotide at the active site in the conjugate decisively affects its activity. The specific internal activity of the immobilized enzyme-cofactor complexes increased by only a limited degree with the bound nucleotide-to-enzyme ratio; thus high efficiency is already attained at nucleotide-to-enzyme ratios near 1. This is in sharp contrast to free-solution systems, in which this ratio is normally at least 10^3 to 10^4. It is interesting to note that the coimmobilized conjugates were actually more efficient (1.5 to 5 times, under the assay conditions used) than corresponding soluble systems. Moreover, the conjugates exhibited increased activity in the presence of excess soluble NAD^+, which allows a catalytic contribution from those bound enzyme subunits to which bound nucleotide is not accessible. Finally, the thermal stability of the immobilized conjugates was greater than that of soluble systems, and the data imply that those enzyme molecules immobilized with bound nucleotide positioned at the active site are actually more stable than those immobilized distant from the nucleotide.

5 SPECIALIZED PROCEDURES, TECHNIQUES, AND APPROACHES

The ever-increasing use of "solid phase" or "multiphase" biochemical systems in affinity chromatography, in the immobilization of enzymes and other macromolecules, and in the solid phase synthesis and degradation of biopolymers has necessitated the development of specialized laboratory procedures and techniques that allow the investigator to properly handle, utilize, or characterize various immobilized conjugates. In a complementary sense, it is also evident that many familiar laboratory operations have benefited enormously—and in many cases have been completely revolutionized by—the availability of "solid phase biochemical" technology. Indeed, techniques virtually unheard of 10 or 15 years ago occasionally provide the only experimental approach for the resolution of important research questions. This section mentions briefly a number of specialized experimental procedures and techniques useful in affinity chromatography procedures. With the aid of the references and the list of reviews, the reader can work more comprehensively through the literature bearing on specific topics of interest. In addition, several cases are cited of the use of affinity chromatography or related procedures in highly sophisticated ways to approach and resolve complex research questions.

1. Quantitation of the amount of ligand bound per unit weight of conjugate (usually milligrams of ligand per gram of support) can be accomplished in a number of ways. For protein or peptide ligands, quantitative amino acid analysis on dried, weighed samples of the conjugate gives excellent results and

is widely used [3, 13, 16, 77]. Several quantitative or semiquantitative color tests for amino- or sulfhydryl-containing conjugates have been described (see reviews and Refs 21, 78–80), and these vary in accuracy, reproducibility, and sensitivity. The conjugate may be "solubilized" to release the ligand for subsequent quantitation by spectrophotometric or other means. A discussion of several such methods for agarose, and their use in conjunction with a color reagent, has been presented by Failla and Santi [81]. A considerably less satisfactory quantitation method is estimation from the difference between starting ligand concentration and that recovered in the washings after coupling to the support, and large errors may be encountered in quantitations based on such calculations. Other techniques such as elemental analysis, titration, or the use of labeled ligands have also been used successfully (see reviews). Recently Werber [82] proposed quantitating agarose conjugates (formed by coupling amines to CNBr-activated agarose) by measuring the amount of *p*-nitrophenol released from *p*-nitrophenyl acetate in the presence of the conjugate. Another interesting approach that should soon become more popular is the quantitation of enzyme ligands by active site titration directly over the conjugate. It is important to keep in mind that all quantitation methods yield data expressed per gram of support, and the method used to dry the conjugate before weighing must be taken into account. Thus shrinking with increasing acetone concentration followed by drying *in vacuo* at 100°C for 24 hr will give a much different volume-to-dry weight ratio for agarose gels than will simple air drying on a laboratory funnel.

2. The ability to carry out direct spectrophotometric examinations of a conjugate *in the immobilized state* would be highly desirable in many applications of immobilized ligands. May and Kuo [13] have described procedures developed in their laboratory for settling a rubredoxin-Sepharose 4B conjugate to a constant density and obtaining spectra directly on the settled gel, using a spectrophotometer equipped with a scattered transmission accessory. Direct spectral examination allowed complete characterization of the iron-binding sites and examination of the redox properties of this immobilized enzyme, for comparison with the soluble species' Koelsch et al. [83] also reported results of direct spectrophotometric examination of settled protein–Sepharose 6B conjugates, but their work was confined to proteins without longwavelength chromophores, and the spectra were obtained simply for quantitation purposes and were not used for examining properties of the immobilized proteins. Routine spectral and fluorescence examinations of conjugates have also been carried out by a number of investigators by suspending conjugates in media such as glycerol, sucrose, glycols, polyacrylamide, or Polyox (see reviews).

3. It is possible to extract quantitative binding information from studies of the binding behavior of species on an affinity column in the presence of

competing, soluble ligand. Thus Dunn and Chaiken [84, 85] developed appropriate mathematical expressions and determined binding constants for staphylococcal nuclease with both a soluble and an immobilized nucleotide ligand from such studies. The data obtained allowed a detailed analysis of the specificity of binding to the column, as well as comparison of the mode of interaction of the soluble and immobilized ligands with the binding site of the enzyme. A similar study with ribonuclease species was recently reported by Chaiken and Taylor [86]. A general quantitative theoretical treatment for affinity chromatography has been developed by Nichol et al. [87] and used to extract binding constants from experimental data by these authors and also by Brinkworth et al. [88].

4. Certainly the most elegant uses of immobilized biomolecules have been in sophisticated mechanistic and structural studies, and it is here that affinity chromatography and other branches of "solid phase" biochemistry converge most impressively. Often complex questions that had been virtually unapproachable experimentally are now being resolved using the new technology associated with heterogeneous biological systems. Among the very numerous examples are the following: the use of specific inhibitor-Sepharose conjugates to confirm directly the ordered lactate dehydrogenase mechanism, to demonstrate the existence of a complex responsible for substrate inhibition, and to clarify the roles of the two halves of the NAD^+ molecule in generating the substrate binding site [89]; the separation of isozymes from each other by cleverly taking advantage of slight differences in their affinities toward substrates and coenzymes [90–92]; the resolution of the catalytic and regulatory subunits of a c-AMP-dependent enzyme on an affinity column, with the attendant production of c-AMP-independent activity [93, 94]; the demonstration that certain residues required for catalytic activity are not required for substrate binding to an enzyme [95]; the demonstration of the transfer of a covalently attached group from one enzyme molecule to a different type of enzyme molecule only after the second enzyme has bound to substrate [96]; the specific isolation of peptides containing tryptophan residues [97]; and the estimation of dissociation constants for the binding of enzymes with cofactors or inhibitors [98].

Finally, in my opinion perhaps the most elegant of all such applications is the work pioneered by Chan [99–101], who used immobilization techniques to explore the question, usually impossible to resolve using conventional solution methods, of whether individual monomers of multimeric enzymes exhibit catalytic activity. The idea here is to covalently attach a multimeric enzyme by, ideally, a bond to only one subunit per molecule, then to wash away the other noncovalently attached material, leaving immobilized subunits. Not only can activity studies on individual monomers then be carried out, but association-dissociation studies using preparations attached to the support by different residues can yield important information about subunit interactions and quaternary structure of enzymes. Interestingly, in such experiments the trick is to use supports with a low degree of functionalization and loading, to minimize the chances of multipoint attachment by adjacent subunits; thus packing as many active groupings or as much ligand as possible onto a support is not always desirable.

6 COMMERCIALLY AVAILABLE MATERIALS

As a practical aid, Table 5.2 lists commercially available materials suitable for use in affinity chromatography procedures. For nonbiological applications, there may be a great many support materials more suitable than those listed here, and the table is confined to materials currently "in vogue" for biological applications. The table is intended to be neither exhaustively detailed nor comprehensive; rather, it is a guide to the major sources of supply in the United States. Some of the large chemical supply houses have recently started to stock some of the more commonly used materials. For clarity, trade names (e.g. Sepharose, Biogel) are usually not specified. In addition, all commercially available immunoadsorbents and immobilized enzymes are omitted from the table. The addresses of suppliers designated are as follows:

1. BioRad Laboratories, Richmond, California 94804. (Trade names for polyacrylamide and agarose are Biogel P and A, respectively.)
2. ICN Pharmaceuticals, Inc., Cleveland, Ohio 44128.
3. LKB Instruments, Inc., Rockville, Maryland 20852.
4. Miles Research Products, Elkhart, Indiana 46514.
5. Pharmacia Fine Chemicals, Inc., Piscataway, New Jersey 08854. (Trade name for agarose is Sepharose. Sephadex is a registered trade name of this company.)
6. Pierce Chemical Company, Rockford, Illinois 61102.
7. PL Biochemicals, Inc., Milwaukee, Wisconsin 53205.

7 REVIEWS AND TECHNICAL INFORMATION

An exceedingly useful source for detailed methodological information on a wide variety of affinity chromatography procedures is *Affinity Techniques*, Volume 34 of the series Methods in Enzymology (W. B. Jakoby and M. Wilchek, Eds., Academic Press, New York, 1974. Also, *Affinity Chromatography* [6] provides exhaustive information and references on many aspects of this area. Numerous less comprehensive summaries and reviews on "affinity," "biospecific," or "ligand-specific" chromatography have appeared and continue to appear in surprisingly diverse sources, and a listing

Table 5.2 Commerically Available Materials

Material	Manufacturer
Support materials	
Agarose (Sepharose, Biogel A)	BioRad, Pharmiacia, PL
CNBr-activated agarose	Pharmacia
Cross-linked Sepharose [with 2,3-dibromo-propanol]	Pharmacia
Glass, controlled pore, various sizes [Corning]	Pierce
Polyacrylamide (Biogel P)	BioRad
Sephadex (cross-linked Dextran)	Pharmacia
Ultrogel	LKB
Spacer-containing supports without specific ligands	
Alkyl- and aryl-agarose, various	Miles, Pharmacia, PL
Aminoalkyl-agarose, various	BioRad, ICN, Miles, Pharmacia, PL
Aminoethyl-succinyl-polyacrylamide derivative	BioRad
Aminoprolyl-succinyl-agarose derivative	BioRad
Carboxyalkyl-agarose, various [free-carboxyl group]	BioRad, ICN, Pharmacia, PL
3,3′-Diaminodipropylamine-agarose derivative	BioRad
3,3′-Diaminodipropylamine-*p*-Aminobenzyl-agarose derivative [terminal arylamine]	ICN
3,3′-Diaminodipropylamine-succinyl-agarose derivative	BioRad
N-Hydroxysuccinimide ester–agarose carboxyl group]	BioRad, Pharmacia
Organomercurial-agarose derivative [various spacer arms]	BioRad, ICN, Miles
Thio-terminal agarose derivatives, various [various spacer arms]	BioRad, ICN, PL
Specially functionalized supports	
Aminoalkyl-agarose, various	BioRad, ICN, Miles, Pharmacia, PL
Bromoacetyl-cellulose	Miles
Carboxyalkyl-agarose, various [free-derivatives [various chain lengths]	BioRad, ICN, Pharmacia, PL
Epoxy-alkyloxy–agarose [conjugate of Sepharose plus 1,4-bis(2,3-epoxypropoxy) butane]	Pharmacia

Table 5.2 (*Continued*)

Material	*Manufacturer*
Glass, controlled pore, derivatized and/or coated [various substances and functional groups available]	Pierce
Glutathione-agarose (Sepharose) disulfide with 2-thiopyridine	Pharmacia
Hydrazide-polyacrylamide derivative	BioRad
N-Hydroxysuccinimide ester–agarose derivatives [various chain lengths]	BioRad, Pharmacia
Organomercurial-agarose derivative [various spacer arms]	BioRad, ICN, Miles
Thio- terminal agarose derivatives, various [various spacer arms]	BioRad, ICN, PL
Specific ligand conjugates	
AMP-, ADP-, and ATP-agarose conjugates [various linkage positions, various phosphate positions]	Pharmacia, PL
Coenzyme A–agarose, various	PL
DNA-agarose	PL
DNA-cellulose	PL
Glutathione-agarose (Sepharose)	Pharmacia
GTP-, UDP-, and UTP-agarose	PL
Inhibitor and substrate analog agarose derivatives	ICN, Miles, PL
Lectin-agarose conjugates, various [for carbohydrate binding]	Miles, Pharmacia, PL
NAD- and NADP-agarose [various linkage positions]	PL
Nucleotide-cellulose conjugates, various	PL
Organomercurial-agarose derivative [various spacer arms]	BioRad, ICN, Miles
Polynucleotide-agarose conjugates, various	Pharmacia, Pl
Multipoint-attached or "neutral" conjugates; miscellaneous	
Hydrazide-agarose derivatives, various ["neutral" gels (see p. 341); multivalently attached derivatives also available]	Miles
Polymer-agarose derivatives, various [multipoint ligand attachment]	Miles
"Blue"-agarose	BioRad, Pharmacia

of some of these follows. The articles by March et al., by Dean and Harvey, by Feinstein, and by Nishikawa provide short introductions to, or updates on, the general area; somewhat more chemistry is discussed in the article by Weetal. More extensive discussions are presented in the articles by May and Zaborsky, by Cuatrecasas, and by May, the latter being considerably more current. Finally, the very excellent work by Porath and Kristiansen is very worthwhile reading for anyone seriously contemplating using the techniques of affinity chromatography.

H. H. Weetal, "Affinity Chromatography," *Sep. Purif. Methods*, **2**, 199–229 (1973).

G. Feinstein, "Affinity Chromatography of Biological Macromolecules," *Naturwissenschaften*, **58**, 389 (1971).

S. C. March, I. Parikh, and P. Cuatrecasas, "Affinity Chromatography—Old Problems and New Approaches," *Adv. Exper. Med. Biol.*, **42**, 3–14 (1974).

A. H. Nishikawa, "Affinity Purification of Enzymes," *Chem. Technol.*, 564–571 (September 1975).

P. D. G. Dean and M. J. Harvey, "Applications of Affinity Chromatography," *Process Biochem.*, 5–10 (September 1975).

J. Porath and T. Kristiansen, "Biospecific Affinity Chromatography and Related Methods," in *The Proteins*, Vol. 1, 3rd ed., H. Neurath and R. L. Hill, Eds., Academic Press, New York, 1975, pp. 95–178.

S. W. May and O. R. Zaborsky, "Ligand-Specific Chromatography," *Sep. Purif. Methods*, **3**, 1–86 (1974).

P. Cuatrecasas, "Selective Adsorbents Based on Biochemical Specificity," in *Biochemical Aspects of Reactions on Solid Supports*, G. R. Stark, Ed., Academic Press, New York, 1971, pp. 79–109.

S. W. May, "Separation Techniques Based on Biological Specificity," in *Recent Advances in Separation Science*, Vol. 5, N. Li, Ed., CRC Press, Cleveland, in press.

Particularly detailed, informative, and high-quality literature on various aspects of affinity chromatography is available on request from Pharmacia. A number of other manufacturers (BioRad, Miles, Pierce, LKB, etc.) also supply very informative material related to the affinity products which they sell. Especially for investigators new to this field, such materials contain helpful hints that can easily make the difference between success and failure in initial experiments.

ACKNOWLEDGMENT

We gratefully acknowledge the support of the National Science Foundation (BMS 74-20830) and the National Institutes of Health (GM 23474) for studies carried out in our laboratory.

References

1. E. Sarkstein, *Biochem. Z.*, **24**, 21 (1910).
2. L. S. Lerman, *Proc. Nat. Acad. Sci., U.S.*, **39**, 232 (1953).
3. R. Axen, J. Porath, and S. Ernback, *Nature*, **214**, 1302 (1967).
4. S. W. May, in *Recent Advances in Separation Science*, Vol. 5, N. Li, Ed., CRC Press, Cleveland, in press.
5. S. W. May and O. R. Zaborsky, *Sep. Purif. Methods*, **3**, 1 (1974).
6. C. R. Lowe and P. D. G. Dean, *Affinity Chromatography*, Wiley, New York, 1974.
7. J. Porath and T. Kristiansen, *The Proteins*, Vol. 1, 3rd ed., H. Neurath and R. L. Hill, Eds., Academic Press, New York, 1975, pp. 95–178.
8. J. Porath, J. C. Janson, and T. Laas, *J. Chromatogr.*, **60**, 167 (1971).
9. J. Porath and L. Sundberg, *Nature*, **238**, 261 (1972).
10. J. Porath, *Methods Enzymol.*, **34**, 13 (1974).
11. I. Parikh, S. March, and P. Cuatrecasas, *Methods Enzymol.*, **34**, 77 (1974).
12. S. C. March, I. Parikh, and P. Cuatrecasas, *Anal. Biochem.*, **60**, 149 (1974).
13. S. W. May and J. Y. Kuo, *J. Biol. Chem.*, **252**, 2390 (1977).
14. J. Porath, R. Axen, and S. Ernback, *Nature*, **215**, 1491 (1967).
15. J. Porath, *Nature*, **218**, 834 (1968).
16. R. Axen and S. Ernback, *Eur. J. Biochem.*, **18**, 351 (1971).
17. R. Axen and P. Vretblad, *Acta Chem. Scand.*, **25**, 2711 (1971).
18. L. Kagedal and S. Akerstrom, *Acta Chem. Scand.*, **24**, 1601 (1970).
19. L. Ahrgren, L. Kagedal, and S. Akerstrom, *Acta Chem. Scand.*, **26**, 285 (1972).
20. G. J. Bartling, H. D. Brown, L. F. Forrester, M. T. Koes, A. N. Mather, and R. D. Stasin, *Biotechnol. Bioeng.*, **14**, 1039 (1972).
21. P. Cuatrecasas, *J. Biol. Chem.*, **245**, 3059 (1970).
22. P. Cuatrecasas and I. Parikh, *Biochemistry*, **11**, 2291 (1972).
23. S. C. March, I. Parikh, and P. Cuatrecasas, *Adv. Exper. Med. Biol.*, **42**, 3 (1974).
24. G. I. Tesser, H. U. Fisch, and R. Schwyzer, *FEBS Lett.*, **23**, 56 (1972).
25. F. Krug, B. Desbuquois, and P. Cuatrecasas, *Nature New Biol.*, **234**, 268 (1971).
26. G. Vaquelin, M. L. Lacombe, J. Hanoune, and A. D. Strosberg, *Biochem. Biophys. Res. Commun.*, **64**, 1076 (1975).
27. G. I. Tesser, H. U. Fisch, and R. Schwyzer, *Helv. Chim. Acta*, **57**, 1718 (1974).
28. M. S. Verlander, J. C. Venter, M. Goodman, N. O. Kaplan, and B. Saks, *Proc. Nat. Acad. Sci., U.S.*, **73**, 1009 (1976).
29. T. Kristiansen, L. Sundberg, and J. Porath, *Biochim. Biophys. Acta*, **184**, 93 (1969).
30. M. Wilchek, *FEBS Lett.*, **33**, 70 (1973).
31. M. Wilchek and T. Miron, *Methods Enzymol.*, **34**, 72 (1974).
32. P. Cuatrecases, *Adv. Enzymol.*, **36**, 29 (1972).
33. W. H. Scouten, *Methods Enzymol.*, **34**, 288 (1974).
34. E. Steers, Jr., P. Cuatrecasas, and H. B. Pollard, *J. Biol. Chem.*, **246**, 196 (1971).
35. R. Cardinand and J. Holgoin, *Biochem. J.*, **127**, 301 (1972).
36. P. O'Carra, S. Barry, and T. Griffin, *Biochem. Soc. Trans.*, **1**, 289 (1973).
37. M. C. Hipwell, J. J. Harvey, and P. D. G. Dean, *FEBS Lett.*, **42**, 355 (1974).
38. Z. Er-el, Y. Ziadenzaig, S. Shaltiel, *Biochem. Biophys. Res. Commun.*, **49**, 383 (1972).

39. S. Shaltiel and Z. Er-el, *Proc. Nat. Acad. Sci., U.S.*, **70**, 778 (1973).
40. B. H. J. Hofstee, *Anal. Biochem.*, **52**, 430 (1973).
41. E. Nystrom and J. Sjovall, *Anal. Lett.*, **6**, 155 (1973).
42. Y. Imai and R. Sato, *J. Biochem.*, **75**, 689 (1974).
43. Z. Er-el and S. Shaltiel, *FEBS Lett.*, **40**, 142 (1974).
44. R. Jost, T. Miron, and M. Wilchek, *Biochim. Biophys. Acta*, **362**, 75 (1974).
45. A. H. Nishikawa and P. Bailon, *Arch. Biochem. Biophys.*, **168**, 576 (1975).
46. P. O'Carra, S. Barry, and T. Griffin, *FEBS Lett.*, **42**, 355 (1974).
47. S. Barry and P. O'Carra, *Biochem. J.*, **135**, 595 (1973).
48. Y. Ashani and I. B. Wilson, *Biochim. Biophys. Acta*, **276**, 317 (1972).
49. P. M. Blumberg and J. L. Strominger, *Proc. Nat. Acad. Sci., U.S.*, **69**, 3751 (1972).
50. S. A. Barker, C. J. Gray, J. C. Ireson, R. C. Parker, and J. V. McClaren, *Biochem. J.*, **139**, 555 (1974).
51. A. Fenslau and K. Wallis, *Biochem. Biophys. Res. Commun.*, **62**, 350 (1975).
52. M. Robert-Gero and J. P. Waller, *Eur. J. Biochem.*, **31**, 315 (1972).
53. R. M. Silverstein, *Thromb. Res.*, **4**, 675 (1974).
54. R. Y. Yon, *Biochem. J.*, **137**, 127 (1974).
55. K. Mosbach, H. Guilford, R. Ohlsson, and M. Scott, *Biochem. J.*, **127**, 625 (1972).
56. K. Mosbach, H. Guilford, P. O. Larsson, R. Ohlsson, and M. Scott, *Biochem. J.*, **125**, 20 (1972).
57. C. R. Lowe, K. Mosbach, and P. D. G. Dean, *Biochem. Biophys. Res. Commun.*, **48**, 1004 (1972).
58. R. Ohlsson, P. Brodelius, and K. Mosbach, *FEBS Lett.*, **25**, 234 (1972).
59. C. R. Lowe and P. D. G. Dean, *FEBS Lett.*, **14**, 313 (1971).
60. C. R. Lowe and P. D. G. Dean, *Biochem. J.*, **133**, 515 (1973).
61. M. Lindberg, P. O. Larsson, and K. Mosbach, *Eur. J. Biochem.*, **40**, 187 (1973).
62. D. B. Craven, M. J. Harvey, and P. D. G. Dean, *FEBS Lett.*, **38**, 320 (1974).
63. P. Brodelius, P. O. Larsson, and K. Mosbach, *Eur. J. Biochem.*, **47**, 81 (1974).
64. D. B. Craven, M. J. Harvey, C. R. Lowe, and P. D. G. Dean, *Eur. J. Biochem.*, **41**, 329 (1974).
65. M. J. Harvey, C. R. Lowe, D. B. Craven, and P. D. G. Dean, *Eur. J. Biochem.*, **41**, 335 (1974).
66. C. R. Lowe, M. J. Harvey, and P. D. G. Dean, *Eur. J. Biochem.*, **41**, 341 (1974).
67. C. R. Lowe, M. J. Harvey, and P. D. G. Dean, *Eur. J. Biochem.*, **41**, 347 (1974).
68. M. J. Harvey, C. R. Lowe, and P. D. G. Dean, *Eur. J. Biochem.*, **41**, 353 (1974).
69. C. R. Lowe, M. J. Harvey, and P. D. G. Dean, *Eur. J. Biochem.*, **42**, 1 (1974).
70. M. J. Comer, D. B. Craven, M. J. Harvey, A. Atkinson, and P. D. G. Dean, *Eur. J. Biochem.*, **55**, 201 (1975).
71. P. Zappelli, A. Rossodivita, G. Prosperi, R. Pappa, and L. Re, *Eur. J. Biochem.*, **62**, 211–215 (1976).
72. S. Barry and P. O'Carra, *Biochem J.*, **135**, 595–607 (1973).
73. K. Das, P. Dunnill, and M. D. Lilly, *Biochim. Biophys. Acta*, **397**, 277 (1975).
74. W. H. Baricos, R. P. Chambers, and W. Cohen, *Enzyme Technol. Digest*, **4**, 39–53 (1975) and references therein.
75. S. W. May and L. M. Landgraff, *Biochem. Biophys. Res. Commun.*, **68**, 786–792 (1976).

76. S. Gestrelius, M. Mansson, and K. Mosbach, *Eur. J. Biochem.*, **57**, 529–535 (1975).
77. O. R. Zaborsky and J. Oglestree, *Biochim. Biophys. Acta*, **289**, 68 (1972).
78. J. K. Imman and H. M. Dintzis, *Biochemistry*, **8**, 4074 (1969).
79. E. Kaiser, R. L. Colescott, C. D. Bossinger, and P. I. Cook, *Anal. Biochem.*, **34**, 595 (1970).
80. D. R. Grassetti and J. F. Murray, *Arch. Biochem. Biophys.*, **119**, 41 (1967).
81. D. Failla and D. V. Santi, *Anal. Biochem.*, **52**, 363 (1973).
82. M. M. Werber, *Anal. Biochem.*, **76**, 177 (1976).
83. R. Koelsch, J. Lasch, I. Marquardt, and H. Hanson, *Anal. Biochem.*, **66**, 556 (1975).
84. B. M. Dunn and I. M. Chaiken, *Proc. Nat. Acad. Sci., U.S.*, **71**, 2382 (1974).
85. B. M. Dunn and I. M. Chaiken, *Biochemistry*, **14**, 2343 (1975).
86. I. M. Chaiken and H. C. Taylor, *J. Biol. Chem.*, **251**, 2044 (1976).
87. L. W. Nichol, A. G. Ogston, D. J. Winzor, and W. H. Sawyer, *Biochem. J.*, **143**, 435 (1974).
88. R. I. Brinkworth, C. J. Masters, and D. J. Winzor, *Biochem. J.*, **151**, 631 (1975).
89. P. O'Carra and S. Barry, *FEBS Lett.*, **21**, 281 (1972).
90. P. O'Carra, S. Barry, and E. Corcoran, *FEBS Lett.*, **43**, 163 (1974).
91. H. Spidmann, R. P. Erickson, and C. J. Epstein, *FEBS Lett.*, **35**, 19 (1973).
92. P. Brodelius and K. Mosbach, *FEBS Lett.*, **35**, 223 (1973).
93. M. Wilchek, Y. Salomon, M. Lowe, and Z. Selinger, *Biochem. Biophys. Res. Commun.*, **45**, 1177 (1971).
94. E. M. Reimann, C. Brostrom, and J. D. Corbin, *Biochem. Biophys. Res. Commun.*, **42**, 187 (1971).
95. K. Kasai and S. Ishii, *J. Biochem.*, **74**, 631 (1973).
96. L. Pass, T. L. Zimmer, and S. G. Laland, *Eur. J. Biochem.*, **47**, 607 (1974).
97. M. Rubinstein, Y. Schechter, and A. Patchornik, *Biochem. Biophys. Res. Commun.*, **70**, 1257 (1976).
98. P. Brodelius and K. Mosbach, *Anal. Biochem.*, **72**, 629 (1976).
99. W. W. C. Chan, *Biochem. Biophys. Res. Commun.*, **41**, 1198 (1970).
100. W. W. C. Chan and K. Mosbach, *Biochemistry*, **15**, 4215 (1976).
101. W. W. C. Chan, *Can. J. Biochem.*, **54**, 521 (1976).

Chapter **VI**

CENTRIFUGING

Charles M. Ambler
Frederick W. Keith, Jr.

---

1 Introduction, Definitions 295

2 Theory 297

Centrifugal Force 297
Sedimentation Theory 298
Hydrostatic Pressure 301
Stresses 303

3 Equipment 303

Bottle Centrifuges 304
Preparative Ultracentrifuges 315
Zonal Centrifuges 317
Analytical Ultracentrifuges 323
Tubular Bowl Centrifuges 328
Disc Bowl Centrifuges 332
Conveyor Discharge Centrifuges 335
Basket Centrifuges 337

4 Uses of Centrifuges 340

5 Operating and Maintenance 340

Balancing 340
Operation 341
Electric Motor and Other Drives 343
Care of Centrifuges 344

Symbols and Nomenclature 345

---

1 INTRODUCTION, DEFINITIONS

There are two basic types of centrifuge:

1. *Solid wall centrifuges*, in which separation or concentration is by sedimentation or flotation; the particles of the dispersed phase migrate either

away from or toward the axis of rotation depending on whether their density is greater or less than that of the continuous liquid phase.

2. *Perforate wall centrifuges*, centrifugal filters in which the particulate phase is supported and retained on a permeable member through which the continuous liquid phase is free to pass.

In at least one industrial model the functions of these two types are combined in a single rotor.

The sedimentation type of centrifuge is used to increase the sedimentation rate and permit the attainment of equilibrium conditions much more quickly than is possible by gravity settling. The field of acceleration in the centrifuge overcomes the diffusion effects from very small particles, even those of molecular dimensions, and promotes compaction to higher solid phase concentrations with sharply defined and stable boundaries. Because of this, sedimentation-type centrifuges under controlled conditions of time and rotational speed are specified for many analytical procedures.

This type of centrifuge is used frequently where filtration is not applicable because the suspended solids are very fine, amorphous, or gelatinous. It is used for the rapid and complete separation of immiscible liquids and the resolution of emulsions formed during extraction procedures. It is particularly useful for the concentration and washing *in situ*, by decantation, of labile systems such as proteins or easily oxidizable inorganic precipitates that are subject to deterioration by exposure to air or other change in environment.

The perforate wall centrifugal filter is used for the removal of surplus liquid from crystalloid and other relatively nondeformable solids. Drainage is more rapid and complete, and washing, when required, is more effective than on a gravity or vacuum filter.

The ultracentrifuge applies a high field of acceleration to the separation and concentration of very small soluble or insoluble particles, down to those of molecular dimensions.

The analytical ultracentrifuge permits observation and accurate measurement of the migration and rate of change of concentration of protein and other relatively large molecules in a similarly high field of acceleration.

Centrifugal force is the force that tends to impel an object or particle away from an axis of rotation.

In the broadest sense, a *centrifuge* is a device in which centrifugal force can be applied to a system. By custom, this term is applied more narrowly—namely, to devices for resolving multicomponent systems.

Centrifuging or *centrifugation* is the operation of performing such a resolution under centrifugal acceleration.

2 THEORY

Centrifugal Force

The scalar velocity of an object rotated at constant speed about an axis is constant. Its vector velocity is constantly changing. This change in vector velocity produces centrifugal acceleration. The force on a particle constrained to move in a circular path results from its acceleration toward the axis of rotation, that is, the rate of change of its velocity away from a linear path tangent to its circle of rotation that it attempts to follow, in accordance with Newton's first law of motion.

In engineering notation, this force is

$$F = ma \tag{6.1}$$

where m is the relative mass of the particle or the difference between its true mass and that of the fluid phase it displaces. In its elemental form this can be demonstrated by whirling around a weight that is attached to a string. Centrifugal force is the outward pull of this weight. Centripetal force is the pull that must be exerted on the operator's end of the string to keep the weight in a circular path.

On the surface of the earth, a is the gravitational acceleration g with a mean value of 980.7 cm/sec^2. In a centrifuge

$$a = \omega^2 r \tag{6.2}$$

where r is the radial distance from the axis of rotation.

For the centrifuge user, a more useful index for comparative purposes is relative centrifugal force (RCF) designated by

$$\text{RCF or } G = \frac{\omega^2 r}{g}, \tag{6.3}$$

the ratio of centrifugal acceleration to gravitational acceleration. Rotative speed is frequently more conveniently measured as revolutions per minute (rpm) and by substitution

$$G = 0.0000111821 r\,(\overline{\text{rpm}})^2 \tag{6.4}$$

In discussing bottle centrifuge tests it is not enough to report results in terms such as "the separation was effected by centrifuging at 3000 rpm in an eight-place head using 50 ml tubes for 20 min." Since the radius and the depth of liquid vary over wide limits in different types of tubes and centrifuges, the work done can be more adequately defined by the following: "the separation was effected by centrifuging in 50-ml tubes for 20 min at a speed that gave a value of 1984 G at the tip and 960 G at the free liquid surface."

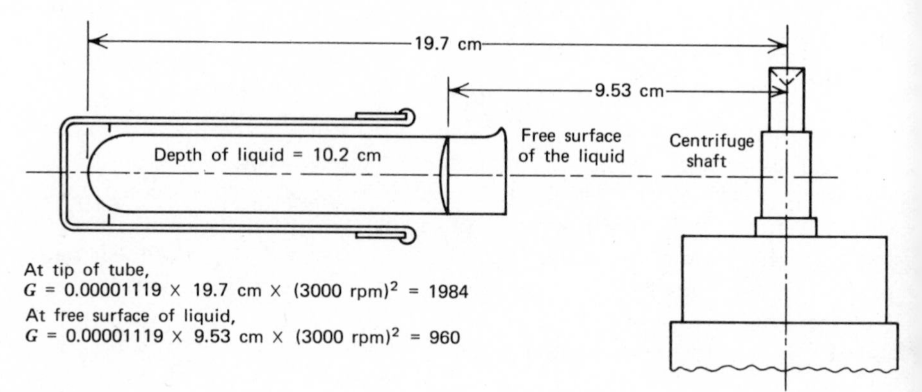

Fig. 6.1. Determination of relative centrifugal force.

By taking the measurements indicated in Fig. 6.1 and applying (6.4), the foregoing values are obtained and the procedure is definitive. By measuring the radius to the tip of the tube in his centrifuge, any other investigator can calculate the equivalent rotational speed of his own centrifuge to obtain the stipulated G value at the tip. The required equivalent liquid depth can be calculated from the expression

$$\text{liquid depth} = \frac{(G \text{ at tip} - G \text{ at surface})}{(G \text{ at tip})}(r \text{ at tip}) \tag{6.5}$$

In this manner the average force in the liquid required to effect the particular separation and the time and force at the tip of the tube required to concentrate the heavy phase to the specified degree are completely defined.

If the axis of rotation is vertical, as it is in the laboratory centrifuge, the effect of gravity acting downward becomes negligible when the value of G exceeds 25. Figure 6.2 demonstrates that under such conditions the resultant vector G' is 25.02 G, an error of less than 0.1 % that becomes progressively less as G increases. It should also be noted that at the specified condition, the angle that the vector G' makes with the horizontal is 2.3°, which decreases to 0.11° at 500 G. The vector effect of gravity produces the same angle between the free surface of the liquid and the axis of rotation, and its influence on the liquid depth is also negligible.

Sedimentation Theory

The force acting on a particle in a centrifugal field (6.1) and (6.2) must be corrected for the mass of the continuous phase fluid it displaces:

$$F = (m_P - m_L)\omega^2 r \tag{6.6}$$

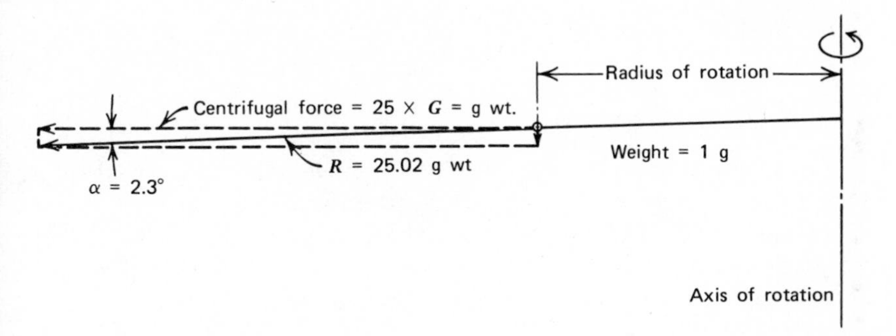

Fig. 6.2. Resultant of gravity and centrifugal force.

If the particle is a sphere

$$F = \frac{\pi}{6} d^3 \Delta\rho\omega^2 r \tag{6.7}$$

where d is the diameter of the particle and $\Delta\rho = \rho_P - \rho_L$, the difference between the density of the particle and that of the fluid in which it is suspended. This is the force that is available to move the particle through the liquid, away from the axis of rotation if the sign of $\Delta\rho$ is positive, and toward the axis of rotation if it is negative.

For a small particle moving not too rapidly (Reynolds no. $dv\rho/\mu \leq$ about 1), the viscous resistance to its motion is expressed by Stokes' law:

$$F = 3\pi\mu dv \tag{6.8}$$

Stokes' law does not apply to larger particles moving more rapidly. For these, the viscous resistance becomes negligible compared to the turbulent resistance, but the controlling particle size in most centrifugal separations falls in or below the Stokes' law range.

When the particle diameter approaches or is less than the mean free path of the molecules of the fluid phase, the particle is subject to Brownian movement because of unbalanced impact of the fluid phase molecules. This random movement may limit the concentration to which small particles will settle in a given field of acceleration, and this has led to increased importance of the ultracentrifuge with its high G values.

Subject to these limitations, the settling velocity, or rising velocity if $\Delta\rho$ is negative, reaches a constant value when the forces creating and those resisting the motion are in equilibrium; thus we have

$$F = \frac{\pi}{6} d^3 \Delta\rho\omega^2 r = 3\pi\mu dv \tag{6.9}$$

and for a spherical particle

$$v_s = \frac{\Delta\rho d^2 \omega^2 r}{18\mu} \tag{6.10}$$

From this basic equation, the velocity of a particle of Stokes equivalent diameter d at any distance from the axis of rotation can be calculated. Since a small particle reaches its limiting terminal velocity very quickly, the distance it will settle can be calculated from its time of exposure to the centrifugal field. In the gravitational field, the terminal sedimenting velocity of the particle is

$$v_g = \frac{\Delta\rho d^2 g}{18\mu} \tag{6.11}$$

The simplest form of centrifuge in which the fluid flow is continuous is a cylinder rotated about its longitudinal axis, provided with end caps, and having means for feeding at one end and discharging at the other. If the thickness s of the liquid layer moving as an annulus is small compared to the radius of the cylinder, then v_s will be approximately constant across this layer and the distance x that a given particle will settle during the time t that the liquid in which it was suspended is in the bowl is

$$x = v_s t = \frac{\Delta\rho d^2 \omega^2 r}{18\mu} \frac{V}{Q} \tag{6.12}$$

where V the volume of a hollow cylinder is $\pi Z(r_2^2 - r_1^2)$ and Q is the volumetric flow rate.

If x is greater than the initial distance of the particle from the outer radius of the moving annular layer, it will be deposited against the bowl wall and removed from the fluid phase. In an ideal system, operating on a homogeneously dispersed feed, when $x = s/2$, one-half of the particles of diameter d will be removed from suspension and one-half will not. This condition is considered to be the "cutoff point" for comparing sedimentation systems. At the cutoff point, by substitution and rearrangement of (6.12), we find

$$Q = \frac{\Delta\rho d^2}{9\mu} \frac{V\omega^2 r}{s} \tag{6.13}$$

Since from (6.11)

$$\frac{\Delta\rho d^2}{9\mu} = \frac{2v_g}{g} \tag{6.14}$$

this may be rewritten as

$$Q = 2v_g \Sigma \tag{6.15}$$

where

$$\Sigma = \frac{V\omega^2 r}{gs} \tag{6.16}$$

In these equations v_g is defined entirely by the parameters of the dispersed system being examined, and Σ contains only parameters of the centrifuge. For centrifuges in which the liquid layer thickness is not small with respect to its radius, the same relationship applies, provided the effective average or integrated values of r and s are used.

This concept of Σ value is a very important one. When r and s are properly developed and the flow patterns are understood for a given type of centrifuge, it permits a direct comparison of the performance of different sizes of the sedimentation centrifuges of various types. The equivalent area of a settling tank operating in the gravitational field is Σ. As with the area of a settling tank, the larger the Σ value of a centrifuge, the greater is its theoretical ability to do useful work. For a given amount of work to be done on a given system,

$$\frac{Q}{\Sigma} = 2v_g = \text{constant} \tag{6.17}$$

and

$$\frac{Q}{\Sigma} = \frac{Q_1}{\Sigma_1} = \cdots \frac{Q_n}{\Sigma_n} \tag{6.18}$$

and this equation sets the basis for comparing different centrifuges.

As a corollary, in a centrifuge, as in a settling tank, sufficient retention time must be provided to permit the settled solids to reach the equilibrium or desired concentration. This depth function of a settling tank or centrifuge may be a limiting factor in many cases, independently of the device's ability to remove particles from the fluid phase.

Hydrostatic Pressure

The hydrostatic pressure exerted on a filled container in a laboratory bottle centrifuge reaches a comparatively high value. In the gravitational field, given a container filled with liquid (e.g., a glass beaker on a laboratory bench), the hydrostatic pressure at a distance h below the liquid-atmosphere interface is

$$p = \rho h \tag{6.19}$$

Since centrifugal force varies as the distance from the axis of rotation, in a container in a centrifugal field, the corresponding integrated form becomes

$$p = \frac{\rho\omega^2(r_2^2 - r_1^2)}{2g} \tag{6.20}$$

As an example of the magnitude of this hydrostatic pressure on the container, consider a glass tube filled with human blood, $\rho = 1.07$ (Fig. 6.3). If the bottom of this tube is 10 cm from the axis of rotation, the liquid column is 5 cm deep, and the rotational speed is 3000 rpm, $\omega = 314$ from (6.20), and the hydrostatic pressure at the bottom of the tube is

$$p = \frac{1.07 \times 314^2 \times (100 - 25)}{2 \times 980.7} = 4034 \text{ g/cm}^2$$

It is obvious that hydrostatic pressures that will rupture glass containers can readily be reached. This internal fluid pressure can be offset to a considerable degree by filling the space between the glass tube and the metal cup with water or some other liquid. For example, consider what happens when the space between the glass tube and the metal cup of Fig. 6.3 is filled with water. The depth of the water annulus to the bottom of the glass tube is 4.5 cm, and the hydrostatic pressure external to the tube would be 3508 g/cm^2.

The net pressure on the bottom of the tube is therefore $4034 - 3508 = 526$ g/cm^2, a pressure that most glassware should withstand satisfactorily. By proper selection of the surrounding liquid and of the depth of the sample in the tube, it is possible to balance the forces so that there is no internal or external pressure on the tube, the ideal condition. Only noncorrosive liquids that will not attack the metal shield should be used. For general purposes glycerine or ethylene glycol is satisfactory. Water should be avoided if the running time is to be long, since much of it may be lost by evaporation.

For operation at high rotational speeds, more elastic containers such as polyethylene or thin-wall polyallomer are used. These deform without rupture sufficiently that the hydrostatic pressure is exerted on the metal shield. As an alternative, containers of aluminum or other compatible rigid material may be used when transparency to observe the sample is not a requirement.

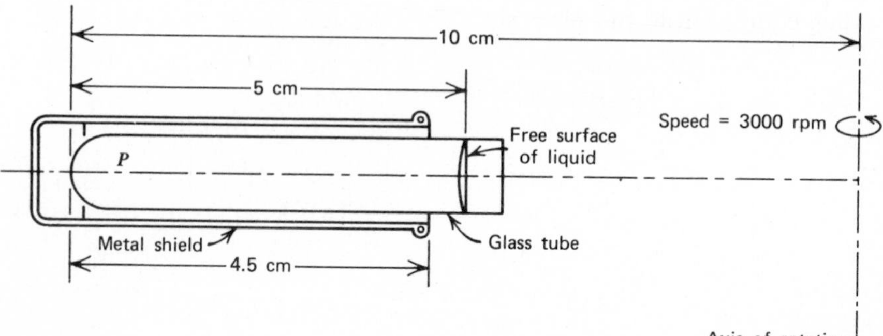

Fig. 6.3. Fluid pressure from centrifugal force.

Stresses

All rotating parts of the centrifuge are subject to the stresses created by centrifugal acceleration acting on them. It is important to realize the limitations imposed by these stresses on the permissible speed of the centrifuge and the hazard that results from operating damaged, corroded, or out-of-balance equipment. In all cases the manufacturer's recommendations of allowable maximum speed under various conditions of loading should be strictly followed.

In the swinging-head bottle centrifuge, the trunnions on the rings that support the metal shields carrying the sample tubes must withstand the total pull exerted by the trunnion rings themselves, the metal shields, the sample tubes, and their contents while under centrifugal acceleration. This pull equals the combined weight of these parts multiplied by the G value, $\omega^2 r/g$, at the radial distance of the center of gravity of the assembly from the axis of rotation.

Since this force increases as the square of the rotative speed, the importance of not exceeding the speed limit established by the designer for given equipment at a specified condition of loading is easily seen. Although most centrifuges have a guard or shield to ensure some degree of protection to the operator in case of breakage, any damage to the equipment usually necessitates extensive and expensive repair.

3 EQUIPMENT

In the laboratory, centrifuges may perform many functions: meeting the specifications of standard analytical procedures, effecting separations during bench-scale experimentation, facilitating small-batch or continuous pilot plant production, or yielding bench-scale or pilot plant data required for scaling equipment to full plant size. The purpose and scale of the work, as well as the type of system being treated, determine the centrifuge to be used and the degree to which theory need be applied to the operation.

The dividing line between bottle centrifuges and some preparatory ultracentrifuges has become indistinct as the rotational speed of the former and the volumetric capacity of the latter have increased. As originally defined by T. Svedberg, the term "ultracentrifuge" referred to a centrifuge, operated at any rotational speed, that permitted optical observation of a system within a transparent cell rotated in a centrifugal field. In modern terminology, the preparative ultracentrifuge covers a range of high-speed laboratory centrifuges that can be divided into analytical and preparative types on the basis that the former include an optical system and the latter do not.

Though recognizing a wide area of overlap, this chapter makes an arbitrary differentiation between bottle centrifuges, for operation up to about 20,000 rpm, and preparative ultracentrifuges, for operation at that speed or higher.

Bottle Centrifuges

For most bottle centrifuges, the basic structure is a motor-driven vertical spindle on which various heads or rotors can be mounted, enclosed in a protective casing with hinged or removable cover. The rotor carries metal containers into which can be fitted tubes and other containers of many shapes and sizes; alternatively, the containers have cavities that can be fed samples to be separated on a batch or continuous basis.

The larger centrifuges are in free-standing cabinets frequently mounted on rollers for portability; the smaller ones are bench-top units.

General-purpose bench-top units (Fig. 6.4) operate at 1800 to 5000 rpm, generating 500 to 3000 *G* in the lower speed range and up to 20,000 rpm with

Fig. 6.4. Bench-top centrifuge with swinging bucket head. Courtesy of Damon/IEC Division, 300 Second Avenue, Needham Heights, Massachusetts 02194.

34,000 G in the high-speed units. They are driven directly from below by a vertical shaft electric motor with speed-regulating control in the base. The interchangeable rotors are mounted on an extension of the motor shaft inside a guard casing with hinged access lid. Optional accessories include timer, tachometer, and manual or automatic braking.

Cabinet models (Fig. 6.5) operate up to 6000 rpm developing up to 8000 G. Attachments permit the use of smaller loads up to 25,000 rpm or 40,000 G. In addition to the more conventional accessories (see above), the cabinet models may be equipped with refrigeration for automatic temperature control ($\pm 1°C$) down to $-10°C$.

Bottle centrifuges can be equipped with three different kinds of rotor: a

Fig. 6.5. Cabinet model refrigerated centrifuge with swinging bucket rotor. Courtesy of DuPont Instruments, Sorvall.

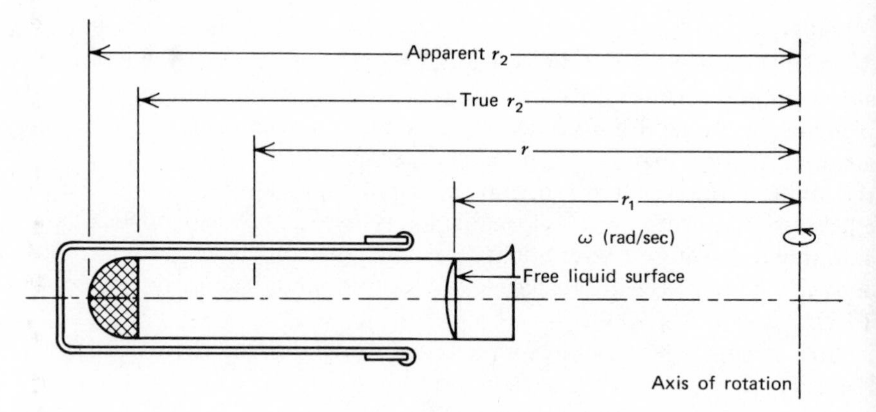

Fig. 6.6. Sedimentation in swinging bucket bottle centrifuge.

swinging bucket (Figs. 6.4 to 6.6), a fixed angle head (Fig. 6.7), or in some, a basket head.

In the swinging bucket type, the sample tubes are placed on elastomeric cushions in metal cups that are carried by trunnion rings set in appropriate slots in the rotor head. Since the center of gravity of each assembly is below the trunnions, the tubes hang vertically while at rest. As the rotor starts to turn, the cups swing out to a nearly horizontal position where they remain as long as the head is rotating. When it is stopped, the cups return to the vertical position. During rotation heavy particles ($\Delta\rho$ positive) travel away from the axis of rotation (Fig. 6.6) and are deposited in the bottom of the tube as a pellet.

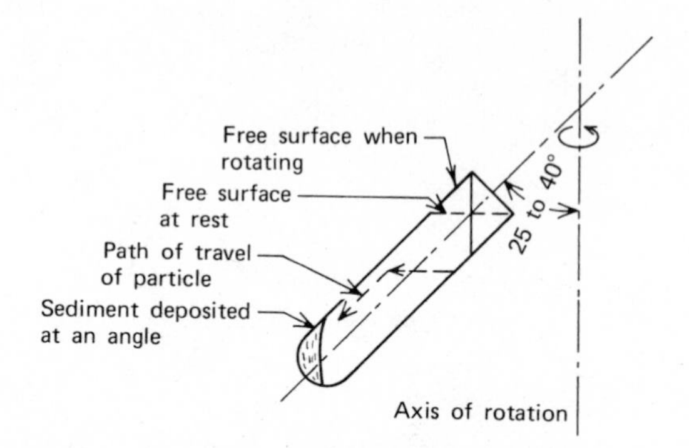

Fig. 6.7. Sedimentation in an angle-type rotor.

This type of rotor has two disadvantages: the path of travel is the full depth of the tube for particles that started near the surface, and hindered settling occurs locally as the solids are concentrated near the bottom. Both increased speed of rotation and time of centrifuging overcome these disadvantages. On the other hand, the long path of travel and the normality of the sedimenting boundary to the axis of the tube are distinct advantages in effecting fractional sedimentation, in applying theory to the bottle centrifuge, and in measuring the position of the liquid-solid interface.

A wide variety of rotors is available to support from 2 to 160 tubes per head. The tubes range in capacity from capillaries holding 175 μl to 1 liter. They include round-bottomed cylindrical or tapered tubes (plain or calibrated), Hoskins tubes tapering to graduated stems, separatory funnels with bottom stopcocks, Babcock bottles with calibrated necks for floating materials such as butterfat, and many others. Plain cylindrical tubes are useful for special sedimentation tests or for isolating a clarified liquid or sedimented solids fraction, pear-shaped tubes with calibrated stems (that may approach capillary size) aid in measuring a small volume of sedimented immiscible liquid or solids, and special tube shapes frequently are specified for particular analytical procedures. For economy and durability as well as solvent and corrosion resistance, the available tube materials include glass, polyethylene, polypropylene, polyallomer, polycarbonate, aluminum, and stainless steel. Many are in bottle form and can be easily capped or corked to prevent evaporation during long-term centrifuging.

The fixed angle head centrifuge is useful when it is not necessary to determine accurately the position of the liquid-solids interface. The sample tubes are inserted into appropriate openings or drilled holes in a streamlined rotor, where they are held at a fixed angle to the axis of rotation, usually in the range of 25 to 40°, both at rest and in rotation. The speeds and corresponding rates of sedimentation possible with this design are somewhat higher than those obtainable with the swinging tube type. The particles travel radially a distance equal to no more than the product of the diameter of the tube and the secant of its angle of inclination (Fig. 6.7), thus greatly reducing the time for sedimentation to a surface. Performance is optimized if the angle selected is the minimum that will permit the solids reaching the tube wall to slide along it to the bottom. Since the solids layer is deposited at an angle, this type of rotor is generally used for liquid-liquid separation or when measurement of the sedimented volume is not required. Only cylindrical or tapered tubes are generally used, with sizes ranging from small capillaries to 250 ml.

In some centrifuge designs a basket-type rotor with either solid or perforate wall is available as an accessory. These are discussed in a later section.

In making comparisons of centrifuging conditions, it is convenient to consider the relative acceleration in the centrifugal field without regard to the physical characteristics of the particle or its ambient fluid. This relative acceleration, $G = \omega^2 r/g$, expresses centrifugal force or acceleration as a function of gravitational acceleration and is dimensionless. For a centrifuge tube in either an angular or a horizontal position, the value of G is less at the free liquid surface than at the bottom of the tube. As an example, in the horizontal tube of Fig. 6.1, if $r_1 = 9.53$ cm and the tube is rotating at 3000 rpm, then at the surface of the liquid

$$G_1 = 0.000011182 \times 9.53 \times 3000^2 = 959 \times \text{gravity}$$

and at the bottom of the tube, where $r_2 = 19.7$ cm, $G_2 = 1984$.

A particle traveling from r_1 to r_2 is subjected to a constantly increasing field varying from G_1 to G_2. This is also a function of the rotational speed of the centrifuge. In addition to these factors, the proportion of the particles sedimented depends on the time of exposure to the centrifugal field. From this it can be seen that comparison of centrifuging conditions requires the proper relation of speed, bottle dimensions, centrifuge dimensions, and centrifuging time. It is almost impossible to duplicate centrifuging conditions completely without using identical equipment unless these variables are related to a Q/Σ function in which Σ combines the physical parameters of the centrifuge as discussed previously.

For sedimentation in a bottle centrifuge, the radius of the surface of the sedimented phase should be considered to be the true value of r_2. Any particle reaching that radius at any time during the run has been removed from the final supernatant liquid. From (6.10) and (6.11) the rate of sedimentation of a particle at radius r between r_1 and r_2 is defined as follows:

$$v_s = \frac{dr}{dt} = v_g \frac{\omega^2 r}{g} \tag{6.21}$$

Separating the variables and integrating for a particle that just reaches r_2 in time t:

$$\int_r^{r_2} \frac{dr}{r} = \int_0^t \frac{v_g \omega^2}{g} \, dt \tag{6.22}$$

which yields the result

$$\ln \frac{r_2}{r} = \frac{v_g \omega^2 t}{g} \tag{6.23}$$

Considering particles of diameter d, one-half are sedimented at the 50% cut-off point in a tube of cross-sectional area A if

$$A(r - r_1) = A(r_2 - r) \tag{6.24}$$

and

$$r_{50\%} = \frac{r_2 + r_1}{2} \tag{6.25}$$

By substitution into (6.23), we have

$$v_g = \frac{g}{\omega^2 t} \ln\left(\frac{2r_2}{r_1 + r_2}\right) \tag{6.26}$$

Since the general development of the Σ function in (6.15) shows that $Q/\Sigma = 2v_g$, for the bottle centrifuge, we have

$$\frac{Q}{\Sigma} = \frac{4.6g}{\omega^2 t} \log\left(\frac{2r_2}{r_1 + r_2}\right) \tag{6.27}$$

Since the rate function Q for a unit batch is V/t, in this case

$$\Sigma = \frac{\omega^2 V}{4.6g \log[2r_2/(r_1 + r_2)]} \tag{6.28}$$

The Q alone has no real meaning as a "flow rate" in a bottle, but it does incorporate the essential time factor. Therefore, by matching Q/Σ through any convenient combination of the parameters of (6.27), centrifuging conditions can be matched between different bottle centrifuges and, for a given suspension, the same degree of sedimentation can be obtained. This approach does not consider abnormal wall effects from tubes of noncylindrical configurations and yields similar values only for the volume of the compacted heavy phase if G at r_2 is the same.

As an example of the calculations, consider a bottle centrifuge turning at 2000 rpm. With a sample volume of 100 ml, $r_1 = 5.0$ cm and, assuming a sedimented cake 1 cm thick, $r_2 = 18.5$ cm; thus

$$\Sigma = \frac{(2\pi \times 2000/60)^2 \times 100}{4.6 \times 980 \times \log[2 \times 18.5/(18.5 + 5.0)]} = 0.00049 \times 10^7 \text{ cm}^2$$

If the centrifuging time is 10 min, (6.27) gives

$$\frac{Q}{\Sigma} = \frac{4.6 \times 980}{(2\pi \times 2000/60)^2 \times 10 \times 60} \cdot \frac{2 \times 18.5}{18.5 + 5.0} = 0.338 \times 10^{-4} \text{ cm/sec}$$

Certain emulsions and suspensions of very fine particles (such as those of molecular dimensions) constitute exceptions to the behavior defined by (6.28). For these, t and $\omega^2 r$ are not interchangeable because a threshold of centrifugal acceleration must be exceeded before breaking of the emulsion or compaction of the fine particles to high concentration occurs, regardless of the time of centrifuging.

Occasionally it is necessary to determine sedimentation results at a very low Q/Σ value, to accomplish a fractional sedimentation, or to obtain data for a Q/Σ versus percentage sedimented curve when some particles are large or of high density. Under these conditions the time required to accelerate and decelerate the centrifuge rotor becomes an important factor. For example, the bottle centrifuge in the example above with two 100-ml tubes requires 20 to 30 sec to approach 2000 rpm. A stopping time (with braking) of less than 10 sec is likely to disturb the surface of the sedimented solids pellet; even longer deceleration time may be necessary with uncompacted solids. The assumption that ω is a linear function of t during acceleration after startup and during deceleration is sufficiently accurate to account for the value of Q/Σ during those periods. Assuming $\omega = kt$, (6.22) becomes

$$\int_r^{r_2} \frac{dr}{r} = \int_0^{t_1} \frac{v_g k_1^2 t^2}{g} dt + \int_{t_1}^{t+t_1} \frac{v_g \omega^2}{g} dt + \int_{t+t_1}^{t_T} \frac{v_g k_2^2 t^2}{g} dt \qquad (6.29)$$

where k is the slope of ω versus t; $k_1 = \omega/t_1$ while starting, and $k_2 = \omega/t_2$ while stopping. The total time of centrifuging is $t_T = t_1 + t + t_2$. This equation resolves to give

$$\frac{Q}{\Sigma} = \frac{4.6g}{\omega^2 t + \frac{1}{3}(k_1^2 t_1^3 + k_2^2 t_2^3)} \log\left(\frac{2r_2}{r_1 + r_2}\right) \qquad (6.30)$$

In general, (6.30) must be used instead of (6.27) only if the centrifuging time is less than 3 min at which point the correction is about 5 % for the time values just given.

If the bottle centrifuge is used for fractional separations, it is necessary to calculate the time needed to sediment all particles of diameter d or larger. In this case (6.22) is integrated between the limits r_1 and r_2 so that

$$\ln \frac{r_2}{r_1} = \frac{v_g \omega^2 t}{g} \qquad (6.31)$$

Most of the particles encountered in centrifugal separation are small enough to settle in accordance with Stokes' law. Equation 6.11, derived from Stokes' law for spherical particles, may be rewritten

$$v_g = \frac{\Delta\rho d^2 g}{4\mu} K \qquad (6.32)$$

where K is a factor relating shape, roughness and particle concentration to the frictional resistance of particles passing through the liquid. For single spheres $K = \frac{2}{9} = 0.222$. If the particle is other than spherical, its size may be defined by its equivalent spherical diameter, which is the diameter of a sphere having the same characteristics as the particle that under the same conditions will settle at the same rate as the irregularly shaped particle. K factors for particles settling in the Stokes' law range have been reported: for cubes, $K = 0.206$ [1]; for discs [2] and tetrahedrons [1], $K = 0.19$; for cylinders [2], $K = 0.03$ to 0.06 and for other conditions, $K = 0.001$ to 0.15 [3, 4]. If (6.31) and (6.32) are combined and the appropriate value of K is inserted, the desired cutoff time for particles of mean diameter d is

$$t_d = 4\mu \ln \frac{r_2}{r_1} \omega^2 d^2 \Delta\rho K \tag{6.33}$$

It is important that the initial concentration of the dispersed phase be low enough to assume "free" or unhindered sedimentation; otherwise corrections must be made to these equations. For precise work this concentration should be less than 1% by volume, but concentrations up to 10 or 15% usually give reasonable approximations. It is possible to obtain a greater proportionate removal of fines of less than the desired cutoff diameter by reslurrying the sedimented solids in a clean portion of the liquid and recentrifuging with the same time limit as before. It is frequently necessary to use dispersing agents to prevent aggregation of the particles during the time required for fractional sedimentation or for determining particle size distribution.

It is convenient to determine the proportion of particles sedimented from a given suspension at several values of Q/Σ by varying centrifuging time or rotational speed or both. For each condition the volume of sediment should be noted and the supernatant suspension carefully decanted. The volume content of the solids in the supernate is obtained by recentrifuging it for an extended period of time until all the solids have been sedimented and their volume has been measured. These data permit construction of a graph of Q/Σ versus percentage sedimented (or unsedimented). It is convenient, and frequently adequate, to construct the Q/Σ curve on the basis of sedimented volume. For greater precision the sedimented weight basis should be used, and for this it is necessary to prepare a weight-to-volume correlation or to dry and weigh the sedimented pellet with correction for the soluble fraction of the liquid phase.

Such a graph appears in Fig. 6.8 for silica particles sedimented from ethylene glycol. On logarithmic probability paper, Q/Σ plots for particles having a normal or Gaussian probability distribution are linear and can therefore be drawn from relatively few points and extrapolated over at least short ranges. On this graph, extrapolation indicates that Q/Σ at the 50% cutpoint is 126×10^{-4} cm/sec. From (6.17) and (6.32), the form for calculating the equivalent spherical diameter becomes

$$d = \left(\frac{2\mu}{g\Delta\rho K} \frac{Q/\Sigma}{E} \right)^{0.5} \tag{6.34}$$

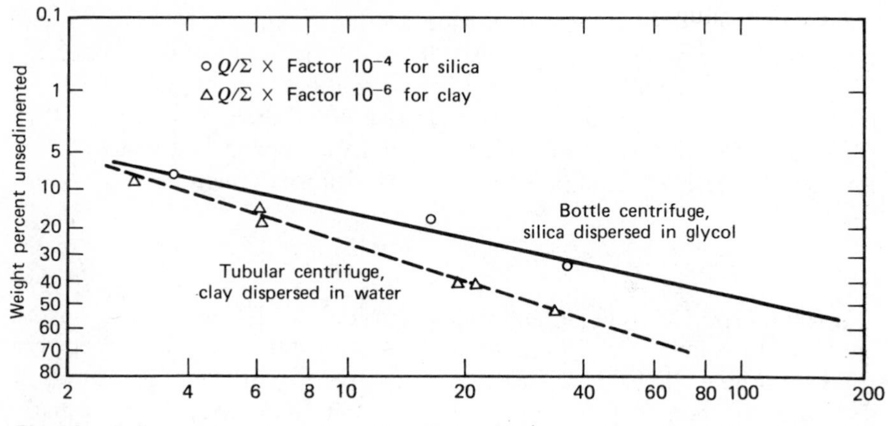

Fig. 6.8. Sedimentation plot: ⊙, $Q/\Sigma \times$ factor 10^{-4} for silica; △, $Q/\Sigma \times$ factor 10^{-6} for clay.

in which E is the fractional estimated efficiency, usually assumed to be 1.0 for the bottle centrifuge at reasonably low suspension concentrations. For $\mu = 0.17$ P, $\Delta\rho = 1.1$ g/ml, and an estimated shape factor $K = 0.17$, the particle size at the 50% cutoff point is 49 μ, which is in good agreement with the supplier's mean value of 54 μ for this grade of silica.

For systems with well-dispersed and discrete particles, the Q/Σ curve permits estimating particle sizes and, particularly, particle size distributions. The width of the distribution is indicated by the slope of the curve, a wide distribution corresponding to a relatively small change in percentage unsedimented for a large change in Q/Σ as in Fig. 6.8. Conversely, a suspension of particles of uniform size gives a curve sloping sharply toward the Q/Σ axis. With appropriate correction factors based on experience, bottle centrifuge Q/Σ curves can be used to estimate the operation of continuous centrifuges on a given material. In theory and in practice the Q/Σ curves from centrifuges of different types operating on the same material vary only by differences in the operating efficiencies of the centrifuge types. Frequently they must also be corrected for agglomeration during the time of exposure to the centrifugal field.

A number of bottle centrifuges and heads are specifically designed for particular purposes ranging from blood plasma separation (from blood bank collections) to microanalyses. Refrigerated centrifuges are useful when working with low melting point solids or with biochemical or other heat-sensitive materials that must be kept in a controlled temperature range to inhibit enzyme or bacterial activity or protein denaturization. Fractional separation based on differential solubility frequently requires controlled low temperatures (e.g., in fractionation of proteins from blood serum).

Heated centrifuges are required for certain analytical procedures, such as the Babcock test for butterfat content of dairy products. Many emulsions are much more easily separated at elevated temperature. Certain systems require a relatively high temperature to maintain one or more of the components in fluid form. For these, the casing or other enclosure of the bottle centrifuge is equipped with thermostatically controlled heating units suitable for continuous operation. As a temporary expedient, hot air or steam may be introduced into the casing, but a penalty may be expected in the form of increased maintenance costs.

A number of specialty rotors are available that permit bottle centrifuges to achieve preparative results previously considered to be in the realm of ultracentrifuges. One bench-top model, described shortly, accepts a zonal rotor that permits the collection of up to 800 ml of sediment with continuous addition of feed and discharge of concentrate. Another rotor for a high-speed benchtop centrifuge is an angle-head type with tubes and accessories designed for continuous addition of feed and discharge of concentrate (Fig. 6.9), at rates of up to 500 ml/min. Separation takes place at about 35,000 *G* maximum, and up to 350 ml of settled solids can be harvested per run.

By introducing the concept of long-path circumferential flow, a refrigerated cabinet model bottle centrifuge can be used to achieve the equivalent of ultracentrifuge performance at only 6000 rpm. The rotor can be operated in various modes and permits the collection of up to 550 ml of sediment. In Fig. 6.10 it is operating as a zonal centrifuge with formation of the gradient, continuous flow of the sample, and recovery of the banded harvest while running at full speed.

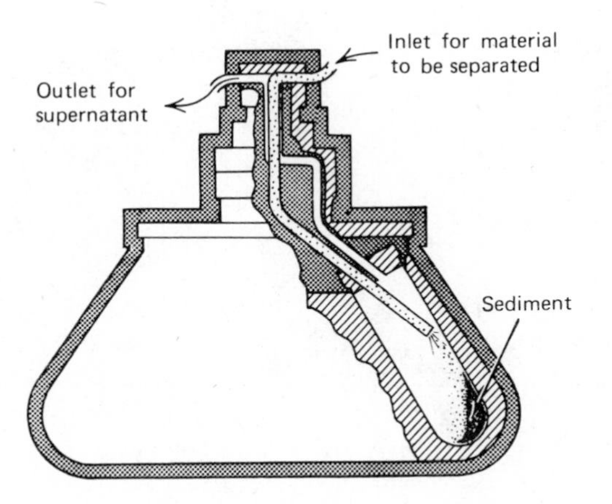

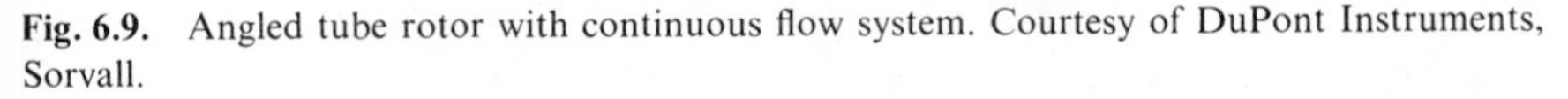
Fig. 6.9. Angled tube rotor with continuous flow system. Courtesy of DuPont Instruments, Sorvall.

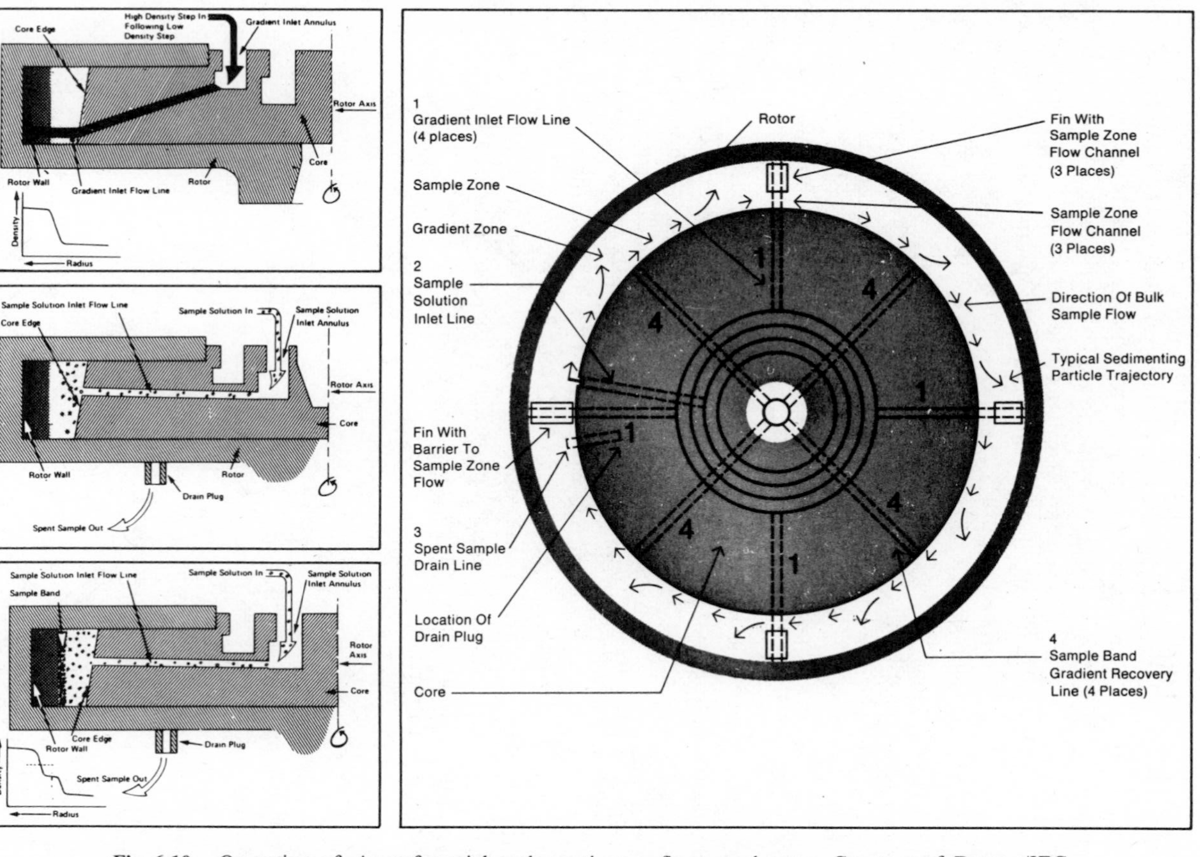

Fig. 6.10. Operation of circumferential path continuous flow zonal rotor. Courtesy of Damon/IEC

Preparative Ultracentrifuges

Following the introduction of the analytical ultracentrifuge with optical system that was capable of processing at most only a few milliliters of sample, increasing interest in the areas of subcellular particles, viruses, and proteins created the need for processing and preparing large quantities by centrifugal separation. The highly instrumented ultracentrifuge developed for analytical work (Fig. 6.18) can be greatly simplified for preparative work, including elimination of the complicated optical systems. Capacities can be much higher, with some reduction in maximum centrifugal acceleration. Many varieties of preparative ultracentrifuges are available, with varying degrees of automated control for specific applications. This section treats their more common features.

Preparative types of ultracentrifuge cover the spectrum of operating speeds from 20,000 rpm, generating about 40,000 *G*, to 75,000 rpm and about 500,000 *G*. The rotor is generally underdriven by electric motor or oil turbine and is surrounded by a high-strength cylindrical casing. The friction of air at atmospheric pressure at these speeds would rapidly raise the temperature of the rotor by an unacceptable amount. A suitable and usually self-contained vacuum system reduces the pressure in the casing surrounding the rotor to less than 0.5 atm in the low-speed models and to as little as 1 μ for very-high-speed rotors. In addition, a refrigeration system provides temperature control in the range of -15 to $+30^\circ$C. Sensors monitor the temperature of rotor, either directly or indirectly, to maintain it within $\pm 1^\circ$C of the set point. This is usually sufficient to inhibit thermal convection and to prevent thermal damage to the sample in the rotor. Electronic controls maintain rotor speed within the required narrow range. Interchangeable printed circuit boards automatically program the centrifuge sequence, reducing both the need for operator attendance and the risk of human error. Electronic ramping circuits control acceleration and deceleration, particularly below 10,000 rpm, to minimize turbulence in the sedimentation chambers. Up to 20 min may be required for this function at each end of the cycle, particularly for centrifuging within a density gradient. With the flexible shaft (self-balancing) drives now in use, "eyeball" balancing of the samples is usually satisfactory.

Many safeguards can be provided for emergency shutdown, including overspeed and no-load interlocks, cases of temperature and vacuum control failure, and excessive unbalance. Among the more sophisticated additions are an optical system for monitoring the progress of sedimentation and an $\omega^2 t$ integrator to sum automatically the product of acceleration and time, including the acceleration and deceleration periods. Preparatory ultracentrifuges are so reliable that the drives are warranteed for several billion revolutions and can be rebuilt with replacement of relatively few parts such as seals.

The proliferation of rotors for preparative work presents an enormous range of capabilities. The simplest basis for classification of the rotors is to consider those operating on a fixed volume of sample as batch and those that accept feed and discharge centrate during rotation as continuous. Perhaps 50 % of preparative work is done in batch rotors, which include angle and swinging bucket types similar to those described earlier and zonal varieties as presented in the next section. The angle rotors hold from four containers of 500-ml capacity each to more than 100 fractional-milliliter tubes. They are used for the preparative fractionation of crude homogenates or cell suspensions to collect the cells and tissue debris as a pellet at the end of the tube [5]. The horizontal or swinging bucket rotor is largely used for density gradient separation or when the volume of the pelleted fraction must be accurately determined.

The simplest centrifugal separation of polydisperse particulates is by selective sedimentation on the basis of size and density from a suspension that is uniformly mixed. Since smaller and lighter particles that were initially near the bottom of the tube also eventually reach the pellet, it contains a mixture of all classes of particles but is richer in the heavier and leaner in the lighter ones than was the original suspension. Repeated reslurrying of the pellet and recentrifuging is necessary to obtain a relatively pure fraction of the heavier solids. If the heterogeneity of the original suspension is due to a range of particle densities, a technique of isopycnic separation improves the efficiency by increasing the density of the fluid phase until it is intermediate to the range of densities of the particles to be separated. Particles denser than the fluid will settle to the bottom, and others will remain suspended or rise to the surface regardless of size.

In another technique, the rotor of Fig. 6.9 can be fed continuously at the outer end of each chamber while rotating and the effluent continuously withdrawn at the inner end [6]. Each tube acts as an elutriator, and particles having sedimentation velocities lower than the flow rate of liquid toward the center are discharged with the effluent, while solids having a higher velocity are retained in the tube. Fractions can be discharged by incrementally changing the flow rate or the rotational speed.

In earlier developments, rotors in the form of tops with undersides shaped to act as turbines were floated on jets of driving air. This approach has been modified with a tiny rotor that amounts at an angle of 18° six plastic tubes, each of 0.17 ml capacity. It operates at 100,000 rpm, giving 160,000 G for achieving rapid separation of small samples.

A useful performance index for comparing preparative rotors [7] is

$$P = \frac{\omega^2}{\log r_2 - \log r_1} \tag{6.35}$$

Some investigators prefer to use this in the reciprocal form $1/P$. The larger the value of P, the higher the performance capability. The time for a particle of known Svedberg value s to transit from r_1 to r_2 is

$$t = \frac{1}{Ps} \tag{6.36}$$

The preparative ultracentrifuge is used in biological research to isolate intracellular fractions to determine their separate functions and reactions to drugs, and to isolate and purify relatively large quantities of viral material from larger quantities of cellular and subcellular matter as in the production of vaccines. Lower speed rotors are used for cellular species and large subcellular organelles such as nuclei, mitochondria, and lyosomes. Intermediate speeds are used for separating subcellular components such as chloroplasts and large viruses. Higher speed rotors are used for the isolation and purification of small subcellular particles such as ribosomes and macromolecules.

Zonal Centrifuges

The concept of zonal centrifugation and the development of the equipment to utilize it have provided a quantum jump in the techniques of isolating and purifying subcellular species down to molecular dimensions. In principle, a density gradient is established normal to the axis of rotation of the rotor, ranging from a maximum at r_2 to a minimum at r_1. The gradient material may be any solution compatible with the system to be separated. Aqueous solutions of low molecular weight solutes such as potassium citrate, sucrose, and cesium chloride are frequently used; the last-named material is particularly useful because of the high density of its saturated solution. The gradient may be formed by introducing a solution of appropriate strength into the rotor and centrifuging for a prolonged period, such as 48 hr, to establish a natural and continuous gradient at equilibrium between centrifugal acceleration and the forces of diffusion. To save time, the gradient material may be introduced in successive layers whose composition changes from low to high density. The higher density solution displaces those of lower density toward the axis of rotation with little intermingling [8, 9].

In the simplest centrifuge for batch bottles with swinging or angle head, the gradient is formed while the tube is at rest, and the sample to be centrifuged is carefully layered on top of it. For more sophisticated operation, the rotors are fed gradient solutions and samples while rotating. Density gradient forming devices are available to develop automatically the desired density profile, even for varying the shape of the gradient from linear to nonlinear. Linear gradients offer the best resolution of such components as proteins, enzymes, and hormones, whereas concave gradients are better for lipoproteins. Stepwise gradients can also be created for the sharp separation of particles of known different densities; thus each fraction forms a layer on the step of density slightly higher than its own.

Rate zonal sedimentation depends on the sedimentation coefficient of each particle and the difference between its density and that of the gradient at each point along the radius in the centrifugal field. The sample is always introduced as a layer on top of the preformed gradient, and particles sediment under centrifugal force at different rates depending on their mass and size (Fig. 6.11). The gradient is selected to ensure that its density is less than that of the heaviest particle, and preferably the run is terminated before the heaviest particles have reached the bottom of the tube.

In the equilibrium or isopycnic banding technique, the particles of the fractions to be separated must have densities between the lowest and highest densities of the gradient column. The sample may be mixed initially with a gradient solution such as cesium chloride which, in time, forms its own gradient in the centrifugal field. The sample may also be layered over the preformed gradient. The run is terminated when the particles of each density have reached their respective equilibrium isopycnic positions.

The use of swinging head or angle-type rotors, particularly the smaller sizes that must be employed at high speeds, severely limits the size of sample that can be processed and the amount of product harvested in each run. This restriction has led to the development of a variety of larger rotors for both

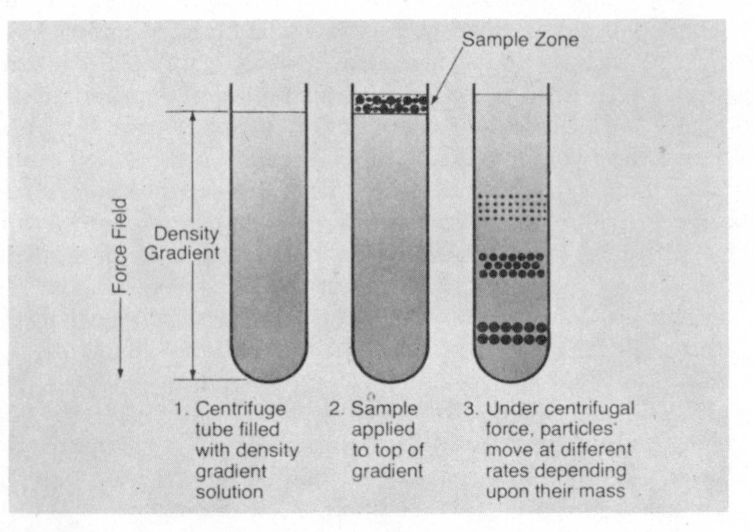

Fig. 6.11. Rate zonal separation in a swinging bucket rotor. Illustration courtesy of Beckman Instruments, Inc.

batch and continuous flow operation. Gradient conditions are stabilized by radial baffles forming sector-shaped chambers within the rotor, thus minimizing wall effects from Coriolis forces. Rotors made of titanium or aluminum may be either cylindrical or, more commonly, bowl shaped as in Fig. 6.12. The latter, with capacities up to 1.7 liters, may be, for example, up to 18 cm i.d. by 8 cm high, with sedimentation path length of about 7 cm; the larger bowls produce about 100,000 *G*, and smaller units may reach 250,000 *G*. These rotors are adapted to batch as well as to continuous flow techniques.

For static loading and unloading, the gradient is introduced with the rotor at rest (Fig. 6.13), starting with the light end and displacing it upward with layers of increasing density. The sample is then layered on top of the gradient. As the rotor is carefully accelerated, the gradient and sample layers assume a position normal to the axis of rotation with little or no intermixing. After the run is completed, the rotor is decelerated to rest with equal care, and the layers reorient themselves so that they can be statically unloaded as cuts of incremental density. The technique is known as reorienting gradient or "reograd."

Fig. 6.12. Bowl-shaped zonal rotor with interchangeable cores. Illustration courtesy of Beckman Instruments, Inc.

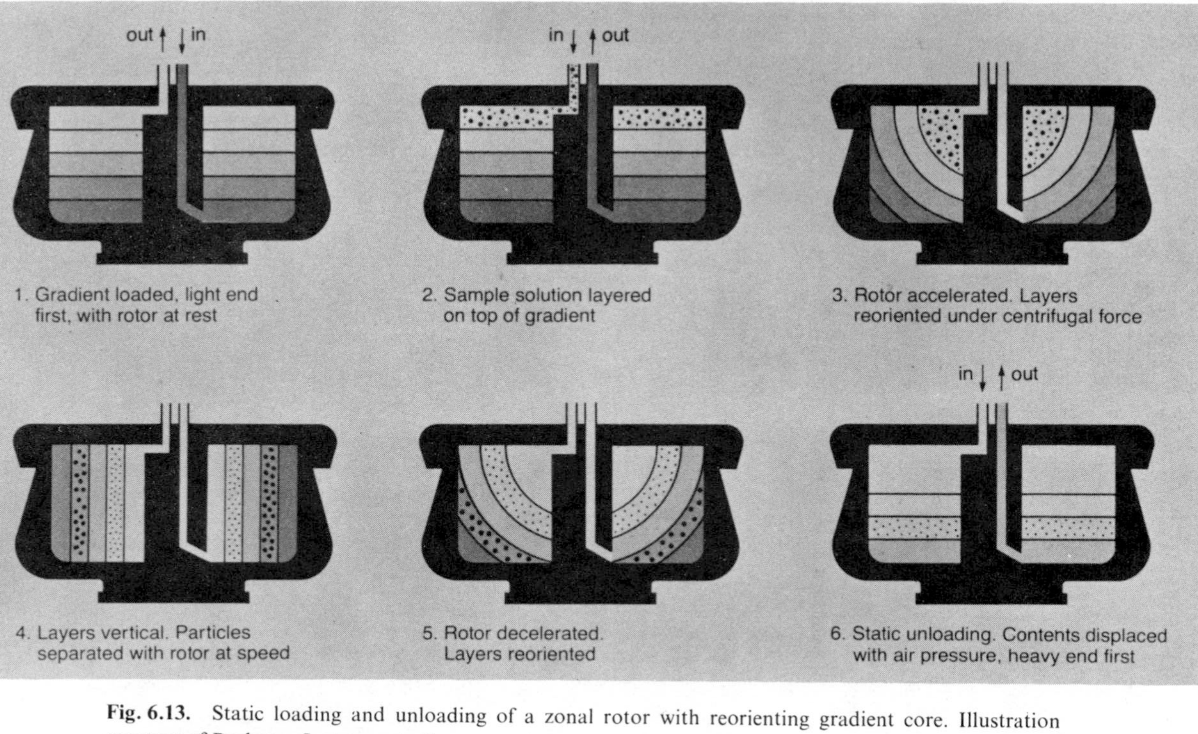

Fig. 6.13. Static loading and unloading of a zonal rotor with reorienting gradient core. Illustration courtesy of Beckman Instruments, Inc.

Since gradients formed in a centrifugal field are more stable than those formed in a gravity field, relatively shallow gradients are feasible when formed during rotation. By the use of a suitable stationary seal, the gradient and sample can be introduced while the rotor is turning at reduced speed (Fig. 6.14). The seal is then removed and acceleration is continued to design speed for the run, after which the seal is replaced for unloading. Sharp separation of the layers during unloading is favored by tapering the core and introducing a dense solution at the bowl wall to displace the fractions at the center.

With more sophisticated seals, continuous feeding of sample and rejection of filtrate at full speed is possible, thus allowing collection of considerably larger quantities of product. Care must be taken not to overload the bands because a maximum band layer thickness of 2 mm is generally considered

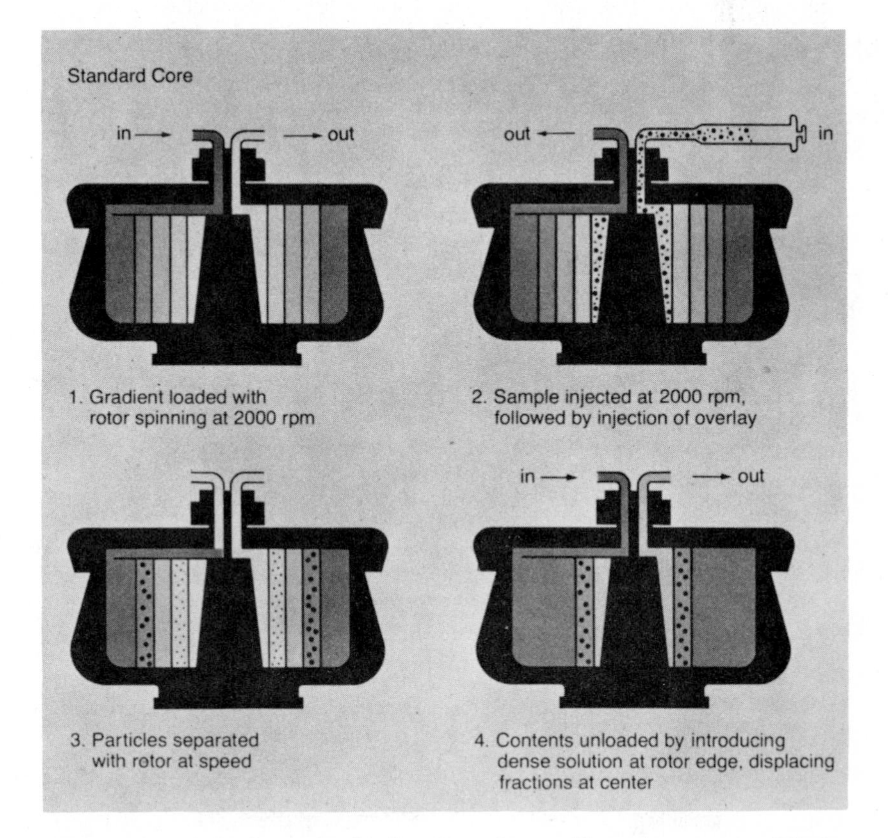

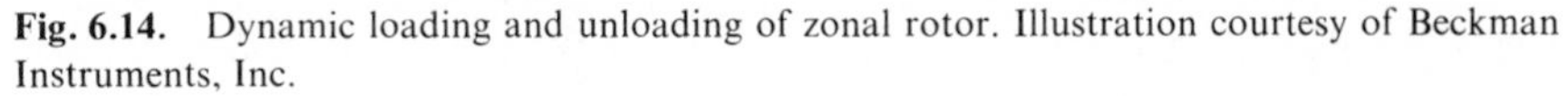

Fig. 6.14. Dynamic loading and unloading of zonal rotor. Illustration courtesy of Beckman Instruments, Inc.

acceptable. Increased loading is obtained with the long tubular rotor of Fig. 6.15. The research and pilot plant model permits flow rates up to 60 liters/hr at 55,000 rpm (150,000 *G*), whereas a larger rotor permits production rates up to 120 liters/hr at 35,000 rpm (90,000 *G*) for full-scale operation. The tubular zonal rotor is usually loaded with gradient and unloaded at rest by the reograd procedure.

Separated bands, varying in density, may be identified and cut points determined by a number of techniques, including solute concentration, refractive index measurement, and spectrophotometric methods. Figure 6.16 shows the separation of subcellular components from rat liver homogenate in a fairly broad density gradient. The separation of cytoplasmic inclusion bodies from nuclear inclusion bodies from the bacteria and cell components of the Tussock moth is illustrated in Fig. 6.17. In this example, a very shallow gradient was needed for high resolution of components with a small difference in banding density. Small beads coded by color with respect to density may be introduced into the rotor to facilitate band density identification.

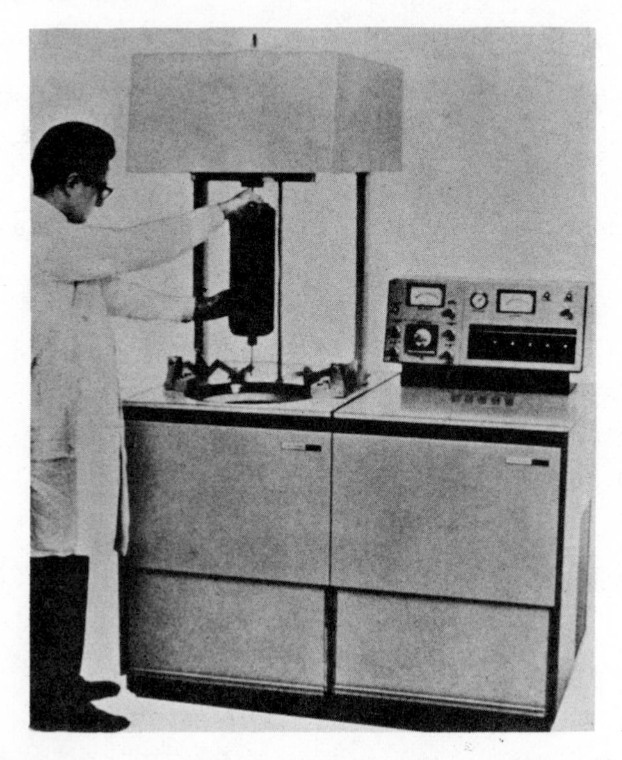

Fig. 6.15. Continuous flow tubular bowl zonal centrifuge. Courtesy of Electro-Nucleonics, Inc.

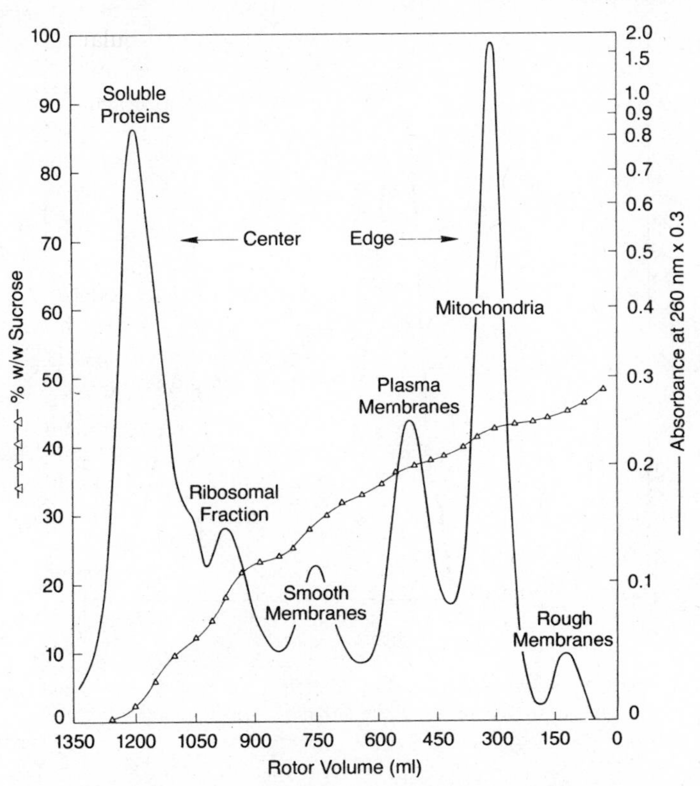

Fig. 6.16. Separation of subcellular components from rat liver homogenate in a broad density gradient. Illustration courtesy of Beckman Instruments, Inc.

Analytical Ultracentrifuges

The analytical ultracentrifuge, originally developed by T. Svedberg and his co-workers, permits continuous observation and recording of sedimentation and diffusion in a sample in a centrifugal field up to 372,200 G in commercially available models (e.g., Fig. 6.18). Details of the theory and design of this centrifuge and the interpretation of results are beyond the scope of this chapter. Reference should be made to such standard works as those by Svedberg and Peterson, Schachman, and Trautman [10–12].

High-strength rotors made from aluminum or titanium forgings are shaped to maximize strength and minimize turbulence of the surrounding atmosphere. They contain two to six cavities for holding the sample observation cells. Suspension from a flexible wire drive shaft provides sufficient freedom for the rotor to be self-balancing and to turn with stability about its own center of mass up to design speed of 68,000 rpm. To ensure safety, the highly

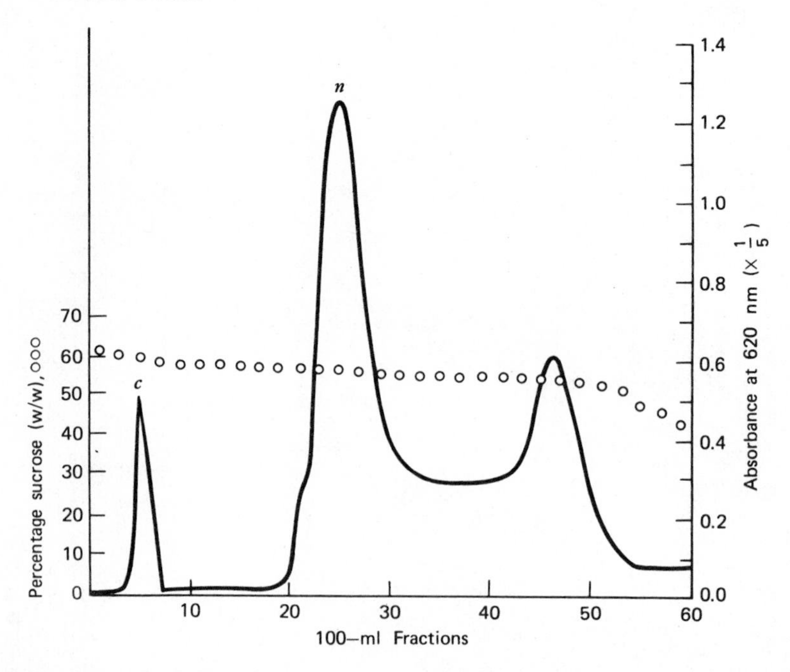

Fig. 6.17. Separation of cytoplasmic inclusion bodies *c* from nuclear inclusion bodies *n* and from cell components of the Tussock moth in a shallow density gradient [8].

stressed rotor is surrounded by a heavy inner steel ring that is free to turn and absorb rotational force. This in turn is encased in a steel vacuum chamber and the whole, including the drive and part of the optical system, is in a steel barricade. The drive is now usually made through gearing from an electric motor, although air turbine and oil turbine drives have also been used. Control of the speed of the drive was originally mechanical and later electronic. Now an electrooptical system uses a light chopper disc to produce a frequency corresponding to the operating speed; by comparing this to a reference signal from the speed selector, rotor speeds are maintained within less than $\pm 0.2\,\%$ of the selected speed within a range of 800 to 80,000 rpm.

To minimize heating of the rotor from windage friction, the chamber is evacuated to about 1 μ absolute by mechanical and oil diffusion vacuum pumps in series. Maintenance of controlled temperature during runs that may last upward of 48 hr is extremely important to avoid thermal convection within the cell. A thermal sensor in the rotor acts through a needle dipping into a mercury pool in the bottom of the casing to provide temperature indication. In turn, this reacts through a bridge circuit to control heating or coolant flow through a liner within the evacuated casing. Rotor temperature

Fig. 6.18. Model E analytical ultracentrifuge. Illustration courtesy of Beckman Instruments, Inc.

at full speed is maintained at 0 to 40°C with an accuracy of ±0.1°C. A high-temperature accessory permits operation to 135°C.

The unique feature of the analytical ultracentrifuge is its provision for continuously observing within the cells the movement of particles down to those of relatively small molecular dimensions. The common methods are Schlieren or Rayleigh fringe interference systems and the absorption of monochromatic or ultraviolet light. Figure 6.19 illustrates typical readings from the several methods. The boundary layer of monodisperse solute molecules is indicated by a change in refractive index gradient within the cell and is shown by Schlieren optics as a peak. The interface of the clarified solvent layer inward of the boundary and the sedimenting zone of higher solute concentration provides a sharp peak, although neither layer shows an internal gradient or any response in the Schlieren system. In a system containing two or more species of particles sedimenting at different rates, the boundary of each would show a peak. This system is normally used with solute concentrations of 0.1 to 2.0 %.

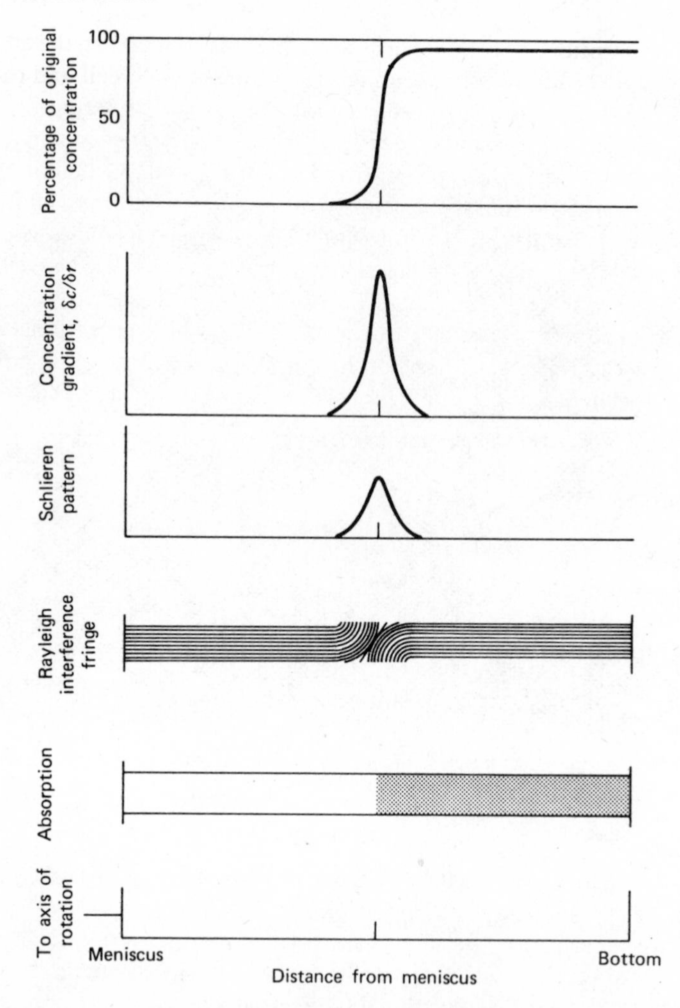

Fig. 4.19. Schematic solute patterns in ultracentrifuge by different optical systems. From *Encyclopedia of Industrial Chemical Analysis*, Vol. 9, Wiley, New York, 1970, p. 184.

The Rayleigh fringe interference system also indicates changes in refractive index along the sedimentation path. It is somewhat more sensitive than Schlieren optics and responds well to solute concentrations in the 0.05 to 0.5 % range. The monochromatic ultraviolet light absorption system, similar to that used by Svedberg originally, has come into wider use with the need for indicating very low concentrations of the order of 0.001 %. It is particularly useful in work with nucleic acids, enzymes, and viruses that show strong ultraviolet absorption. Several devices are available to increase the efficiency of this method. A photoelectric scanner can monitor each cell and can convert variations in transmission intensity into integral and derivative curves directly. A monochromatic light source operated at preselected wavelengths can optimize the system for the absorption peak of a particular species. Another accessory automates the scanner to move from cell to cell in a pattern to permit the continuous monitoring of all the cells in the rotor.

The cells are usually sector shaped to minimize wall effects and include single and double sectors, provision for a synthetic boundary or mechanical partition, and others for special purposes. For most applications the cell mean radius from the axis of rotation has been standardized at 65 mm, with a sample radial depth of 14 mm. The liquid thickness through which light is transmitted ranges from 1.5 to 30 mm in cells of aluminum, Kel-F, and Epon with windows of quartz or synthetic sapphire. Cell capacity is in the range of 0.1 to 2.1 ml.

The basic equation in analytical ultracentrifugation balances sedimentation in the centrifugal field against the forces of diffusion:

$$c\omega^2 xM(1 - V_* \rho_*) = -RT\frac{dc}{dx} \tag{6.37}$$

The Svedberg sedimentation constant is defined by

$$\begin{aligned} s &= \frac{1}{\omega^2 x}\frac{dx}{dt} \\ &= \frac{\ln(x_2/x_1)}{\omega^2(t_2 - t_1)} \end{aligned} \tag{6.38}$$

These equations lead to two methods of calculating molecular weights based on sedimentation equilibrium:

$$M = \frac{2RT\ln(c_2/c_1)}{(1 - V_*\rho_*)\omega^2(x_2^2 - x_1^2)} \tag{6.39}$$

or on sedimentation velocity

$$M = \frac{RT}{D(1 - V_*\rho_*)}s \tag{6.40}$$

where c is the concentration of sedimenting material, R the gas constant, T the absolute temperature, D the diffusion coefficient for the particular species, M the molecular weight, V_* the partial specific volume of solute, ρ_* the density of dispersion, and x_2 and x_1 are the radial positions of a particle or the interface at times t_2 and t_1, respectively; s, usually expressed as s_{20°, the Svedberg constant, is the settling velocity in a unit field of force in a fluid of the density and viscosity of water at 20°C expressed in units of 10^{-13} cm/sec.

The range of applications of the analytical ultracentrifuge is almost unlimited and is constantly expanding, since it permits examination of even highly labile species in their natural environment. The measurement of sedimentation velocity and diffusion equilibrium helps to determine molecular weight and shape, sample purity, and particle size distribution in polydisperse systems. Solvation of molecules and monomer-polymer equilibrium can be studied. Sedimentation equilibrium methods, particularly with the introduction of density gradient techniques, permit measurement of concentration distributions. Many proteins, viruses, hormones, and other subcellular particles including DNA and RNA are being studied by this technique.

Tubular Bowl Centrifuges

Laboratory-size continuous flow centrifuges are available for the clarification or separation of larger amounts of material than are feasible with the bottle centrifuge. Previous sections have covered several centrifuges that are really laboratory instruments, up to the larger tubular zonal centrifuge used for commercial recovery of viruses at high purity. The next three sections include several small industrial centrifuges. These are used not only on a bench scale but also to obtain data from which to predict the performance of much larger commercial models of similar types.

If the aim of the work is primarily to obtain data for scaling up, preselection of the type of centrifuge to be used ultimately should be made in time to use the same type for test work. The science and art of centrifugation has not yet advanced to the point of accurately comparing centrifuges of different basic flow patterns. Many of the factors leading to a choice between tubular, perforate and imperforate basket, and conveyor discharge centrifuges have been reviewed in the current literature [13, 14].

The laboratory tubular centrifuge (Fig. 6.20) is a flexible unit for continuous treatment of a liquid-solid suspension or a liquid-liquid mixture in which the solid content is low and the particle size is medium to fine. It is convenient for handling batches of 4 liters or more or for pilot plant production at feed rates up to 4 liters/min, depending on the system. It consists of a relatively long, hollow cylindrical bowl of small diameter suspended from a flexible spindle at the top and guided at the bottom by a loose-fitting bushing. With standard motor drive it operates at 22,000 rpm. The turbine-driven design can be run at any speed up to 50,000 rpm at which $G = 62{,}400$. Inside the bowl is an accelerating device whose function is to ensure rotation of the liquid at bowl speed.

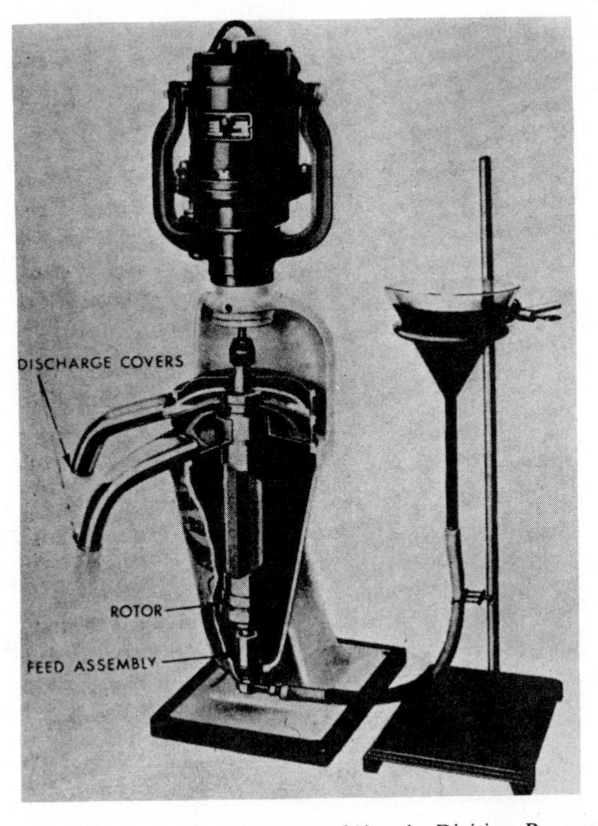

Fig. 6.20. Laboratory supercentrifuge. Courtesy of Sharples Division, Pennwalt Corporation.

Feed is jetted into the bottom of the bowl through a hole in the guide bushing and is thrown outward to the wall, where it forms a layer of thickness controlled by selection of effluent dams. Clarification or separation is effected during the time the feed passes up the length of the bowl. In a clarifier bowl, solid particles of density greater than the liquid are deposited as a packed cake on the bowl wall, while the clarified liquid is discharged by overflowing a dam at the top of the bowl. When the cake has built up sufficiently to interfere with clarification, the bowl is stopped and the cake is recovered.

In a separator bowl for resolving liquid-liquid mixtures, the heavier liquid settles to the bowl wall to form a continuous layer on which floats a layer of the lighter liquid. Unresolved emulsion remains at the interface until broken; premature discharge of interface material is prevented by an annular dam near the top of the bowl. Light liquid continuously overflows the inner circumference of the annular dam and is discharged into a cover through radial holes in the bowl. Heavy liquid passes outside the annular dam and is discharged over the inner circumference of a second dam, which maintains the heavy liquid layer at the desired depth inside the bowl. The heavy liquid is discharged into a separate cover from the light effluent. Heavy solids introduced with the emulsion are also retained as a cake in the separator bowl.

Development of Σ theory for tubular centrifuges has been reported [13, 14], and only its application is covered. In a continuous centrifuge the term Q denotes the volumetric flow rate through the bowl and provides the necessary time parameter. The separating capacity of a clarifier bowl is related to its physical dimensions by

$$\Sigma = \frac{2\pi Z\omega^2}{g}\left(\frac{3}{4}r_2^2 + \frac{1}{4}r_1^2\right) \tag{6.41}$$

where Z = effective length of the bowl,
r_1 = inner radius of liquid layer = radius of the liquid discharge dam,
r_2 = outer radius of liquid layer = inner radius of the bowl wall.

As cake builds up at the bowl wall, r_2 is reduced and holdup time is also reduced, serving to decrease clarification efficiency. Until the bowl is too full of cake, the loss in efficiency can be temporarily offset by a reduction in the throughput Q.

The Σ value for any clarifier bowl can be readily computed from (6.41) if the bowl dimensions and operating speed are known. The data for a bowl with a specific overflow diameter are, for example: Z = 18.7 cm, r_1 = 0.71 cm, r_2 = 2.22 cm, at 20,000 rpm.

$$\Sigma = \frac{2\pi \times 18.7 \times (2\pi \times 20{,}000/60)^2}{980}(0.75 \times 2.22^2 + 0.25 \times 0.71^2)$$

$$= 0.19 \times 10^7 \text{ cm}^2$$

At 50,000 rpm, as attained in a turbine-driven model, Σ becomes 1.17×10^7 cm^2.

If the bottle centrifuge example calculated earlier is assumed to give satisfactory clarification on a particular material, the theoretical throughput rate can be calculated for the laboratory tubular centrifuge to give the same degree of clarification. Assume that the tubular device operates at 20,000 rpm with $\Sigma = 0.19 \times 10^7$ cm^2 as previously and that $Q/\Sigma = 0.338 \times 10^{-4}$ cm/sec from the bottle centrifuge; then Q for the tubular bowl is theoretically $0.338 \times 10^{-4} \times 0.19 \times 10^7 = 63.5$ cm^3/sec. In practice, the actual tubular centrifuge feed rate for this selected degree of clarity may vary from 20 % to nearly 100 % of theoretical depending on the system. Hard, nonaggregated solids result in high efficiency, but flocculent, agglomerated feeds may suffer considerable dispersion as a result of the shear forces imposed during acceleration to the speed of the bowl. For scaling up centrifuging data from a laboratory-sized tubular, tests should be made at approximately the same rotational speed as for the larger unit to avoid this kind of variation in efficiency factor.

The clarifier bowl of the tubular centrifuge is readily used to determine Q/Σ curves for estimating particle size distribution or mean particle size if the shape factor K can be estimated and the centrifuge efficiency for the system is known [15]. Also, by comparison with bottle centrifuge Q/Σ data, the efficiency of the tubular bowl for the particular system can be approximated. Figure 6.8 gives a set of Q/Σ data for the laboratory tubular centrifuge on a suspension of clay particles in water with Tamol NNO as a dispersant. For this system the centrifugal efficiency is estimated to be 80 %, and K can be estimated at 0.18; the density difference is 1.64 g/cm^3 and the viscosity 0.89 cP. Since Q/Σ at 50 % sedimented is 31.7×10^{-6} cm/sec, the calculated mean equivalent spherical diameter of the clay particles is 0.49 μ from (6.41). A particle size distribution curve provided by the manufacturer shows a size of 0.55 μ at the 50 % weight point, a reasonable correlation.

In a separator bowl, Σ values can be calculated for each layer; that is, r_1 and r_2 correspond to the inner and outer radii, respectively, of each layer. This requires knowledge of the radius of the interface, which can be approximated by (6.20) from known overflow dam sizes because the pressures exerted at the interface by the light and heavy phases will be equal. In simplified form we have

$$\frac{\rho_{\text{heavy phase}}}{\rho_{\text{light phase}}} = \frac{r_I^2 - r_L^2}{r_I^2 - r_H^2} \tag{6.42}$$

where r_I = radial distance interface from axis,
r_L = radial distance light phase discharge from axis,
r_H = radial distance heavy phase discharge from axis.

More exact data are found by a study of the interface mark usually left on the accelerator device during an experimental run. The Σ value of the phases can be varied inversely by adjusting the diameter of the discharge dam.

Many variations are possible in tubular centrifuges to meet special conditions. Work with solvents requires a vapor-tight housing; for operation at elevated temperatures, the frame can be jacketed. Light solids that float on a heavier liquid can be discharged, as in the case of wax from lubricating oil. Highly viscous fluids such as soap or soapstock can be discharged in a frame hopper. Refrigerated frames are used for biochemical operations requiring low temperatures. Completely closed bowls can be used for treatment at high centrifugal force of 250-ml batches.

Disc Bowl Centrifuges

Disc bowl centrifuges are units for effecting on a continuous basis (*a*) the clarification of suspensions in which the particles are fine and the concentrations relatively low, (*b*) the separation of liquid mixtures with or without a solids content, or (*c*) the concentration into a relatively small stream of the solids existing at low concentrations in the feed. The first two functions are similar to those of the tubular centrifuge, although in commercial sizes the latter are generally smaller, less expensive, and of lower capacity than disc units. Function *c* is unique to the disc centrifuge and provides its largest field of application.

The distinctive feature of the design is a stack of concentric metal cones separated by spacers and so arranged in the bowl that feed must pass through the narrow spaces between discs before leaving the bowl. The purpose of the discs is to reduce the settling distance from the full depth of liquid in the bowl to the radial distance of separation of adjacent conical surfaces. The angle of the cones with the vertical axis is large enough to permit solids or liquid layers deposited on their surfaces to move to the inner or outer edge of the stack under the centrifugal field. In smaller sizes the bowl is underdriven and is supported on a spindle (Fig. 6.21). The feed is usually introduced at the top center of a clarifier bowl and passes down through the center post and under the disc stack, to the periphery of the bowl. Here it is distributed up the height of the stack and passes through the parallel disc spaces. Clarified feed is discharged over a dam or through orifices whose radial location can be varied to maintain the proper submergence of the disc stack.

Heavy solids are deposited on the underside of the discs and after aggregating, they slide down the surface to the periphery of the disc and then settle to the outer wall of the bowl. When the total volume of solids to be collected is very small, the bowl wall is solid and cake builds up in the space outside the discs. When cake builds into the disc stack, however, the stack loses efficiency and the bowl must be cleaned. When the proportion of solids is large, the bowl wall is sloped, allowing the solids to slide to discharge ports around the periphery, where they may be continuously discharged from the bowl in a stream of liquid. The volume of this solids concentrate stream is controlled by the number of discharge ports and the size of the nozzles used in the ports. Two limitations on the discharge ports may prevent obtaining a solids stream as concentrated as is desired: (1) since an even number of ports should be used to maintain bowl cake symmetry, two ports is the minimum number; and (2) the nozzles should be at least twice the diameter of the largest solids particle present. In addition, the solids must flow or be extruded through the nozzles at the pressures available, and their angle of repose must be low enough to prevent the cake between discharge ports from building up into

Fig. 6.21. Westfalia laboratory disc centrifuge. Courtesy of Centrico, Inc.

the disc stack. Some commercial-size disc machines utilize valved ports to permit greater concentration of solids before discharge, but these are not available in machines for laboratory use.

The Σ value for a disc machine involves the disc angle and the number of discs as well as the parameters previously introduced; the literature covers the derivation of the formula and the factors affecting it [13, 14]. For a clarifier bowl the Σ value is given by

$$\Sigma = \frac{2\pi n\omega^2(r_2^3 - r_1^3)}{3gC} \tan\theta \tag{6.43}$$

where n = number of spaces between discs,
r_1 = radius of inner periphery of disc slope,
r_2 = radius of outer periphery of disc slope,
C = factor for fluid flow pattern between discs,
θ = angle of the disc from the vertical.

The disc centrifuge of Fig. 6.21 is a flexible laboratory unit in that a solid-wall clarifier bowl, a nozzle discharge solids concentrator, and a liquid-liquid separator bowl can be mounted interchangeably on the same frame. The design characteristics of the nozzle discharge bowl are as follows: $n = 25$, $r_1 = 2.1$ cm, $r_2 = 3.6$ cm, $\theta = 45°$, and rotational speed is 12,000 rpm. With these data it is possible to calculate a value, thus

$$\Sigma = \frac{2\pi \times 25(2\pi \times 12{,}000/60)^2(3.6^3 - 2.1^3)}{3 \times 980 \times 1.0 \times \tan 45°} = 0.32 \times 10^7 \text{ cm}^2$$

Again using the system previously computed for the bottle centrifuge example with $Q/\Sigma = 0.338 \times 10^{-4}$ cm/sec for satisfactory clarity, the theoretical throughput for this disc machine clarifier bowl is $Q = 0.338 \times 10^{-4} \times 0.32 \times 10^7 = 107 \text{ cm}^3/\text{sec}$. Experimental tests have shown that the efficiency of a disc machine as a clarifier seldom exceeds 55 %, C in (6.43), and is considerably lower for solids subject to dispersion by shearing stresses or for those that do not aggregate readily. The practical feed rate for the desired clarity in this example is then probably not more than 60 cm^3/sec.

Nozzle discharge in disc centrifuges may be either "peripheral" (i.e., at the full radius of the bowl) or "return," in which solids are withdrawn at the full radius of the bowl but are returned toward the center of the bowl in an internal channel and discharged at some radius R, where $r_2 > R > r_1$. There is little difference in the two types except that the return nozzle gives a smaller flow rate for a given nozzle size because of the lower pressure head at the smaller radius. The laboratory machine discussed previously uses return nozzles with a minimum orifice size of 0.3 mm. Two such nozzles give a flow rate of the order of 3 cm^3/sec; therefore the maximum concentration ratio is roughly 20:1 for the example given.

In some solid-wall clarifier bowls the disc stack is replaced by two concentric chambers, the outer having the radius of the bowl wall and the inner one being of smaller radius. Feed enters the inner chamber, then the outer chamber, before discharge. Because of the depth of the liquid layer, the efficiency of this type of bowl is lower than that of a disc stack or tubular centrifuge. The Σ value may be estimated as the sum of the Σ values for the two chambers, where each is computed from (6.41).

The design of the liquid-liquid separator differs in two respects from that of a clarifier bowl. Since each layer must be clarified, the feed enters the disc stack from the bottom through a ring of holes piercing the discs and forming feed channels vertically through the stack. The liquid streams are discharged at the top of the bowl into separate covers. Solids, if any are present, form a cake at the bowl wall. For maximum efficiency, the interface should be maintained at the radius of the center line of the holes by proper selection of the heavy phase discharge dam. In disc machines designed for a specific separating problem, the radius of the location of the disc holes is selected to give the required separation in the phase representing the more difficult problem. In an all-purpose laboratory machine the holes are usually located at from one-half to two-thirds of the disc width. In the centrifuge of Fig. 6.21 the holes are halfway out; other dimensions for the separator bowl are $n = 25$, inner radius of disc $= 2.4$ cm, radius of the center line of the holes $= 3.5$ cm, outer radius of disc $= 4.6$ cm, $\theta = 45°$, and speed is 12,000 rpm.

In evaluating separator bowls, a Σ value is computed separately for each layer by (6.43); the light phase is assumed to fill the stack inward from the centerline of the holes, and the heavy layer extends outward from that radius. Thus for the dimensions of the bowl just given, we have for the light layer, $r_1 = 2.4$ cm and $r_2 = 3.5$ cm, so that $\Sigma_L = 24.6 \times 10^5$ cm^2; and for the heavy layer, $r_1 = 3.5$ cm and $r_2 = 4.6$ cm, so that $\Sigma_H = 45.9 \times 10^5$ cm^2. Throughput data are determined for the phase of greater importance. In general, the total capacity for the separator above is reported not to exceed 70 ml/sec.

Conveyor Discharge Centrifuges

The conveyor discharge centrifuge of Fig. 6.22 sediments solids from a continuous flow of feed slurry and discharges these solids continuously by

Fig. 6.22. The P-660 continuous conveyor discharge centrifuge, 6-in. diameter. Courtesy of Sharples Division, Pennwalt Corporation.

means of an internal helix that is caused to rotate at a speed different from, and usually lower than, that of the rotor itself (Fig. 6.23). It develops up to about 3000 G at the bowl wall and is usually limited to the removal and dewatering of solids above the 1 to 2 μ range. Many of the smaller particles report to the effluent which, in some applications, is further clarified on tubular or disc centrifuges or on a polishing filter. The concentration of solids in the feed may vary over wide limits without affecting the general performance, provided the bulk solids handling capacity of the conveyor is not exceeded. The smallest units with internal rotor dimensions of $r_2 = 7.6$ cm and overall length of 26 to 33 cm are more often used for pilot plant or small-scale industrial service or for obtaining data for scaling to commercial sizes than for bench-top work.

Originally, rotors were shaped like truncated cones, and the solids were conveyed "up" the cone, out of the liquid layer, and over a dry beach section for drainage before discharge at the small end. The modern design adds a cylindrical section to the large end of the cone, for increased retention time under centrifugal acceleration and correspondingly improved solids capture. The centrifuged liquid phase discharges at the end away from the cone through overflow ports that are adjustable to control the free liquid surface and, consequently, the pond depth and dry beach length within the bowl. The feed is usually introduced through the conveyor hub to the junction of the conical and cylindrical sections to maximize the clarification length.

Although it represents a great oversimplification of the flow pattern within this type of rotor, (6.41) for the Σ value of a tubular bowl supplies an excellent approximation for comparing at equal performance the capacities of

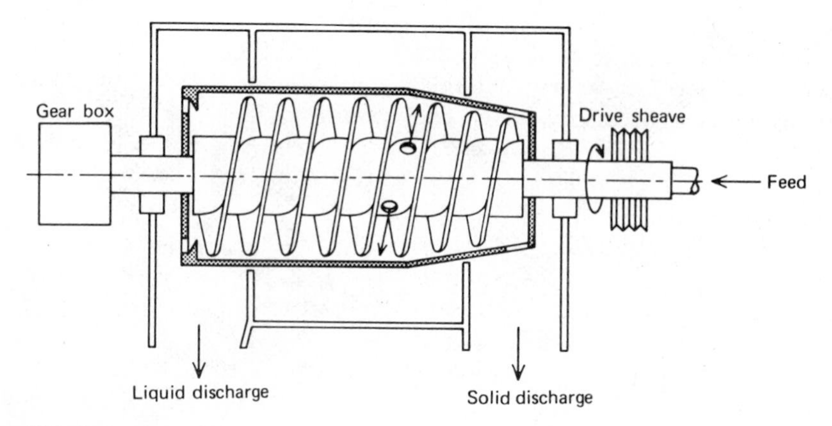

Fig. 6.23. Operating schematic of horizontal conveyor discharge centrifuge. Courtesy of Sharples Division, Pennwalt Corporation.

conveyor discharge centrifuges of different sizes and rotational speeds. The effective length Z is the mean distance between the feed ports in the conveyor and the liquid overflow weirs.

The critical point in the conveyance of solids is their transit across the dry beach inward of the free liquid surface. By raising the pond level to approximately the radius of the solids discharge, this critical zone is eliminated and a much wider variety of solids can be satisfactorily discharged from the rotor, including cellular matter such as yeast and biomass from aerobic treatment of wastewater. Raising the pond level also provides increased retention time for compaction of sedimented solids and minimizes turbulence. Sometimes flow rate for a given proportion of solids captured and dryness of discharged solids can be substantially improved by treating the feed inside the rotor with suitable agglomerating or flocculating agents such as polyelectrolytes. Reducing the differential speed between rotor and the conveyor may also help scrolling, as well as reducing associated turbulence, but with a corresponding loss in compacted solids handling capacity. When centrifuging easily conveyed, relatively granular solids, the length of dry beach available for drainage can be increased by lowering the pond level.

The dimensions of the pilot plant unit illustrated are Z (effective clarifying length) $= 17.9$ cm, $r_2 = 7.6$ cm, $r_1 = 6.5$ cm, and speed is 6000 rpm ($\omega = 628$). The Σ value from (6.41) is 0.246×10^7 cm^2. If it is assumed that the solids used in the previous examples are scrollable, then for $Q/\Sigma = 0.338 \times 10^{-4}$ cm/sec, the theoretical capacity is 83 ml/sec. Centrifuges of this type have efficiencies ranging from 65 % (on clay) down to 15 % on solids that are difficult to scroll; thus the practical value on this application is probably about 40 ml/sec, provided solids loading is not a critical factor.

Basket Centrifuges

For bottle centrifuges of some designs, interchangeable perforate or imperforate basket rotors with suitable filtrate or effluent collection chambers are available as accessory heads. The cylindrical rotor with a lip ring at the top and a solid bottom is mounted on top of the drive spindle. Heavier duty pilot plant basket centrifuges in sizes upward from 12-in. rotor diameters are available as separate units (Fig. 6.24).

In the perforate basket, solids are supported and retained on a porous medium through which the liquid fraction is free to pass under centrifugal acceleration. This type is used to separate from mother liquor the crystalline and fibrous solids that form a porous cake. Washing of the formed and drained cake with an appropriate solvent is usually efficient. The bowl shell has relatively large perforations to permit free escape of liquid. The basket is enclosed in a collecting chamber with a drain for continuous discharge of the filtrate. The solids are supported on a porous liner consisting of a fine mesh

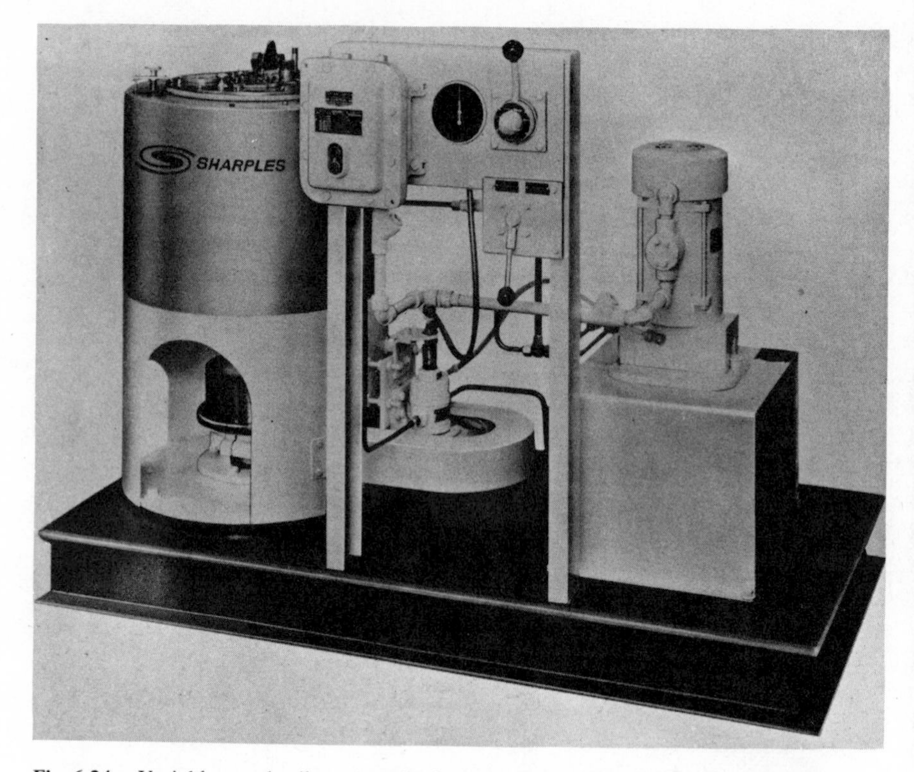

Fig. 6.24. Variable-speed pilot plant basket centrifuge. Courtesy of Sharples Division, Pennwalt Corporation.

filter screen or cloth, or filter paper supported in turn by a coarser mesh screen to furnish transverse drainage to the holes in the bowl shell. The porosity and other characteristics of the cake determine the time required to load, rinse, and drain to the required dryness. Laboratory perforate basket centrifugal filters range from 5 to 14 in. diameter, with a Z/D ratio of from 0.5 to 0.8. Most of the data required to scale to commercial centrifugal filters can be obtained from 12 or 14 in. units.

In general, the flow rate through a porous cake is

$$Q = \frac{\pi \rho_L \omega^2 Z(r_2^2 - r_1^2)}{\mu \alpha \ln(r_2/r_c)} \tag{6.44}$$

where α is specific resistance and r_c refers to cake surface. If the cake is compressible, the exponent of ω will be less than 2 and that of $[(r_2^2 - r_1^2)/\ln(r_2/r_c)]$ may be greater than 1.0. Prediction of the amount of liquid adhering to the solids after a finite spin time is impossible without experimental data. As time approaches infinity, a useful relationship for the volume ratio of residual liquid to solids is

$$S = k / \; d^{0.5}\left(\frac{\omega^2 r}{g}\right)^{0.5} \rho_L^{0.25} \tag{6.45}$$

Solid wall rotors are used for the accumulation of solids batchwise from a continuous flow of feed slurry with continuous discharge of the lip ring overflow. Suspensions containing very fine particles or gummy or gelatinous solids cannot be filtered readily because of rapid clogging of the filter medium or lack of porosity in the cake itself. A solid-wall rotor may be satisfactory for recovering such solids but does not give as dry a cake as the centrifugal filter. The size of the lip at the top of the bowl determines the volume available for holdup; in some models an adjustable skimmer dipping into the liquid surface is used to remove the clarified supernate, permitting variation of the holdup volume. Feed accelerators or radially disposed baffles are desirable to keep the liquid rotating at bowl speed and to minimize disturbance of the settled solids during deceleration. Feed is introduced through a tube extending to the bottom of the bowl from which it is directed to the liquid layer. Since solids sediment through the liquid layer as it moves toward the top of the bowl, the pattern is much the same as for the clarifier bowl of the tubular centrifuge. The same theory applies, but the Z dimension must be reduced by the vertical distance required to accelerate the incoming feed to rotor speed.

The advantage of solid-wall basket over the tubular bowl centrifuge is its larger solids holding capacity and easy access for removing the accumulated solids. It operates at much lower acceleration level with G about 1800 and its capacity is restricted accordingly.

The dimensions of two sizes of laboratory solid wall basket centrifuges are given in Table 6.1.

Table 6.1

Dimension	*Small Size*	*Large Size*
Vertical height Z (in.)	3	6
Inner radius of lip r_1 (in.)	2.5	4.1
Bowl wall radius r_2 (in.)	4	7
G at r_2	1390	1800
Σ (cm^2)	5.7×10^5	25.5×10^5

Referring again to the example in which satisfactory clarity is obtained at $Q/\Sigma = 0.338 \times 10^{-4}$ cm/sec, the theoretical throughputs of the smaller and larger baskets are 19.3 and 86.2 ml/sec, respectively. The efficiency factor for solid basket centrifuges is in the range of 50 to 80 %, and these rates would be correspondingly reduced in practice.

4 USES OF CENTRIFUGES

Many of the techniques of preparative and analytical chemistry are based on centrifuges. The use of these devices is often a matter of convenience and saving in time, but in other cases, such as the concentration of nonfilterable material, their use is a necessity. Their application under carefully controlled conditions is specified for certain Standards of the American Society for Testing and Materials and other official procedures in the field of petroleum, food, and general analysis, clinical medicine, public health, and particle size determination.

In practice the techniques involved in these tests are frequently applied to commercial operation. Pilot plant and laboratory tests provide basic data for the transition to full-scale operation.

5 OPERATING AND MAINTENANCE

Balancing

Every rotating body tends to vibrate at certain speeds regardless of how carefully it has been balanced. These critical speeds occur when the rotational speed reaches the naturally occurring frequency of the assembly, and they are accentuated by stress distortion, errors in loading, and normal manufacturing tolerances.

Commercial centrifuges are either large in diameter with respect to height, like the laboratory bottle centrifuge, or small in diameter with respect to length, like the tubular bowl centrifuge. Both are inherently stable forms.

To permit centrifuges to pass freely through their critical speeds, some degree of flexibility in their mounting must be provided. In addition, this flexibility permits the rotating body to find its desired axis of rotation about its center of mass, which may differ from its geometric axis. The energy absorbed by a fixed mount without such flexibility would be greater than the driving power normally provided, and the assembly would not reach its rated speed. Various systems are used to provide this required flexibility. In small centrifuges the entire machine may be mounted on rubber, but only rotating parts may be spring or rubber mounted in larger sizes. In the suspended bowl tubular SuperCentrifuge, flexibility is provided by cushioned mounting of the drive and flexibility of the spindle, even though the frame is rigidly mounted on the laboratory bench.

The disc centrifuge is usually underdriven, the weight being carried on a fixed thrust bearing at the lower end of the drive spindle. The required flexibility is provided by a radial bearing that is either spring or rubber cushioned and is located immediately below the bowl assembly. In smaller sizes disc centrifuges are also rigidly mounted on the working surface. Larger disc centrifuges are usually mounted on flexible base cushions. Since centrifuges of the conveyor discharge type are commonly designed to be operated below their critical speed, rubber base cushions provide the required absorption. Continuous centrifuges are essentially self-balancing under normal operating conditions. Certain precautions, however, must be observed in loading the laboratory bottle centrifuge. It is adequate to fill tubes with a capacity of 20 ml or less to the same height by visual inspection. Tubes carrying similar loads should be placed 180° apart in the rotor head, and the annulus around each should be filled with water or other suitable liquid if the centrifuge is to be operated at a speed of 2000 rpm or higher. Tubes of 50-ml capacity or more should be balanced more carefully, and those to be placed opposite each other in the rotor should be balanced to within 0.25 gram. The recommended procedure is to place the glass tube in the metal shield and use the trunnion ring to support them on the balance, one assembly on each balance pan, for loading the glass tubes with equal weights of sample. With larger tubes it is always advisable to use a supporting fluid. The loaded and balanced assemblies should be placed in the rotor head in such a manner that each balanced pair is at opposite ends of a common diameter.

Similar precautions should be observed in loading angle centrifuge heads; in the case of the preparatory head of the high-speed ultracentrifuge, extreme care should be exercised in balancing diametrically opposite tubes.

Operation

After the rotor head is properly loaded, the centrifuge cover should be closed and latched. For centrifuges equipped with temperature control devices, the preliminary adjustment should have been made previously, and final adjustment is performed at this time. With the rheostat or speed at zero or lowest speed position, the power is turned on. The speed control handle can then be advanced slowly until the desired speed is attained.

There are two reasons for this procedure:

1. To limit the current inrush to the motor at the start.
2. To prevent the swinging tubes from oscillating violently and possibly expelling their contents or even breaking.

Larger centrifuges of this type are frequently equipped with a built-in relay to prevent their being started unless the controller handle is in the zero position. They are also frequently supplied with a tachometer to provide visual indication of the operating speed.

Although many centrifuge rotor assemblies can be operated at the full speed of the centrifuge, others cannot because of stress considerations, mentioned earlier, and the manufacturer's recommendation of permissible speeds at various loading should be carefully adhered to if damage to the equipment is to be avoided.

The manner in which the centrifuge should be stopped depends on the nature of the sedimented phase. Considerable agitation at the interface occurs when the acceleration or deceleration is rapid. This usually does no harm during starting. During stopping, however, considerable remixing may occur, particularly if the sediment is a relatively light flocculent material that does not pack well. In such a case it is necessary to stop the centrifuge very slowly. Some centrifuges are equipped with belt tension releases, and other devices take as much drag off the shaft as possible during this period. Such precautions are unnecessary if the sediment packs down as a hard layer, and the centrifuge may be braked to a stop in these cases.

In the basket type of centrifuge a perforate basket should be lined with an appropriate filter medium supported on a coarser screen. Unless it is equipped with locking rings or grooves to seal the filter medium in place, it should be rotated at a slow speed and thoroughly wetted with some of the slurry to set the filter medium in place. The speed can then be increased and the suspension added at a rate that will cause it to overflow the lip ring. The solids are retained in the basket and the filtrate passes to the outer casing, from which it can be caught in a container placed under the spout. When the sample has been loaded or the basket approximately filled with solids, the feed is stopped and the collected solids are allowed to drain at maximum centrifugal force. When washing of the cake is desired, the wash medium should be distributed as accurately as possible over the entire surface of the cake. The wash may be introduced in any convenient manner (a perforated spray pipe, manually controlled single spray or nozzle, etc.). In small centrifuges a wash bottle can be used.

For maximum washing efficiency the rinse should be started before the mother liquor has left the cake. Otherwise the cake may crack and offer a convenient channel for the passage of the wash liquor without useful work. The formation of a smooth cake of uniform thickness throughout the basket promotes the efficiency of the rinsing operation. This can usually be accomplished by control of the centrifuge speed and manipulation of the slurry feed during the loading operation. After spinning off the mother or rinse liquor, the centrifuge is stopped and the basket unloaded.

A solid basket with vanes is first brought up to the desired speed; then the feed slurry slowly poured in, as close to the bottom as possible. The bowl gradually fills, and the clarified supernatant liquor overflows the lip ring into the outer chamber. For successful operation, the feed should enter the bottom of the bowl at such a rate that the desired separation has been completed when the liquid reaches the top and starts to overflow. At the completion of the run, the centrifuge is stopped and the bowl is unloaded.

Similar considerations apply to the operation of the disc-type centrifuge except that the assembly makes provision for the introduction of the feed stream to the proper point in the rotor. The tubular bowl centrifuge is also always brought to the desired operating speed before the feed is introduced. Its operation differs in one very important particular. In the basket and disc centrifuges the size of the feed nozzle is not critical. The tubular bowl is fed from the bottom, however, and since there is no mechanical connection between the feed nozzle and the rotor, the feed must leave the nozzle with sufficient upward velocity to jet into the accelerating blades with which this type of bowl is equipped. In general, a feed nozzle size must be selected that will provide a standing jet at least 15 cm high at the desired operating rate.

Centrifuges of the tubular bowl type used for the separation of immiscible liquids are equipped with "ring dams" for direct control of the depth of the heavy phase liquid within the bowl, and for indirect control of the depth of the light phase liquid, since the total liquid depth is a fixed value of the bowl itself. The larger the opening of the ring dam, the shallower the depth of the heavy phase liquid layer and the lower its respective Σ value. Since the total Σ value of the bowl remains constant for a given rotational speed, this means that larger ring dams give greater Σ value to the portion of the bowl that is working on the lighter phase. When the feed mixture or emulsion contains less than 50 % by volume of the heavier liquid phase, the bowl should be primed with enough heavy phase liquid to seal off the outer annulus before continuous separation is attempted.

Disc-type centrifugal separators are equipped with ring dam control, and it is essential that the adjustment be made so that the interface falls within the holes in the disc stack for maximum efficiency. The conveyor discharge centrifuge should also be brought to operating speed before feed is introduced.

Electric Motor and Other Drives

Most centrifuges are now driven by electric motors. The principal exception is that certain models of the tubular bowl SuperCentrifuge are driven by air or steam turbine at speeds up to 50,000 rpm. For the latter, air or steam at 40 psig is required, and speed may be adjusted to any lower limit by controlling the air or steam pressure to a corresponding lower value.

Speed regulation during startup and control during operation are desirable for most motor-driven laboratory centrifuges. For centrifuges requiring not more than 0.75 horsepower, universal series or compound wound motors are usually supplied. The speed of these may be regulated by a variable rheostat connected on one side of the line or by a Variac transformer for alternating current only. In either case the speed regulation is accomplished by reducing line voltage. These universal motors can be used with direct current or any cycle alternating current. Since their speed is a function of the applied voltage, care should be taken that the line voltage corresponds to the nameplate reading of the motor.

For direct current motors larger than 0.75 horsepower, rheostat control of the speed is also used. Either the brush-shifting repulsion type, in which speed is regulated by adjusting the brushes, or the capacitor type with autotransformer control in the primary circuit is satisfactory for single-phase motors in sizes larger than 0.75 horsepower. Many continuous centrifuges that are normally driven at some fixed speed can be regulated to some extent by varying the ratio of the driving to the driven pulley, or by interposing a variable-speed belt drive between the two. The use of polyphase alternating current motors on laboratory or pilot plant centrifuges is comparatively infrequent. When speed adjustment of this type is required, the same method is applicable.

Care of Centrifuges

Most centrifuge troubles are connected with (1) improper lubrication; (2) unbalance of the rotating parts due to mechanical damage or distortion, corrosion, unequal loading, or inadequate cleaning of the rotor; (3) poor condition of the mounting and drive, particularly with respect to bearings, bushings, and rubber cushions; or (4) poor condition of the motor, particularly with respect to bearings, brushes, and commutator. Lubrication troubles arise from too much as well as too little lubrication. If bearings run dry or hot, wear and damage to adjacent parts from overheating may result, and replacement of parts may be required. Overlubrication may result in seepage of excess oil or grease into the motor, onto the surrounding working area, or even into the centrifuged product. Oil on motor parts collects dust that may lead to grounding and causes deterioration on motor windings. It is always advisable to follow exactly the manufacturer's recommendations on lubrication.

The proper balancing of the working load has already been discussed. Even minor unbalance in the rotating system may interfere with quantitative results or cause sufficient vibration to damage the centrifuge. When damage has occurred, parts should be replaced, or the equipment must be returned to the supplier for repair and rebalancing.

Continuous centrifuge rotors generally need thorough cleaning after each use. Even a one-gram undistributed mass of dirt in a tubular bowl results in an eccentric load of more than 60 kg on the rotor when it is revolving at full speed. The bushings, bearings, and belts of all centrifuges should be inspected periodically for wear and corrosion, particularly if the centrifuge is used only intermittently, since the oil film on the bearings may become contaminated with dust or condensed moisture. These parts should be replaced periodically before they wear out completely or fail. It is false economy to risk a valuable experiment in the hope that worn parts will last "just one more run."

Centrifuges equipped with rubber base pads, cushions, or vibration isolators should be inspected periodically for deterioration from oil, solvent, or age. When such conditions are found, the worn parts should be replaced promptly. Many laboratory centrifuge motors are of the universal type with brushes and commutators. Commutators should have smooth, cylindrical surfaces, free from scratches and grooves. If they become somewhat roughened or scratched, fine sandpaper no. 0 or 00, but not emery cloth, may be used to smooth and polish them while the centrifuge runs at low speed. If the commutator is markedly cut or grooved, the armature should be polished in a lathe. Only the brushes recommended by the manufacturer should be used; replacement brushes should be properly seated with a brush-seating stone while the centrifuge runs at low speed. All motors not of the commutator type should undergo periodic checks for loose connections.

The most important factor in maintenance of centrifuges is often overlooked: they must be kept clean. Broken glass, spilled samples, and surplus lubricant and dust should be removed as part of a regular preventive maintenance program. When glass breakage occurs in a bottle centrifuge, the rubber cushions should be removed from the metal shields and thoroughly cleaned because embedded glass particles may create point stress on succeeding glass tubes and cause them to fail. The highly stressed parts of most centrifuges are of metal construction, and corrosion should be avoided. When this is impossible, the stressed parts should be inspected after each use and replaced whenever weakness is suspected.

SYMBOLS AND NOMENCLATURE

A area
a acceleration
C factor for fluid flow pattern between discs
c concentration of sedimenting material
D centrifuge rotor diameter, or diffusion coefficient

d	diameter of particle or drop
F	force
G	relative centrifugal acceleration $(\omega^2 r/g)$
g	acceleration of gravity
h	height
I	subscript referring to interface
K	shape factor
k	constant or slope
L	subscript referring to liquid phase
M	molecular weight
m	mass (m_P, mass of particle; m_L, mass of liquid it displaces)
n	number of disc spaces in disc stack
P	performance index
P	subscript referring to particle (solid or drop)
p	pressure
Q	flow rate (volumetric)
r	radius
r_1	radius, free liquid surface or interface
r_2	radius, bowl wall or boundary layer
r_c	radius of cake surface
R	gas constant
S	volume liquid/volume of solid
s	radial distance, or sedimentation velocity in the Svedberg system
T	absolute temperature
T	subscript referring to total centrifuging time
t	time
V	volume
v	velocity
v_s	velocity of sedimentation
v_g	velocity in gravitational field
x	distance
Z	distance parallel to axis of rotation, as in effective sedimentation length of cylindrical bowl
α	specific cake resistance
$\Delta\rho$	difference in density, as of particle minus fluid
θ	half of included disc angle

μ absolute viscosity
ρ density
Σ equivalent area of sedimentation centrifuge
ω angular velocity (rads/sec)

References

1. E. S. Pettyjohn and E. B. Christiansen, *Chem. Eng. Prog.*, **44**, 157 (1948).
2. J. H. Perry, Ed., *Chemical Engineer's Handbook*, 4th ed., McGraw-Hill, New York, 1963, pp. 5–59.
3. G. H. Neale and W. K. Nader, *Am. Inst. Chem. Eng. J.*, **20**, 531 (1974).
4. F. C. Bond, *Chem. Eng.*, **61**, 195 (1954).
5. R. J. Casciato, *Am. Lab.*, **53** (July 1969).
6. C. R. McEwan et al., *Anal. Biochem.*, **23**, 369 (1968).
7. E. G. Pickels, *Methods Med. Res.*, **5**, 107 (1952).
8. G. B. Cline, *Progress in Separation and Purification*, E. S. Perry and C. F. van Oss, Eds., Wiley, New York, 1971, pp. 299–306.
9. O. M. Griffith, *Techniques of Preparative, Zonal, and Continuous Flow Ultracentrifugation*, Spinco Division of Beckman Instruments, Inc., Palo Alto, California, 1975.
10. T. Svedberg and K. O. Pedersen, *The Ultracentrifuge*, Oxford University Press, New York, 1940.
11. H. K. Schachman, *Ultracentrifugation in Biochemistry*, Academic Press, New York, 1959.
12. R. Trautman, "Ultracentrifugation," in *Instrumental Methods in Experimental Biology*, D. W. Newman, Ed., Macmillan, New York, 1964, pp. 211–297.
13. C. M. Ambler, "Centrifuges" in *Chemical Engineers' Handbook*, 4th ed., J. H. Perry, Ed., McGraw-Hill, New York, 1963, pp. 19/86–100.
14. A. C. Lavanchy and F. W. Keith, Jr., "Centrifugal Separation," in *Encyclopedia of Chemical Technology*, Vol. 4, 2nd ed., A. Standen, Ed., Wiley-Interscience, New York, 1956, pp. 563–606.
15. T. Lee and C. W. Weber, *Anal. Chem.*, **39**, 620 (1967).

IV

Chapter **VII**

FILTRATION AS A LABORATORY TOOL

C. S. Oulman
E. R. Baumann

1 Introduction 349
General 349
Aims of Filtration 350
Materials To Be Filtered 351
2 Theory 353
Limitations of Theory 353
Flow Through Porous Media 353
Cake Filtration 358
Granular Bed Filtration 360
Chemical Filter Aids 364
3 Apparatus 365
Types of Filter 365
Pressure and Area 365
4 Filter Media 367
General 367
Filter Paper 368
Membranes 374
Woven Filter Media 388
Perforated Metals 408
Loose Granular Materials 410
Sources of Information 421

1 INTRODUCTION

General

There are many chemical procedures in which solids must be separated from a liquid, or vice versa. Filtration, one frequently employed process for effecting the separation, can be used either to improve the clarity of a liquid by removing suspended solids from it or to remove excess liquids from solids. Usually, however, both results cannot be obtained in the same filtration.

Filtration may be defined as the process in which particles are separated from a liquid by passing the liquid through a permeable material. The *filter medium* is the permeable material that separates particles from the liquid passing through it. The device used to carry out the filtration process is called a *filter*. It consists of the filter medium and the associated hardware that supports the filter medium in the flow pathway.

Most of the filter media used for laboratory procedures can be classified in five categories: papers, membranes, loose materials, woven fabrics, and rigid, porous media. Filter paper is the filter medium most often used in laboratory procedures. Membranes are often used when the particles suspended in the liquid are too small to be trapped by filter paper. Loose materials such as diatomaceous earth are sometimes added to a suspension that is difficult to filter, to speed up the filtration while maintaining filtrate clarity. When woven fabrics and rigid, porous media are the materials used as filter media, they can be reused. Frequently woven fabrics and rigid, porous media act as a support for another material such as loose material that serves as the filter medium. The other material used as a filter medium may even consist of the particles being removed from the liquid as the particles build up into a filter cake.

The solid-liquid material to be filtered is variously referred to as the *suspension, slurry, liquor, sludge, dispersion, prefilt, feed,* or *influent*. Although the terms "suspension" and "sludge" would not be used interchangeably in describing a given material, all the terms designate a material containing particles and liquid that may be subjected to filtration. The particles retained on the filter may be referred to in general as a *residue* or, if accumulated with some thickness in the form of a more or less homogeneous deposit, as a *cake*. The liquid passing out of the filter is called the *filtrate* or, when large quantities are produced, the *effluent*.

Aims of Filtration

There are really only two aims in employing filtration as part of a chemical procedure. One is to remove and recover solids from a liquid, and the other is to remove and recover the liquid from a solid. In part, the kind of filter that is used and the filtration procedure that is needed depend on the specific aim of the filtration.

The most common aim of filtration is that of obtaining liquid free of solids. In many applications the objectionable materials are large particles of dirt, fibers, metallic scale, or other extraneous matter with which the liquid was contaminated. Usually only a small amount of impurity is present, and it is easy to remove. A similar situation arises when adsorbents or decolorizing agents such as activated carbon have been used in treating a liquid.

In other applications it is much more difficult to obtain the desired clarity of filtrate. When the particle size of the suspended matter is small and the suspended matter concentration is low, it may be difficult or impossible to clarify the liquid by direct filtration. The addition of a filter aid such as diatomaceous earth to the suspension may be helpful in trapping the solids without unduly restricting the passage of liquid.

The other aim of filtration (i.e., to free solids of liquid) arises in a variety of applications in preparative and analytical chemistry. At the least complicated level filtration may be a simple, economical, and convenient preliminary step in the drying of many materials. It is also a common method of separating a precipitate from solution. Filtering techniques are also convenient for washing or leaching solids to remove impurities that are more soluble than the solids retained on the filter.

Filtration apparatus also can be used in transferring solids from one liquid to another. Crystals from aqueous suspension can be drained as dry as possible by filtration, then treated with acetone or another water-miscible liquid. The acetone replacement technique of changing from an aqueous to a nonaqueous liquid, or vice versa, is conveniently accomplished when the solids are supported on a filter.

Materials to be Filtered

Both the particles and the liquids in which they are suspended, as well as the way that they interact, contribute to the behavior of materials being filtered. The particles can be classified according to size, shape, rigidity, and concentration, and liquids according to viscosity, density, surface tension, dielectric constant, and other intrinsic properties. Classification of the mode of interaction between particles and liquids in which they are suspended may be according to the stability or lack of stability of the suspension.

In general it can be said that small particles are more difficult to filter than large ones. The filter media needed to trap small particles have inherently more resistance to flow than those needed to trap large particles, and filter cakes formed from small particles contain void spaces that are smaller and less permeable to flow than are those of large particles.

Mixtures of particle sizes respond differently depending on the properties of the particle size distribution itself. A continuous particle size distribution ranging from large to small particles may be filtered easily on a filter medium that is only as fine as it must be to retain the larger particles in the distribution. The finer particles will be trapped within the voids between solids that are deposited on the filter medium. A bimodal particle size distribution may be much harder to filter because clarity of filtrate is controlled by the finer range in particle sizes. The filter medium needed to retain particles of that size range may be quickly clogged by the accumulation of what could be a large mass of coarser solids from the other end of the particle size distribution.

Particle shape is related to surface area per unit volume and for a given volume of particle, the one with the greatest surface area will offer the greatest resistance to flow. On the other hand, some irregularly shaped particles with reentrant surfaces form much more porous cakes than do those formed by compact spherical particles, with the result that the permeability is increased. Still other particles with flat, platelike structures have a relatively high surface area–volume ratio, but they nest together in a filter cake tightly enough to become almost impermeable.

All other things being equal, rigid particles are the easiest to filter and compressible particles are the hardest to filter. As particles compress, they may close off virtually all the interstices that carry away liquid. Generally, the difficulty of filtration increases with particle concentration, but it is often difficult to filter a very dilute suspension, too. When the concentration is high, the filter cake builds up quickly and becomes resistant to flow. When the concentration is low, some solids tend to penetrate into or through the filter medium instead of bridging over the passageways; as a result, the filter medium either is clogged quickly or fails to provide the desired clarity.

The most noticeable property of liquids that affects filtration is viscosity. To a first approximation, the resistance to filtration is directly proportional to the viscosity of the liquid. Because hot water is less viscous than cold water, many filtration operations are speeded up when carried out at a high rather than a low temperature.

Other properties of the liquid such as density have a much more subtle effect because the values for density vary over a narrower range than do those for viscosity. Still other properties of liquids such as surface tension and dielectric constant have effects that are only intuitively obvious, and then only with respect to the way the liquid will potentially interact with particles of varying composition. Surface tension affects the strength of the molecular attraction that the liquid has for molecules in the solid surface. Since the dielectric constants of the liquid and the molecules at the surface of the particles affect the accumulation of electrical charge that can build up at the interface, and that, in turn, affects the attraction that the molecules of the liquid will have for the molecules at the solid surface, the dielectric constant is an important property of the liquid.

Finally, the way that the liquid does interact with the solid, or vice versa, is a factor that must be considered in classifying the material to be filtered. Some suspensions are unstable; that is, left to themselves, they will separate into a solid phase and a liquid phase. Other suspensions are stable and as a practical matter would not separate, at least not over a relatively long period of time.

Unstable suspensions are inherently easier to filter than stable suspensions. The object, in fact, of many of the pretreatment procedures leading up to filtration is to form an unstable suspension. Viewed in this way, digestion of a precipitate to increase the particle size of large crystals at the expense of small crystals has the effect of decreasing the amount of water that is associated with the particles at the liquid-solid interface. The addition of a small amount of gelatin or some other hydrocolloid around which particles can form an envelope squeezes out water that would otherwise be associated with particle surfaces. The addition of an electrolyte may increase the ionic concentration in the vicinity of a hydrocolloid surface and "salt out" solids by separating the colloid from the water that was immobilized in a surrounding envelope around an extended hydrocolloid. Because of the strong attraction of water molecules for other water molecules, suspensions may be destabilized by reducing the attraction of water for the molecules at the surface of the particles—the water then squeezes out the particles of suspended solids.

2 THEORY

Limitations of Theory

The bench chemist, who is not likely to make much use of filtration equations, may be predisposed to consider filtration theory as a form of mathematical gymnastics. Mathematical models, however, do provide a means of showing apparent relationships between variables in a process. To the extent that models help people visualize what is going on, they may be valuable decision-making tools in the selection of apparatus and techniques for a particular filtration application.

The main deficiency in current filtration theory is its preoccupation with resistance to flow, almost to the exclusion of considerations of filtrate quality. It is fairly easy, and at least possible, to estimate the resistance to flow of a clean filter medium. It is usually not possible to estimate with comparable accuracy what the resistance will be as the filter begins to trap solids. And except in the most trivial cases, it is impossible to predict what filtrate quality will be obtained without making a trial filtration. Filtration is still more art than science, but the astute worker can hardly afford to ignore such guidance as can be provided by theory as it exists today.

Flow Through Porous Media

Filtration, as noted earlier, is a process in which particles are separated from a liquid by passing the liquid through a permeable material. The resistance to flow through a porous medium has been the subject of research for many years. The simplest expression relating filtration variables to each other can be stated as follows: the energy lost in filtration is proportional to the rate of flow per unit area. Symbolically, this is

$$H = R\left(\frac{Q}{A}\right) \tag{7.1}$$

where H is the head loss across the filter, Q/A is the flow rate per unit area, and R is a proportionality constant for the system called resistance.

A simple model that has been used to describe flow through a porous medium assumes that the interstices conducting flow through the porous medium correspond to an assembly of conduits. The head loss through a conduit is proportional to the velocity head ($u^2/2g$). The velocity head is the height to which liquid would rise if projected upward against gravity with the stated velocity—that is, if its kinetic energy were converted to potential energy. It is observed that the relationship between head loss and velocity head is

$$H = f\left(\frac{L}{d}\right)\frac{u^2}{2g} \tag{7.2}$$

where f is a friction factor, L is the length of the conduit, and d is its diameter. This equation indicates that the head loss is directly proportional to the length of the conduit and inversely proportional to its diameter. The friction factor can be related to a dimensionless combination of variables called the Reynolds number:

$$N_{\mathrm{RE}} = \frac{\rho d u}{\mu} \tag{7.3}$$

where N_{RE} is the Reynolds number, ρ the liquid density, u the liquid velocity, and μ the liquid viscosity. Up to Reynolds numbers of about 2000, the friction factor is inversely proportional to the Reynolds number:

$$f = \frac{64}{N_{\mathrm{RE}}} \tag{7.4}$$

Substituting (7.4) in (7.2) and simplifying, we have

$$H = \frac{32\mu L u}{\rho d^2 g} \tag{7.5}$$

This is a form of Poiseuille's equation, and it is useful in describing the flow of fluids through small-diameter conduits under conditions described as laminar flow.

This model, although relatively simplistic in approach, provides much information that is useful relating to filtration. It tells us that head loss will be directly proportional to the viscosity of the fluid and the rate of fluid flow. If we can assume that the length of the flow pathways is proportional to the thickness of the porous medium, the head loss is directly proportional to the thickness of the porous medium, or in the case of cake filtration, the thickness of the filter cake.

Furthermore, although the density of a liquid does not vary to any great extent, we find that the head loss is inversely proportional to the density of the liquid. For a cylinder, the contact area of the liquid with the conduit is $\pi L d$ and the volume of the flow pathway is $\pi L d^2/4$. The ratio of contact area to the volume of the flow pathway S is therefore $4/d$. If S is substituted for $4/d$ in (7.5).

$$H = \frac{2\mu S^2 L u}{\rho g} \tag{7.6}$$

it is apparent that head loss is directly proportional to the square of the contact area to volume of flow pathway ratio for the conduit.

If a porous medium consists of spherical particles with a diameter D that are arranged in the filter medium with a porosity ε, where porosity is the ratio of void volume to total volume of voids and particles, S will be

$$S = \frac{(1 - \varepsilon)\pi D^2}{\varepsilon(\pi D^3/6)} \tag{7.7}$$

Also, if the approach velocity v of the liquid to the filter medium is used in place of the velocity in the flow pathways of the filter medium, u, we have

$$v = \varepsilon u \tag{7.8}$$

Substituting (7.7) and (7.8) in (7.6) gives

$$H = \frac{(1 - \varepsilon)^2 \mu L v}{18 g \varepsilon^3 \rho D^2} \tag{7.9}$$

which tells us that head loss is very sensitive to both particle diameter and porosity. A tenfold decrease in particle size will cause a hundredfold increase in head loss. This in part points up the value of digesting a precipitate to increase particle size before filtration.

Furthermore, spheres normally arrange themselves in a porous medium so as to have a porosity of about 0.4 at the most with a random packing. If the small spheres in a suspension to be filtered are flocculated with gelatin to produce more irregular aggregates, a more porous filter cake will be formed and the head loss, which is directly proportional to the porosity factor $(1 - \varepsilon)^2/\varepsilon^3$, will be decreased. A change in filter media (or filter cake) porosity will have the following relative effects on head loss:

Porosity, ε	0.1	0.2	0.3	0.4†	0.5†	0.9*
$\dfrac{(1-\varepsilon)^2}{\varepsilon^3} =$	810	80	18.2	5.6	2	0.0137
Relative head loss, $\varepsilon = 0.9,^* = 1$	59,049	5832	1323	409	146	1
Relative head loss, $\varepsilon = 0.4,^† = 1$	145	14.3	3.25	1	0.36	0.0024

The porosity factor is 5.6 when the porosity is 0.4, it is 2.0 when the porosity is 0.5, and it is 0.7 when the porosity is 0.6. Flocculating the particles to ensure that the cake porosity is increased from 0.4 to 0.5 effectively cuts the head loss by nearly two-thirds, and increasing the porosity still further to 0.6 cuts the head loss by nearly two-thirds again.

Deformation of the solids in a filter cake can affect both the porosity and contact area in a cake. A cake consisting of a cubical arrangement of spherical particles has a porosity of about 0.48. If the thickness of such a cake is compressed to 90 % of its initial thickness by deforming the spheres into oblate spheres without increasing the filter area, the volume of particles in the cake will remain the same and the total volume of particles and voids will decrease to about 80 % of the initial void volume. The new porosity will be 0.42. Based on the decreased porosity and thickness of cake, the head loss will increase to 1.6 times the initial head loss. Since the surface area of the oblate spheres will be greater than that of the original spherical particles, the actual head loss increase would be even greater than the indicated value. Additional deformation to 80 % of the initial volume would decrease the porosity to about 0.35 and increase the head loss to at least 3.3 times the initial value, again neglecting the increase in surface area.

Among the parameters that have been used to describe a particle size distribution are the effective size and the uniformity coefficient. The effective size of a distribution is the largest diameter of the smallest 10 % by weight of the particles, that is,

$$D_E = D_{10\%} \tag{7.10}$$

The uniformity coefficient is the ratio of the diameters of the largest and smallest particles of the half of the distribution that is just larger than the

* Approximate porosity of diatomite filter media.
† Approximate porosity of coal and sand filter media.

effective size, that is,

$$U = \frac{D_{60\%}}{D_{10\%}} \tag{7.11}$$

For a granular material whose particle size is log normally distributed as a function of the percentage by weight smaller than a given size, the value of $\log D_{10\%}$ will be 1.282 standard deviations less than the average value of $\log D$, and the value of $\log D_{60\%}$ will be 0.253 standard deviation greater than the average value of $\log D$. The standard deviation s is therefore

$$s = \frac{\log U}{1.535} \tag{7.12}$$

and the mean particle size $D_{50\%}$ is

$$D_{50\%} = \text{antilog}(\log D_{10\%} + 1.282s) \tag{7.13}$$

A granular material with an effective size of 1 μ and a uniformity coefficient of 5, would have a mean particle size of 3.84 μ. The surface area to volume ratio S^* for a granular material with a porosity of 0.4 would be

$$S^* = \frac{6(1 - \varepsilon)}{D} = \frac{3.6}{D} \tag{7.14}$$

For the distribution just described, the surface area–volume ratio would be 1.64 M^2/cm^3. The particle diameter of a unisize material with the same porosity and surface area–volume ratio would be 2.19 μ. This would be the same as the $D_{30\%}$ diameter in the granular material previously described. Both materials would have the same head loss. It is apparent, therefore, that the smaller particles control the head loss in a granular material that is log normally distributed by weight.

Summarizing, the theory relating to flow through porous media tells us that head loss is directly proportional to the filtration rate, the thickness of the filter cake, and the viscosity of the liquid being filtered when laminar flow conditions exist in the filter media or filter cake. The head loss is inversely proportional to the density of the liquid being filtered, but that probably is not of any great importance because the density of liquids subjected to filtration does not change very much anyway. Head loss is very sensitive to the particle size of the solids deposited in the cake, being inversely proportional to the square of the particle diameter. Head loss is also sensitive to factors affecting the porosity of the cake. Well-flocculated solids or solids having a high porosity in the cake will have a lower head loss than that of more poorly flocculated solids. Compression of the cake will greatly increase the head loss.

The benefits to be derived from adding a filter aid like diatomaceous earth to a suspension of compressible solids is easily explained in terms of the preceding theoretical development of flow through porous media. The diatomaceous earth removed with the compressible solids will add to the surface area of the solids in contact with the flowing liquid and the length of the flow path, therefore it will add components to the head loss that is generated during the filtration process. This can be more than compensated for by the contribution the filter aid can make to maintaining the porosity of the cake, since a material like diatomaceous earth is relatively noncompressible. Furthermore, diatomaceous earth has a relatively high porosity (about 0.90); thus to the extent that the filter aid is the dominant material in the cake, the head loss will be relatively low. If, however, too little diatomaceous earth is used, point-to-point contact of diatomaceous earth will not be obtained, the cake will still be compressible, and porosity will be lost. The head loss with too little filter aid will be greater than it would be if no filter aid were used at all, because of the added thickness of the cake. When sufficient filter aid is added to establish a relatively noncompressible cake, filtration is improved by adding more filter aid. If extremely high amounts of filter aid are added, a point of diminishing returns is reached, and the filter aid merely adds to the cake thickness without providing any compensating benefits in terms of increased cake porosity. Since such high dosages of filter aid are usually much greater than those a practical person would add to speed up filtration, this is not a matter for great concern.

Cake Filtration

When a suspension is filtered, a deposit of solids forms that varies in resistance in proportion to the thickness of the cake that builds up and also to the resistance of the filter itself. Returning to (7.1); the resistance term may be separated into two parts, one relating to the filter cloth R_F and the other to the cake that builds up R_C:

$$R = R_F + R_C \tag{7.15}$$

The total head loss H therefore has two parts—the head loss required to overcome the resistance of the filter cloth H_F and the head loss needed to overcome the resistance of the filter cake H_C:

$$H = H_F + H_C \tag{7.16}$$

For a given filter cloth, all the terms in (7.9) except for the viscosity and density of the liquid, the approach velocity, and the head loss are constant and are collectively equal to the resistance of the filter cloth:

$$H_F = \frac{R_F \mu v}{\rho} \tag{7.17}$$

For a filter cake, the average specific cake resistance α is dependent on cake thickness:

$$\alpha = \frac{(1 - \varepsilon)^2}{18g\varepsilon^3 D^2} \tag{7.18}$$

Substituting (7.18) in (7.9) for H_C gives

$$H_C = \frac{\alpha L \mu v}{\rho} \tag{7.19}$$

Combining (7.16), (7.17), and (7.19) and solving for v, we obtain

$$v = \frac{\rho H/\mu}{\alpha L + R_F} \tag{7.20}$$

If $v = Q/A$ and V is the volume of filtrate collected in a given time of filtration θ, we write

$$\frac{dV}{d\theta} = Q \tag{7.21}$$

and

$$\frac{dV}{d\theta} = \frac{\rho HA/\mu}{\alpha L + R_F} \tag{7.22}$$

If the volume of cake deposited per unit volume of filtrate is m, then

$$L = \frac{mV}{A} \tag{7.23}$$

and

$$\frac{dV}{d\theta} = \frac{\rho HA/\mu}{\alpha mV/A + R_F} \tag{7.24}$$

If this equation is integrated for a constant head loss, we have

$$\int_0^\theta d\theta = \frac{\mu}{\rho HA^2} \int_0^V \alpha mV\,dV + R_F A\,dV \tag{7.25}$$

The result of the integration will be:

$$\frac{\theta}{V} = \left(\frac{\mu m}{2\rho HA^2}\right)V + \frac{\mu R_F}{\rho HA} \tag{7.26}$$

Equation 7.26 is the equation of a straight line. If the line is extrapolated to the point where $\theta/V = 0$, for filtration test results obtained in a constant head filtration, the intercepted volume $-2AR_F/\alpha m$ represents the volume of filtrate that would have produced a cake having the same head loss as that of the filter medium support.

Seen in this way, the time of filtration varies with the square of the volume of liquid to be filtered, making allowance for the resistance of the filter medium and its support. Since the resistance of the filter medium is generally very small compared to that of the filter cake, the second term in (7.26) can be ignored. When the volume of filter cake solids per unit volume of filtrate m is high, this is usually the case. However when m is low, the solids formed on the filter medium may penetrate the voids in the filter medium, partially blinding them and making the filter medium much more resistant to flow. This does not happen at higher concentrations to as great an extent because some reorientation of particles arriving at the filter medium support would be needed to accomplish the blinding action. At higher concentrations the reorientation of particles is less likely to happen because the particles forming the cake tend to get in the way of one another. Bridging over of openings into the voids of the filter medium rather than blinding of the openings seems to prevail at higher concentrations of solids in a suspension.

Again, these statements of theory explain in part why filter aid may be helpful when filtering either very dilute suspensions or suspensions containing particles that have an inherently small particle size. A filter aid like diatomaceous earth increases the concentration of solids forming a cake, thereby preventing the blinding of the filter medium. Used in the correct manner, neither the filter medium nor the filter aid will be very resistant to flow and the filtration will take place rapidly.

Granular Bed Filtration

Unlike cake filtration, which may be used either for the recovery of suspended solids or for the clarification of the liquid, granular bed filtration is suitable only for the clarification of the liquid. Because of the limited capacity of the filter medium for the storage of suspended solids that have been removed, this kind of filter finds its primary application in polishing a liquid that has been pretreated by other processes that have a greater capacity for handling solids (e.g., chemical coagulation followed by settling).

A granular filter bed consists of up to several feet of inert granular material through which a liquid containing suspended solids may be passed; thus suspended solids are removed by the granular material, and the liquid becomes clarified. The granular materials used in the filters are ordinarily sand or sandlike materials, and particle diameters range from 0.2 to 3 or 4 mm. Some suspended solids trapped in a granular bed filter may be too large to pass between the grains of the filter medium, but in most properly designed filters, most of the suspended solids are small enough to pass through the void spaces when the filter medium is clean. Therefore some mechanism other than simple straining must be responsible for suspended solids removal. As the filter removes more and more solids, the size of the void spaces through which the liquid must pass becomes progressively smaller, and more of the removal of suspended solids may ultimately be accomplished by simple straining.

For any removal process other than simple straining, there are two steps that must be completed to bring about clarification—namely, particle transport and particle attachment. The transport step is needed to move suspended solids from the bulk of the liquid flowing past the grains of the filter medium to the surface of the grains where removal can occur. The attachment step is needed to retain the suspended solids on the surface of the grains of the filter medium; this prevents the solids from being swept away and out of the filter.

Several phenomena may either aid or hinder the transport of solids to the grains of the filter medium, including electrostatic forces, gravitational forces, and inertial forces. Attachment may be brought about by chemical or physical forces. The great difficulty of removing some suspended solids from grains of the filter medium during cleaning operations indicates that the energy of mutual attraction of the molecules in the suspended solids and in the grains of the filter medium is reasonably high once adhesion has taken place.

As suspended solids are removed by a granular bed filter, deposits are formed within the bed that restrict flow in the void spaces between the grains of the filter medium. These deposits can form anywhere along the pathways of flow through the filter bed; under ordinary circumstances, however, the downflow filter begins trapping suspended solids near the top surface of the filter and suspended solids are removed at progressively lower depths in the bed during the run.

As void spaces become clogged at a given level within the filter, the velocity of the liquid in the remaining void spaces must increase to carry the same flow. The increased velocity increases the drag forces between the liquid and the granular medium, with the result that the head loss increases during the filter run. The increased velocity also influences the deposition of solids. For example, the effectiveness of inertial forces for bringing suspended solids in contact with grains of the filter medium may be increased. On the other hand, increased velocity may carry particles past the face of a particle before gravitational forces can move the particles out of the streamlines of flow to the surface of the granular medium where attachment can take place.

At very low filtration rates the approach velocity is so low that virtually all the solids are deposited at or near the top surface of the granular bed filter. The run length under these circumstances is very short because the solids collected at the surface become the primary filter. If the deposit that forms is compressible, the void spaces close up as the run progresses and the resistance to flow becomes very great. This is essentially the same as cake filtration, discussed in the previous section; that is, the granular bed filter becomes a filter medium support when a surface deposit is formed, and the filter cake formed becomes the filtering medium.

At increasing filtration rates solids are driven more deeply into the filter bed, and although the head loss through the clean filter is increased by the higher velocity, the rate of head loss development caused by solids deposition is decreased and the filter run can be continued for a longer time. If the rate of flow is increased beyond the optimum flow rate, the head loss of the clean filter continues to increase but the rate of head loss development does not decrease enough to compensate for the initial head loss, thus shortening the run length.

The initial head loss through a granular bed filter can be predicted with reasonably good accuracy with (7.9) for a given flow rate on the basis of a sieve analysis of the filter medium, the physical properties of the liquid to be filtered, and the depth, porosity, and flow rate for the filter.

Although much effort has been expended on formulating mathematical models for the rate of head loss development in a granular bed filter, we can make now only the crudest of generalizations relating to an operating filter. However if a material balance (input − output = accumulation) is written for a differential element of a granular bed filter of thickness dz, cross-sectional area A, operated at a filtration rate per unit area of $Q/A = v$:

$$vAC\,d\theta - vA\left[C + \left(\frac{\partial C}{\partial z}\right)dz\right]d\theta = A\,dz\left(\frac{\partial \Omega}{\partial \theta}\right)d\theta + A\,dz\left(\frac{\partial C}{\partial \theta}\right)d\theta \quad (7.27)$$

or:

$$-v\left(\frac{\partial C}{\partial z}\right) = \frac{\partial \Omega}{\partial \theta} + \frac{\partial C}{\partial \theta} \quad (7.28)$$

where C is amount of suspended solids per unit filter volume and Ω is the accumulation of solids in the filter per unit filter volume. In a nonflow process where no liquid is being added or removed, the first term of (7.28) is zero and the material balance becomes

$$\frac{\partial \Omega}{\partial \theta} = -\frac{\partial C}{\partial \theta} \quad (7.29)$$

Under nonflow conditions, particles would be depleted from the suspension by either sedimentation or adsorption. These processes are probably important in the removal of suspended solids in a flow process also, but the inertia of the suspended solids as they are swept along with the liquid in a flow process probably makes attachment of the particle much more difficult. In a steady flow process, the third term ($\partial C/\partial \theta$) in (7.28) becomes negligible if the volume of the filter is small compared with the volume of liquid that is processed. The material balance then becomes

$$-v\left(\frac{\partial C}{\partial z}\right) = \frac{\partial \Omega}{\partial \theta} \tag{7.30}$$

If the rate of mass transfer in filtration r_F is expressed as

$$r_F = K(C - C^*) \tag{7.31}$$

where K is an overall mass transfer coefficient and C^* is a function of Ω and can be described as the concentration of suspended solids in equilibrium with the solids that have already been deposited, it is generally assumed that C^* is nearly zero.

The mass transfer coefficient does not appear to be constant, however. Therefore the rate of mass transfer changes throughout the filter run. When the particles of the filter medium are clean, there is little tendency for particles of suspended solids to adhere to the medium, and the rate of mass transfer is very low. As deposits form on the particles of the filter medium, additional suspended solids adhere more strongly and the rate of mass transfer increases. As the thickness of the deposit increases, however, the flow pathway in the vicinity of the deposit becomes smaller and the velocity of the liquid increases. Suspended solids carried by the liquid also move with a higher velocity, and their rate of deposition is diminished. Up to a point, the deposition of solids in the filter increases the rate of mass transfer, but beyond that point the rate decreases.

The rate of head loss development is proportional to the rate of solids deposition in the granular bed filter. A linear increase in head loss development indicates that solids are being removed in a way that serves to reduce void space throughout the depth of the filter. A sharp exponential rate of increase in head loss with increasing filtrate volume indicates that most of the solids are being removed near the top of the filter, probably as a cake of compressible solids. When this happens, it may be possible to obtain depth removal either by increasing the flow rate to drive solids more deeply into the filter bed or by increasing the particle size of the granular medium. With either of these alternatives, it may be necessary to use a chemical such as alum or an organic polymer as a filter aid to improve particle attachment.

Chemical Filter Aids

Often a difficult filtration may be helped along by the addition of a filter aid. Filter aids can be either inert materials that increase the rigidity, porosity, and permeability of a filter cake, or chemicals that react with the suspended solids to produce aggregates having improved filtering characteristics. Chemical filter aids may also cause particles of solids depositing in a filter bed to adhere more strongly to the particles of the filter medium.

Chemicals used as filter aids often are water-soluble polymers sold as flocculating agents. Some of these materials used at higher dosages, however, are also effective as dispersing agents. Water-soluble polymers can be divided into three distinct categories—natural products, modified natural products, and synthetic products. Water-soluble polymers derived from vegetable or animal sources include starches, dextrins, natural gums, alginates, casein, and gelatin. Compounds produced by modification of the chemical structure of a low-priced natural product like starch may make it a suitable flocculent equivalent to a more costly synthetic material. Although inherently more expensive than either natural or modified natural products, synthetic polymers are rapidly displacing the other materials because lower dosages may do an equally effective job.

Water-soluble polymers exhibit three kinds of electrochemical behavior: nonionic, anionic, and cationic. Polymers that do not ionize in water are classified as nonionic polymers, and most water-soluble polymers fall into this category. Anionic polymers ionize in aqueous solution to produce a polymeric species bearing a negative charge. This is the second largest category of water-soluble polymer. Only a few commercially available water-soluble polymers exhibit cationic behavior. These include acrylamide copolymers, polyethyleneimine, and derivatives of casein, starch, and guar gum.

Although at quite low concentrations the presence of a water-soluble polymer such as gelatin may "protect" a finely divided mineral suspension from the precipitating action of simple electrolytes, still lower concentrations of gelatin are effective in causing the rapid flocculation of clay. This can be explained in the following way. When the number concentration of gelatin particles is greater than that of the clay particles, the gelatin forms an envelope around the clay particles that protects them from aggregation—the protected particles have no tendency to adhere when they collide with one another because of their strong attraction for the water molecules that surround them. If the numbers of the two kinds of particle are about the same, neither can form a complete envelope around the other, and the tendency for mutual adsorption is satisfied by flocculation, which results in an extended network branching in all directions and growing until the aggregated particles can no longer remain in suspension.

3 APPARATUS

Types of Filters

Laboratory filters can be classified into a number of categories including screens and strainers, funnels, leaf filter and filter sticks, membrane filter holders, and cartridges. Sieves, screens, and strainers employ either woven fabric or perforated metal as a filter medium. Their usual application is for dry sieving of granular materials, but they may also be used for the separation of noncaking materials from liquids.

Funnels of various kinds are used as supports for filter paper and are the most frequently employed type of filtration apparatus. When a circle of filter paper is folded in half, then folded in half again, and opened into the form of a cone, the angle at the apex is 60°. The paper cone will be fully supported in an ordinary 60° funnel, and this is the usual filter–filter medium combination used for gravity filtration. For vacuum filtration the filter paper is supported flat on the bottom of a Büchner funnel or any one of a variety of vacuum filters that are available.

Leaf filters, filter sticks, and porous-bottom filter crucibles are devices for vacuum filtration. Filter sticks and porous-bottom filter crucibles employ fritted glass or other porous materials as a filter medium. Leaf filters may be covered with woven fabric or perforated metal.

Membrane filter holders are a kind of vacuum filter funnel that has been fabricated to support a membrane filter disc. They are usually obtained from the manufacturer to match the size of membrane filter disc being used. Membrane filter holders may be obtained in a variety of materials including glass, plastics, and stainless steel.

Cartridge-type filters are generally proprietary devices containing a filter element with a long service life. The cartridge ordinarily would be used in processing the water or other solvents required in laboratory work. Materials used as filter elements in a cartridge-type filter include paper, string, and fibrous pads. The filter element is designed for a single-use application—when it becomes dirty, it is removed from the filter housing and replaced with a new filter element.

Pressure and Area

Based on filtration theory, only a few factors can be controlled in the design of apparatus that will improve filtration. Among these few factors are the control of filtration pressure and filter area. In the use of an ordinary 60° funnel and filter paper, the filtration pressure is supplied by the leg of filtrate that fills the stem of the funnel, and the filter area is established by the area of contact between the liquid being filtered and the cone of filter paper. The limiting area is determined in part by the size of filter paper that is used and the amount of liquid being filtered.

The area of filter paper in contact with the liquid can be increased by using a larger funnel and proportionately larger filter paper and when large volumes of suspensions must be filtered; this is the usual means of meeting the need. Larger areas of contact can also be provided by the use of prefolded filter paper. The accordion fold that is used gives about twice the filter area and eliminates the triple thickness of paper that is inherent in the manner of folding filter paper to produce a cone. Uniform support of the filter paper and the negative pressure developed by liquid filling the stem of the funnel is lost, however, when folded filter paper is used.

The pressure applied to the filter could be increased by extending the length of the stem of the funnel, but the benefits of doing this would be offset by practical limitations relating to the volume of filtrate needed in filling the stem and the unwieldy shape of the funnel that would result from extending the length of the stem. A more practical means of increasing the negative pressure is to apply a vacuum to the tip of the stem by means of a vacuum filter flask, collecting the filtrate in the filter flask that also serves as a receiver. In this way the applied pressure is increased from the 4 to 5 in. that can be produced by the liquid collecting in the stem of the filter to as much as about 20 ft, which can be produced by the vacuum pump.

A 60° funnel is not a very effective filter apparatus to use in a vacuum filter system. The unsupported part of the filter paper cone over the filter stem is highly susceptible to being ruptured, and the largest portion of filter paper is in contact with the walls of the funnel, thereby providing a capillarylike flow zone for filtrate with a long flow path to the stem. The Büchner funnel and other vacuum filter funnels overcome these defects by providing support over the whole circle of filter paper while controlling the unsupported area to small holes that are distributed over the whole area. By this means, the distance that filtrate must travel through the filter paper to the receiver is minimized, and the filter paper has enough wet strength to bridge the small holes leading to the receiver.

The ability of a vacuum filter to handle larger volumes increases with the square of its diameter. Being of heavy construction, however, these filter funnels become unwieldy when they are much more than a foot in diameter. Laboratory vacuum filters are therefore limited by both pressure and area in their capacity.

If still larger capacity is needed, one must resort to pressure filters or to vacuum filter systems employing multiple leaf elements as the filter surface. With a pressure filter, the capacity is limited more or less by the ability to support the filter medium without damage. In ultrafiltration the pressures used may range up to 200 to 300 psi, and the membranes used are capable of holding back viruses, bacteria, clays, and materials of similar size.

4 FILTER MEDIA

General

The term "filter medium" generally includes all permeable materials that will separate particles from the fluid in which they are suspended. Thus the term often describes materials that also serve as a filter septum or supporting medium when loose materials are added to a suspension to form a more porous filter cake. Filter media in common use include filter papers, membranes, woven fabrics, loose materials (granular and fibrous), and rigid filter media and prefabricated filter elements.

The selection of a filter medium for solids-fluids separation in a particular chemical laboratory depends on the purpose of the filtration, the quantity of material to be filtered, and the apparatus available for assembly of the filter. The purpose of the filtration is important because it must be ascertained whether the interest is in the recovery of the fluid or in the recovery of the solids. If the interest is in the recovery of the solids in a "pure" state, it is not usually satisfactory to introduce loose filter materials to speed up the filtration if the presence of such materials in the resulting cake would contaminate the product. On the other hand, if the interest is in the recovery of the clarified fluid, the use of loose filter materials to improve filter capacity would be desirable as long as the loose materials did not introduce soluble materials into the fluid. Thus a filter medium must have certain characteristics that are related to its use in a specific solid-fluid separation. As a first requirement, the medium should preferably be inert to the liquid to be filtered, being insoluble therein and undergoing little physical change (e.g., adsorption, swelling, shrinking, or distortion) that might affect its permeability characteristics. The filter medium is usually selected to provide the desired quality of filtrate (maximum retention of suspended particles) while at the same time ensuring a maximum rate of filtration. The finer the pore openings in a filter medium, the greater will be the particle retention and the smaller will be the volume throughout.

The presence of inert loose materials in a filter cake is not detrimental in all cases. For example, suppose there is interest in separating two different suspended solids from suspension in a fluid. Using loose materials as filter aids, the "three" suspended solids can be removed and dewatered in the form of a filter cake. Then a solvent for one of the solids to be recovered can be used to wash the filter cake and leave the other solids present in the cake. The loose materials then serve to retain the other solid within the filter cake matrix.

When small quantities (1 to 2000 ml) of suspension are to be filtered, the filtration normally is conducted under gravity conditions or under vacuum (20 to 25 in. Hg). When larger volumes of suspension are to be filtered, it is customary to conduct the filtration under pressure conditions. Pressures of 0 to 100 psi are commonly used in precoat-type filtrations using loose materials as a filter medium collected on a filter septum. Pressures of 200 to 300 psi are used across ultrafiltration membranes, and pressures to 1000 psi and up are used across reverse osmosis membranes. When vacuum filtration is used, dissolved gases in the fluid can come out of solution and clog the filter media. In all cases the filter medium selected must be strong enough when exposed to the fluid under the pressure involved to retain the integrity of the filter medium. Some open-textured media are unsuited for pressure filtrations, since fine particles may enter the pores and, by plugging, make the filtering surface almost impermeable. In many such cases, it is possible to use pore openings larger than many of the particles to be removed and to depend on bridging of the particles (or loose materials) over the openings to form a porous cake. In such cakes a bridging cake will be more likely to form as the concentration of the solids in the fluids is increased, and refiltration of the filtrate through its filter cake can effectively remove particles that pass through the media early in the original filtration.

It should be obvious that different types of apparatus are required with different types of filter media and suspensions. The medium used must be compatible with both the suspension to be filtered and the filter apparatus to be employed.

Filter Paper

One of the most versatile and effective filter media for use in laboratory filtrations is filter paper. Filter papers are made primarily from bleached cellulose, either from cotton linter or wood pulp derivation or from glass fibers. Suitable blends of pure or mixed fibers of various types can produce filter papers that demonstrate unique combinations of filtering properties. The final filter paper is usually checked to assure its compliance with a manufacturing standard using the following ASTM standard tests:

Basis weight	ASTM D646-67	Rapidity	ASTM D981-56
Thickness	ASTM D645-67	Water adsorption	ASTM D824-67
Wet strength	ASTM D774-67	Ash in paper	ASTM D586-63
Air resistance	ASTM D726-58	Retention	ASTM D981-56

Several definitions are important in discussing filter papers.

1. *Ashless Filter Paper.* Filters specified for use in quantitative analysis should have a low ash content remaining after burning off the filter paper.

The ash may be composed of various chemical residues, metallic matter from machinery used in the paper manufacture, or mineral matter in the pulp. Ashless filter paper should contain less than 0.010 % ash.

2. *Ash-Low Filter Paper.* Filter papers specified for use in qualitative analysis may bear a higher ash content than those used for quantitative analysis. However qualitative analysis filter paper should contain less than 0.04 % ash after combustion.

3. *Basis Weight.* The paper industry defines the basis weight of filter paper as the average weight for a given number of paper sheets cut to a specified size. Since filter papers for laboratory use are cut in comparatively small sizes, frequently the basis weight is expressed in grams per square meter for 100 sheets of filter paper.

4. *Filter Paper.* Porous absorbent paper without glue or filter additives is made of soft, expandable, natural or synthetic, organic or inorganic fibers, which may be partly impregnated with synthetic resins to obtain special properties.

5. *Hardened Filter Papers.* These materials are specially treated with resins to provide the property of exceptionally high wet strength.

6. *Ribbon Papers.* Specially manufactured for quantitative chemical analysis, these papers are produced in series of retention grades that are intended for use in retaining different size particles from coarse to very fine.

7. *Surface Property.* Filter papers can be produced smooth or creped depending on the need for flow and retention. For the same type of paper, the creping process improves permeability but reduces retention because of disruption of the physical arrangement of the paper fibers. The filters can be obtained as plain filter discs, or they can be purchased prefolded to provide increased flow capacity.

Most filter paper suppliers produce several grades of laboratory use filter papers (general purpose, low-ash qualitative, and ashless quantitative filter paper) which are available in a series of flow retention characteristics to meet specific filtration requirements. Schleicher and Schuell (Keene, New Hampshire 03431) uses a diagram (Fig. 7.1) to give a qualitative picture of the general applicability of each of their best known analytical and general-purpose filter papers. The most popular grades are listed along the walls of the cylinder, related to their retention efficiency in separating a uniformly used precipitate from distilled water. Table 7.1 lists the physical properties of the filter papers used in the formulation of this retention scale. Similarly, Table 7.2 gives typical properties of a number of popular filter papers manufactured by Eaton-Dikeman (E-D: Mount Holly Springs, Pennsylvania 17065). Retention efficiency, which may be expressed differently by each paper manufacturer, is expressed on a numbering scale; that is a 1 in Table 7.2 denotes excellent

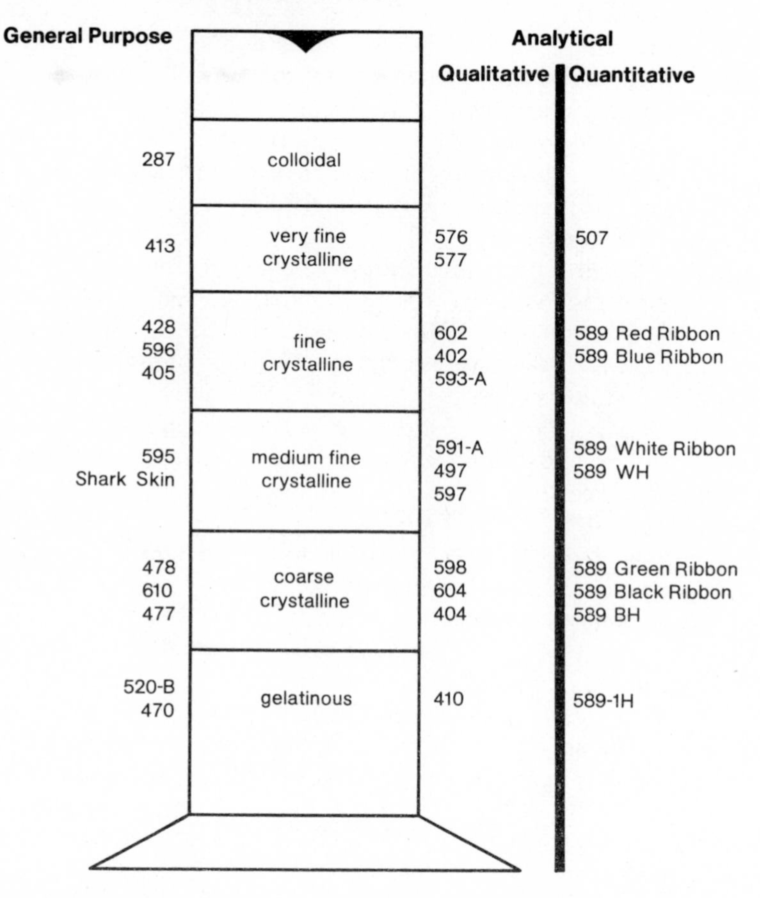

Fig. 7.1. Retention scale for selecting proper grade of Schleicher and Schuell filter paper for retaining specific particles. Courtesy of Schleicher & Schuell, Inc.

small particle retention, a 6 denotes medium particle retention, and 2, 3, 4, and 5 indicate intermediate values. These numbers are based on Coulter counter tests of particle sizes that pass through the papers. E-D does not place an arbitrary or empirical rating on its papers, since the absolute retention of a given paper grade can be affected by too many process factors to permit specification of a meaningful rating. The permeability of a filter paper is a measure of the flow rate of a pure liquid through the paper under a standard applied pressure. In practice, the flow rate does not remain constant but falls off as the solids removed plug the pores in the paper. E-D uses a qualitative instead of a quantitative measure of permeability in Table 7.2,

Property	Grade	Basis Weight (g/m²)	Surface	Filtration Speed[b] (sec)	Precipitates Retained	Absorbancy[c] (sec)	Wet Strength	Thickness (mm)	Ash[d] (%)
Ashless for quantitative analysis	589 Black ribbon	69	Rough	15.4	Coarse	3.6	Low	0.193	0.005
	589 Green ribbon	146	Rough	24.8	Coarse	2.0	Low	0.342	0.010
	589 White ribbon	81	Rough	48.0	Med. fine	6.0	Low	0.200	0.005
	589 Blue ribbon	81	Rough	147.2	Fine	10.5	Low	0.180	0.005
	589 Red ribbon	81	Smooth	150.6	Fine	51.1	High	0.139	0.007
Hardened ashless	589-1H	60	Smooth	6.6	Gelatinous	2.0	High	0.180	0.007
	589-BH	72	Smooth	8.7	Coarse	3.3	High	0.218	0.006
	589-WH	81	Smooth	44.5	Med. fine	26.8	High	0.195	0.005
	507	80	Very smooth	205.2	Very fine	55.0	High	0.101	0.007
Ash-low for qualitative analysis	604	74	Rough	14.0	Coarse	4.1	Low	0.198	0.035
	598	148	Rough	20.8	Coarse	3.0	Low	0.350	0.110
	597	83	Smooth	34.1	Med. fine	9.6	Low	0.193	0.023
	591-A	83	Smooth	48.1	Med. fine	6.4	Low	0.175	0.024
	593-A	179	Smooth	63.0	Fine	2.5	Low	0.340	0.026
	602	83	Smooth	125.5	Fine	25.9	Low	0.165	0.039
Hardened ash-low	410	62	Smooth	6.6	Gelatinous	4.3	High	0.177	0.043
	404	74	Smooth	8.4	Coarse	2.8	High	0.198	0.028
	497	83	Smooth	36.9	Med. fine	28.6	Very high	0.180	0.070
	402	83	Smooth	121.0	Fine	42.2	Very high	0.177	0.070
	577	82	Very smooth	95.2	Very fine	77.3	Very high	0.129	0.004
	576	81	Very smooth	99.0	Very fine	34.9	High	0.114	0.070
General purpose	470	303	Smooth	7.6	Gelatinous	1.0	Low	0.886	
	520-B	152	Rough	7.8	Gelatinous	53.7	High	0.401	
	477	70	Rough	7.9	Coarse	7.2	High	0.228	
	610	62	Smooth	9.7	Coarse	7.2	High	0.149	
	478	155	Rough	15.7	Coarse	3.7	Very high	0.492	

Table 7.1 (*Continued*)

Property	Grade	Basis Weight (g/m^2)	Surface	Filtration Speed[b] (sec)	Precipitates Retained	Absorbancy[c] (sec)	Wet Strength	Thickness (mm)	Ash[d] (%)
	Shark skin	39	Rough	17.8	Med. fine	42.5	High	0.134	
	595	63	Smooth	20.0	Med. fine	18.6	Very high	0.137	
	405	220	Rough	48.9	Fine	4.7	Very high	0.457	
	596	54	Smooth	51.8	Fine	22.2	High	0.147	
	428	122	Smooth	52.9	Fine	3.7	High	0.337	
	413	137	Smooth	80.4	Very fine	10.8	Very high	0.289	
Special analysis	10 Ruled	NA	Smooth black with blue lines	1.0	Coarse	None	Very high	0.119	
	9 Grid	62	Smooth	6.1	Coarse	11.6	High	0.162	
	8 Ruled	62	Smooth	8.4	Coarse	4.4	High	0.154	
	508	200	Rough black	60.0	Fine	2.5	High	0.459	
	551	90	Smooth black	66.2	Fine	14.2	High	0.193	
	287	150	Rough	48.4	Colloidal	4.0	High	0.279	
Glass fiber	24	80	Smooth	14.8	Finest	10.4	Low	0.810	
	25	73	Smooth	17.3	Finest	2.6	Low	0.294	
	29	60	Smooth	13.4	Finest	2.0	Low	0.246	
	30	59.5	Smooth	16.2	Finest	2.5	Low	0.241	

[a] Courtesy of Schleicher & Schuell, Inc.

[b] Filtration speed (rapidity) is the time taken for 10 ml of distilled water to pass through a 110-mm folded circle, according to ASTM method D981-56 (TAPPI T471M-47).

[c] The time of absorption is the time required for the paper to completely absorb a specified quantity of water according to ASTM method D824-67 (TAPPI T432LS-64).

[d] Percentage of ash is determined by measuring the remaining residue after complete combustion of the paper at 950 ± 25°C, according

Table 7.2 Typical Properties of E-D Filter Papers[a]

	Unit Weight			Thickness				
Grade	Paper System[b]	lb/yd²	g/m²	in.	μ	Surface	Perme-ability	Relative Retention Value
E-D 613	20	0.13	70	.006	150	Smooth	Slow	2
E-D 950	25	0.16	88	.007	180	Smooth	Very slow	1
E-D 939	39	0.25	137	.024	610	Smooth	Super fast	6
E-D 623	70	0.45	246	.024	610	Smooth	Medium	3
E-D 627	75	0.49	264	.026	660	Smooth	Slow	2
E-D 301	140	0.91	492	.050	1270	Smooth	Slow	1
E-D 992	12	0.08	42	.006	152	Creped	Medium	3
E-D 953	15	0.10	53	.007	180	Creped	Fast	4
E-D 615	20	0.13	70	.009	230	Creped	Fast	3
E-D 617	35	0.23	123	.018	450	Creped	Fast	4
E-D 634	54	0.35	190	.028	710	Creped	Very fast	5
E-D 633	70	0.45	246	.039	990	Creped	Very fast	5
E-D 654	84	0.54	296	.042	1070	Creped	Slow	2

[a] Courtesy of Eaton-Dikeman.
[b] Weight in pounds of 500 sheets of 20 × 20 in. paper.

where the paper system unit weight of a paper is the weight in pounds of 500 sheets of 20 × 20 in. paper.

Laboratory filter papers are usually available in boxes (100 sheets) of discs with diameters of 5.5, 7, 9, 11, 12.5, 15, and 18.5 cm. Special papers are also available in 20, 25, 33, 40, 45, and 50 cm diameters.

Glass fiber filters are generally made from microfibers of borosilicate glass without the addition of any organic binders. Therefore these fibers resist clogging, provide more retention with a faster flow rate, and can be used at high temperatures and still retain superior wet strength. They possess high and low adsorption characteristics and are resistant to chemical attack. Since they are weakened by folding, they should always be used flat. They find particular use as a prefilter ahead of membrane filters to prevent clogging of the membranes.

It is evident that the success of a laboratory filtration depends on the proper choice of filter paper for a given application. Factors to be considered include the volume, temperature, and viscosity of liquid to be filtered, the size and permeability of the particles, the filtration apparatus and method used, and the degree of clarification required. Among the special types of filter paper available for special types of laboratory analysis are the following:

1. Filter paper impregnated with activated carbon for adsorption of color and odors in liquids.
2. Keiselguhr paper impregnated with diatomaceous earth for separating heavy liquids with low turbidities resulting from the formation of semi-colloids.
3. Phase-separating paper, which is a silicone-impregnated paper for quick and complete separation of immiscible aqueous solution from organic solvents. These "throw-away" separation papers are hydrophobic (i.e., they are pervious only to the organic phase) and are excellent, therefore, for water–organic liquid separations.

Membranes

One of the newest developments in the area of filter media is in membrane technology. Membrane filters are distinctive in that they can remove from gases or liquids all particulate matter that is larger than the membrane pores. To ensure the accomplishment of this, the membrane filter and filter apparatus must be properly chosen. Membrane filters are very thin plastic films with holes or pores of precise size. The number of pores in a filter medium is so large that about 80 % of the volume of a membrane consists of the pore volume. Practically, the size of the pore determines what size particles the filter will retain. Membrane filters can be used either to retain particles for analysis or to produce clean filtrates. In some cases membrane filters can be used to selectively remove particles above a given particle size by their selective removal in a series of filtrations through membranes of progressively smaller pore sizes. A key advantage of membranes is that they consist of a single plastic sheet and are not made up of individual fibers, which can work loose during filtration and contaminate the filtrate.

Membrane filters are produced commercially in a number of sizes, materials, and purposes by at least two companies, the Millipore Corporation, Bedford, Massachusetts 01730, and the Gelman Instrument Company, 600 South Wagner Road, Ann Arbor, Michigan 48106. Products of the former company are generally referred to as Millipore membrane filters, and Gelman products are designated by the brand names Metricel and Acropor (membrane filters). Millipore filters are manufactured in a standard range of pore sizes, filter diameters, and membrane materials.

The range of pore sizes and the uniformity of pore sizes in a typical Millipore membrane filter are listed below:

Filter Pore Sizes (μ)	*Maximum Rigid Particle that will Penetrate Filter* (μ)
14	17
10	12
8	9.4
7	9.0
5*	6.2
3*	3.9
2	2.5
1.2*	1.5
1.0	1.1
0.8*	0.85
0.65	0.68
0.60	0.65
0.45*	0.47
0.30*	0.32
0.22	0.24
0.20*	0.25
0.10	0.108
0.05	0.053
0.025	0.028

* Sizes available also from the Gelman Instrument Company.

Tables 7.4 and 7.6 give specific pore sizes available in the various filter materials.

The pore size that must be selected for a given filtration is, of course, a function of the size of the particle or particles to be removed. The largest particles will be retained on the surface of the filter, and some of the smaller particles may be trapped inside the filter structure. The primary function of membranes, however, is the surface removal of the suspended particles. A human hair has a diameter of about 70 to 80 μ, yeast cells about 3 to 4 μ, and bacteria 0.5 to 2 μ. An approximate pore size reference guide can be set down as follows:

Pore Size (μ)	*Particle Retained*
0.2	All bacteria
0.45	All coliform group bacteria
0.8	All airborne particles for analysis
1.2	All nonliving particles considered dangerous in fluids intravenously fed
5	All significant cells from body fluids for analysis

Membrane filters are manufactured in circular discs of standard size. The principal factor in the selection of a disc size for use in a given filtration is the amount of liquid that must be filtered. The membrane disc size required is related to volume throughput approximately as follows:

Membrane Diameter (mm)	*Volume Throughput*
13	Syringe volumes
24, 25	Syringe volumes
47	100–300 ml
90	500 ml to 2 liters
142	5–20 liters
293	More than 20 liters

Clearly doubling the diameter of the membrane should permit an increase in the volume throughput by a factor of 4 times under similar operating conditions. Use of a larger pore size (5 to 10 μ) and a low suspended solid concentration (5 to 10 mg/liter) should be expected to produce larger volumes of throughput than small pore sizes (0.2 μ) and higher suspended solids concentrations ($>$100 to 200 mg/liter).

Membrane filters are made of a number of different materials to ensure chemical resistance to a wide range of solutions that must be filtered in a chemical laboratory. The Millipore Corporation describes their membrane materials of construction as follows:

1. *MF-Millipore (mixed esters of cellulose).* These filters are autoclavable, standard membrane filters composed of pure, biologically inert mixtures of cellulose acetate and cellulose nitrate. This filter is not attacked by dilute acids and alkalies, aliphatic and aromatic hydrocarbons, and nonpolar liquids. It is recommended for all analytical, ultracleaning, and sterilizing applications below 75°C, except those involving ketones, esters, ester-alcohols, nitroparaffins, or strong acids and alkalies. Incineration at 820°C yields ash of about 0.045% of initial weight.
2. *Fluoropore (PTFE).* Fluoropore membranes are made of polytetrafluoroethylene (PTFE) bonded to a high-density polyethylene net to increase ease of handling. This filter type is recommended for all strong solvents, acids, and bases, or chemically active materials generally. The major exception is aromatic hydrocarbons at temperatures above 80°C. Fluoropore filters can be steam sterilized in place (121°C at 15 psi) and used at temperatures up to 130°C; above that temperature the backing structure begins to soften. The filter material is completely hydrophobic, therefore will remove small quantities of free (not bound) water from liquids being filtered. These filters are not

practical for filtering aqueous solutions unless the filter is first saturated with a compatible liquid that will wet it (e.g., methyl alcohol).

3. *Duralon (nylon).* Duralon filters are made of nylon and are resistant to chemical attack by ketones, esters, and many other liquids that dissolve standard filters. They are not resistant, however, to concentrated acids and alkalies, methyl alcohol, and ethyl alcohol. They are strong and flexible and will tolerate significant reverse-flow surges without failure. They are stable at temperatures to 75°C but are not recommended for residual ash analysis or autoclaving.

4. *Celotate (cellulose acetate).* This membrane material is recommended for filtering the lower molecular weight alcohols such as methanol and ethanol and is particularly useful in low-nitrogen background analysis, since it yields very low residual ash on incineration. It is useful at temperatures up to 75°C but is not autoclavable.

5. *Mitex (Teflon).* The filter membrane is made of pure Teflon (PTFE) and is unaffected by almost all liquids, including organic solvents, concentrated acids and alkalies, rocket propellants, and cryogenic liquids. The membranes are stable at temperatures in excess of 260°C and down to −100°C, and they are biologically and chemically inert. They are hydrophobic, but unlike the Fluoropore membranes they have no backing material. They are, however, available in larger pore sizes. They are fabricated by a process that forms a continuous mat of microfibers fused together at each contact to prevent dislodgement into the filtrate.

6. *Polyvic (polyvinyl chloride).* Polyvic membranes exhibit properties of strength, flexibility, and resistance to low molecular weight alcohol and moderately strong acids and alkalies. Temperature tolerance is limited to 65°C, above which the material softens. It can be readily heat-sealed to form envelopes, cylinders, or specialized shapes. The membranes are neither autoclavable nor recommended for residual ash analysis.

Table 7.3 is a guide to the proper selection of Millipore filters and certain other equipment with regard to the chemical compatibility with a wide variety of liquids or gases. Recommendations are based on static exposure for 72 h at 25°C and 1.0 atm pressure. Operating conditions at moderate temperatures will not change the recommendations, but high liquid temperatures may do so in some cases.

Millipore membrane filters range from about 90 to 170 μ ($\pm 10\%$) in thickness, depending on the material type. They are normally white (reflective index is 94 %) except for selected pore sizes of the MF type, which are also available in black to afford optical contrast with certain contaminant materials. Some types have an imprinted 3.1-mm grid-marked surface to facilitate counting of contaminants retained on the surface. Since the filters are formed of completely homogeneous polymeric materials, they contain no detachable

Table 7.3 Chemical Compatibility (**R** = recommended, N = not recommended) of Millipore Membrane Filters[a]

	Disc Filters						Cartridge							O-Rings					Other[c]				
	MF-Millipore	*Duralon*	*Polyvic*	*Fluoropore*[b]	*Celotate*	*Mitex*	*Millitube*	*Fluorotube*[b]	*Polytube*	*Duratube*	*Microtube*	*Lifegard*	*Milligard*	*Teflon*	*Viton-A*	*Buna-N*	*Silicon*	*E-P*	*Pumps*	*PVC-142*	*Sterifil*	*Swinnex*	*Inline-47*
Acetic acid, glacial	N	N	R	R	N	R	N	R	N	N	N	N	N	R	N	N	N	N	N	R	N	R	R
Acetic acid, 5%	R	R	R	R	N	R	R	R	R	R	N	R	R	R	R	N	R	R	R	R	R	R	R
Acetone	N	R	N	R	N	R	N	R	N	N	R	N	N	R	N	N	N	R	N	N	N	R	R
Ammonium hydroxide, 6*N*	R	N	R	R	N	R	N	R	N	N	N	N	N	R	N	N	R	R	N	R	N	R	R
Amyl acetate	N	R	N	R	N	R	N	R	N	N	R	N	N	R	N	N	N	R	N	N	N	R	R
Amyl alcohol	N	R	R	R	R	R	N	R	N	N	R	N	N	R	N	N	N	R	R	R	R	R	R
Benzene	R	R	N	R	R	R	N	R	N	N	R	N	N	R	R	N	N	N	R	N	N	N	N
Benzyl alcohol, 4%	R	R	R	R	R	R	R	R	R	R	R	R	R	R	R	R	R	R	R	R	R	R	R
Boric acid	R	R	R	R	R	R	R	R	R	R	R	R	R	R	R	R	R	R	R	R	R	R	R
Brine (seawater)	R	R	R	R	R	R	R	R	R	R	R	R	R	R	R	R	R	R	R	R	R	R	R
Butyl alcohol	R	N	R	R	R	R	N	R	N	N	R	N	N	R	R	R	N	N	R	R	R	R	R
Butyl acetate	N	R	N	R	N	R	N	R	N	N	R	N	N	R	N	N	N	N	N	N	N	R	R
Carbon tetrachloride	R	N	N	R	R	R	N	R	N	N	R	N	N	R	R	N	N	N	R	N	N	R	R
Cellosolve (ethyl)	N	R	N	R	N	R	N	R	N	N	R	N	N	R	N	N	N	N	N	N	N	R	R
Chloroform	R	N	N	R	N	R	N	R	N	N	R	N	N	R	R	N	N	N	N	N	N	R	R
Cyclohexanone	N	R	N	R	N	R	N	R	N	N	R	N	N	R	N	N	N	N	N	N	N	R	R
Developers (photo)	N	R	R	R	N	R	N	R	N	N	R	N	N	R	R	R	N	N	R	R	N	R	R
Diethyl acetamide	N	R	N	R	N	R	N	R	N	N	N	N	N	R	N	N	R	N	N	N	N	R	R
Dimethylformamide	N	R	N	R	N	R	N	R	N	N	N	N	N	R	N	N	R	N	N	N	N	R	R
Dioxane	N	N	N	R	N	R	N	R	N	N	R	N	N	R	N	N	N	N	N	N	N	R	R
Ethyl alcohol	N	N	R	R	R	R	N	R	N	N	R	N	N	R	N	R	R	R	N	R	R	R	R
Ethers	R	R	R	R	R	R	N	R	N	N	R	N	N	R	N	N	N	N	N	N	R	R	R
Ethyl acetate	N	R	N	R	N	R	N	R	N	N	R	N	N	R	N	N	N	N	N	N	N	R	R
Ethylene glycol	N	R	R	R	N	R	N	R	N	N	R	N	N	R	R	R	R	R	R	R	N	R	R
Formaldehyde	N	N	R	R	N	R	N	R	N	N	N	N	N	R	N	N	N	N	N	R	N	R	R
Freon TF or PCA	R	R	R	R	R	R	N	R	N	N	R	N	N	R	N	R	N	N	R	R	N	R	R
Gasoline	R	R	R	R	R	R	R	R	R	R	R	R	R	R	R	R	N	N	R	R	N	R	R
Glycerine (glycerol)	R	R	R	R	R	R	R	R	R	R	R	R	R	R	R	R	R	R	R	R	R	R	R
Hexane	R	R	R	R	R	R	R	R	R	R	R	R	R	R	R	R	N	N	R	R	R	R	R
Helium	R	R	R	R	R	R	R	R	R	R	R	R	R	R	R	R	R	R	—	—	—	R	R
Hydrochloric acid, conc.	N	N	R	R	N	R	N	R	N	N	N	N	N	R	R	N	N	N	N	N	N	R	R
Hydrogen (gas)	R	R	R	R	R	R	R	R	R	R	R	R	R	R	R	R	N	R	—	—	—	R	R
Hydrofluoric acid	N	N	R	R	N	R	N	R	N	N	N	N	N	R	R	N	N	N	N	N	N	N	N
Hypo (photo)	R	R	R	R	R	R	R	R	R	R	R	R	R	R	R	R	R	R	R	R	R	R	R
Isobutyl alcohol	R	N	R	R	R	R	N	R	N	N	R	N	N	R	R	N	R	R	R	R	R	R	R
Isopropyl acetate	N	R	N	R	N	R	N	R	N	N	R	N	N	R	N	N	N	N	N	N	N	R	R
Isopropyl alcohol	N	N	R	R	R	R	N	R	N	N	R	N	N	R	R	N	R	R	R	R	R	R	R
Kerosene	R	R	R	R	R	R	N	R	N	N	R	N	N	R	R	R	N	N	R	R	R	R	R

le 7.3 (*Continued*)

	Disc Filters						*Cartridge*							*O-Rings*					*Other*[c]					
	MF-Millipore	*Duralon*	*Polyvic*	*Fluoropore*[b]	*Celotate*	*Mitex*	*Millitube*	*Fluorotube*[b]	*Polytube*	*Duratube*	*Microtube*	*Lifegard*	*Milligard*	*Teflon*	*Viton-A*	*Buna-N*	*Silicon*	*E-P*	*Pumps*	*PVC-142*	*Sterifil*	*Swinnex*	*Inline-47*	*PL Housing*
:hyl alcohol	N	N	R	R	R	R	N	R	N	N	R	N	N	R	N	R	R	R	N	R	R	R	R	R
hylene chloride	N	N	N	R	R	R	N	R	N	N	R	N	N	R	N	N	N	N	N	N	N	R	R	N
hyl ethyl ketone	N	R	N	R	N	R	N	R	N	N	R	N	N	R	N	N	N	R	N	N	N	R	R	N
hyl isobutyl ketone	N	R	N	R	N	R	N	R	N	N	R	N	N	R	N	N	N	N	N	N	N	R	R	N
eral spirits	R	R	R	R	R	R	R	R	R	R	R	R	R	R	R	R	N	N	R	R	R	R	R	R
ic acid, conc.	N	N	R	R	N	R	N	R	N	N	N	N	N	R	R	N	N	N	N	N	N	R	R	N
obenzene	N	R	N	R	N	R	N	R	N	N	N	N	N	R	N	N	N	N	N	N	N	N	N	N
ogen (gas)	R	R	R	R	R	R	R	R	R	R	R	R	R	R	R	R	R	R	—	—	—	R	R	R
oleum base oils	R	R	R	R	R	R	R	R	R	R	R	R	R	R	R	R	N	N	R	R	R	R	R	R
tane	R	R	R	R	R	R	R	R	R	R	R	R	R	R	R	R	N	N	R	R	R	R	R	R
:hloroethylene	R	N	N	R	N	R	N	R	N	N	R	N	N	R	R	R	N	N	N	N	N	R	R	N
oleum ether	R	R	R	R	R	R	R	R	R	R	R	R	R	R	R	R	R	R	R	R	R	R	R	R
nol, 0.5 %	R	N	R	R	N	R	N	R	N	N	R	R	N	R	R	R	R	R	R	R	R	R	R	R
dine	N	R	R	R	N	R	N	R	N	N	N	N	N	R	N	N	R	N	N	R	N	R	R	N
:one oils	R	R	R	R	R	R	R	R	R	R	R	R	R	R	R	R	N	R	R	R	R	R	R	R
ium hydroxide, conc.	N	N	R	R	N	R	N	R	N	N	N	N	N	R	N	N	R	R	N	R	N	R	R	N
uric acid, conc.	N	N	R	R	N	R	N	R	N	N	N	N	N	R	R	N	N	N	N	N	N	R	R	N
uene	R	R	N	R	R	R	N	R	N	N	R	N	N	R	R	N	N	N	R	N	N	N	N	N
hloroethane	R	N	N	R	N	R	N	R	N	N	R	N	N	R	R	N	N	N	N	N	N	R	R	N
hloroethylene	R	N	N	R	N	R	N	R	N	N	R	N	N	R	R	N	N	N	N	N	N	R	R	N
ene	R	N	N	R	R	R	N	R	N	N	R	N	N	R	R	N	N	N	R	N	N	N	N	N

ourtesy of the Millipore Corporation.

ontain hydrophobic PTFE membranes that are not wetted by aqueous solutions. With pore sizes of 1.0 μ ess, their water intrusion pressure is too high for practical filtration of aqueous solutions unless the filter rst saturated with a compatible wetting liquid such as methyl alcohol.

ecommendations apply only to materials of construction other than O-rings. If the liquid in question is npatible with filter holder construction materials, determine O-ring compatibility by reference to the holder cription in the Millipore Corporation catalog, where standard and available alternative O-rings are listed.

fibers or other particles of construction material that can break loose and contaminate the filtrate.

Table 7.4 summarizes the specifications applicable to Millipore membrane filters of various material types and pore sizes. Standard MF-Millipore and Duralon filters are available in all disc sizes. Fluoropore, Celotate, Mitex, and Polyvic filters are available in all sizes except the 24-mm size. If membrane filters are to be used in laboratory filtrations, it is important to be able

Table 7.4 Millipore Membrane Filter Specifications[a]

Filter Type	Pore Size (μ)	*Flow Rate*[b]		Porosity (%)	Autoclavable	Refractive Index	Bubble Point[e]		Water Extractables (w
		Water[c]	*Air*[b]				kg/cm^2	psi	
MF-Millipore (mixed cellulose acetate and nitrate)									
SC	8.0	630	65	84	Yes	1.515	0.28	4	6.0 %
SM	5.0	400	32	84	Yes	1.495	0.42	6	6.0 %
SS	3.0	296	30	83	Yes	1.495	0.70	10	6.0 %
RA	1.2	222	20	82	Yes	1.512	0.84	12	5.0 %
AA	0.80	157	16	82	Yes	1.510	1.12	16	4.0 %
AA (black)	0.80	157	16	82	No	N.A.	1.12	16	N.A.
DA	0.65	111	9	81	Yes	1.510	1.34	19	3.0 %
HA	0.45	38.5	4	79	Yes	1.510	2.32	33	2.5 %
HA (black)	0.45	38.5	4	79	No	N.A.	2.32	33	N.A.
PH	0.30	29.6	3	77	Yes	1.510	2.81	40	2.0 %
GS	0.22	15.6	2	75	Yes	1.510	3.87	55	2.0 %
VC	0.10	1.5	0.6	74	Yes	1.500	17.6	250	1.5 %
VM	0.05	0.74	0.5	72	Yes	1.500	26.4	375	1.5 %
VS	0.025	0.15	0.2	70	Yes	1.500	35.2	500	1.5 %
Duralon (nylon)									
NC	14.0	756	75	68	No	See	0.18	2.5	1.5 %
NS	7.0	452	50	65	No	note[f]	0.28	4.0	2.5 %
NR	1.0	118	29	63	No		0.56	8.0	2.5 %
Fluoropore (PTFE)									
FA	1.0	90	16	85	Yes	See	0.21	3	—
FH	0.5	40	8	85	Yes	note[g]	0.49	7	—
FG	0.2	15	3	70	Yes		0.91	13	—
Celotate (cellulose acetate)									
EA	1.0	178	7	74	No	1.47	0.98	14	4.0 %
EH	0.5	49.6	4.5	73	No	1.47	1.97	28	2.5 %
EG	0.2	15.6	2.2	71	No	1.47	3.87	55	2.0 %
Mitex (Teflon)									
LC	10.0	126	14	68	Yes	N.A.	0.04	0.5	—
LS	5.0	51.9	9	60	Yes	N.A.	0.06	0.9	—
Polyvic (polyvinyl chloride)									
BS	2.0	231	19	79	No	1.528	0.28	4	3.2 %
BD	0.6	33.3	3	73	No	1.528	0.70	10	3.2 %

[a] Courtesy of the Millipore Corporation.

[b] Flow rates are based on measurement with clean water and air and represent *initial* flow rates for a liquic 1 cP viscosity at the start of filtration, before filter plugging is detectable. Actual initial flow rates may v from the average values given here. Variability depends on filter type and is roughly proportional to pore s

[c] Water flow rates are milliliters per minute per square centimeter of filtration area, at 25°C with a differen pressure of 52 cm Hg (10 psi). *Flow rates for Fluoropore and Polyvic filters are based on methanol instead of wa*

[d] Air flow rates are liters per minute per square centimeter of filtration area, at 20°C with a differential press of 52 cm Hg (10 psi) and exit pressure of 1 atm (14.7 psia).

[e] Bubble point pressure is the differential pressure required to force air through the pores of a water-wet fi (except methanol-wet for Fluoropore, Mitex, and Polyvic filters).

[f] Duralon filters cannot be made completely transparent with immersion oil.

[g] Crystalline and amorphous regions of Fluoropore filters have differing refractive indexes, and it is theref not possible to obtain uniform clearing.

to estimate the initial filtration rate (Table 7.4) under a variety of applied pressures. In Fig. 7.2 the filtration rate appears as a function of differential pressure across the membrane *only* using clean water (or methanol). Figure 7.2 does not include pressure drop due to the filtration system inlet and outlet or of the membrane septum support.

Gelman Instrument Company has developed a broad line of unsupported membrane filters made from cellulose triacetate, mixed cellulose esters, cellulose acetate, regenerated cellulose, polyvinylchloride, and so on, marketed under the "Metricel" name. The membranes designated "unsupported" do not include a supporting substrate. All Metricel filters have void volumes in excess of 75 %. Basic membrane quality tests include the following:

1. Precise flow rate tests using both air and liquid (similar to results in Fig. 7.2).
2. Mean and maximum pore size of the membrane lot determined by the bubble point test method (ASTM Standard Procedure D-2499-66T).

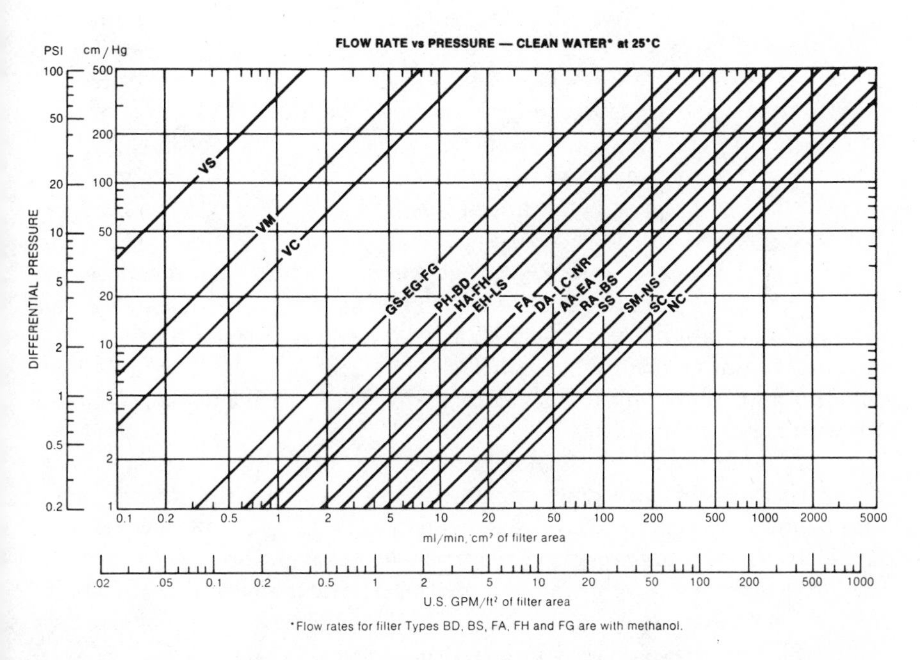

Fig. 7.2. Flow rate versus pressure drop for clean water (or methanol) at 25°C. Flow rate correction for viscosity: for a liquid having a viscosity significantly different from that of water (1 CP), divide the water flow rate by the viscosity of the liquid in centipoises to obtain the approximate initial flow rate for the liquid in question (viscosity of methanol is 0.55 CP at 25°C). Courtesy of the Millipore Corporation.

3. Membranes used for bacterial filtrations must meet rigid bacterial removal specifications.
4. The 0.45 μ filter gives 100% removal of *Serratia marcescens.*
5. The 0.2 μ filter gives 100 % removal of *Pseudomonas diminuta.*

The Metricel filter types are described briefly as follows:

1. *Type GA Triacetate Metricel.* This membrane is best adapted for biological and pharmaceutical filtrations. It is pure white, packaged dry, and has a matte, pattern-free surface. The material has high thermal stability, which permits autoclaving for sterilization. It is compatible for use with ethyl alcohol. A small amount of wetting agent (starchlike cellulose) is incorporated in the membrane to enhance wettability in water and aqueous solutions.
2. *Type TCM Metricel.* This membrane is identical to the type GA membrane but is manufactured without inclusion of detergents, surfactants, or wetting agents. It does, however, exhibit only slight resistance to wetting by water.
3. *Type GN Metricel.* This membrane is made from mixed esters of cellulose, a blend of nitrocellulose and cellulose acetate. It is not as heat stable as the type GA membrane but can be autoclaved or sterilized by means of ethylene oxide. It can adsorb proteins, in contrast to the type GA membranes, which do not. The filters are available in black or green and are used for bacteriological counting and for microscopic analyses of multicolored contaminants. The green membranes are suitable for routine use where eye fatigue is a problem.
4. *Type GN-6 Filters with Hydrophobic Edge.* These special membranes designed for food and drug sterility testing are made with a hydrophobic edge around the membrane perimeter. Thus the outer ring is water repellent, and aqueous solutions readily pass through the effective filtering area without penetrating or adhering to the treated rim. The membranes are available both with and without grids.
5. *Alpha Metricel.* This membrane is solvent resistant, with the same pore structure and filtration characteristics as the classical cellulosic membrane filters. It is composed of regenerated cellulose and is pure white. It is unaffected by the most common organic solvents and boiling acetone. It is used routinely for filtering methyl ethyl ketone, methyl isobutyl butone, methyl Cellosolve acetate, and dimethylformamide. Exposure to water causes the membrane to swell and reduces filtration rates.
6. *Vinyl Metricel.* This membrane is made of polyvinyl chloride to increase its resistance to concentrated acids and alkalies. It has a very low (1 %) moisture pickup and is well suited to gravimetric analysis of air particles.

It cannot be autoclaved for sterilization, but sterilization using ethylene oxide is possible. It is not suitable for operations requiring a low ash content.

7. *DM Metricel.* This membrane is made of a copolymer of acrylonitrite and polyvinyl chloride to provide outstanding resistance to acids and alkalies. Moisture pickup is less than 1.3 %, making it suitable for gravimetric determinations requiring constant weight and a low ash content. The filters are smooth, and less brittle and easier to handle than cellulosic filters.

The Gelman membrane filters coated onto a nylon web are referred to as Acropor type filters. The membrane coating is the same in chemical composition as DM Metricel membranes, a copolymer of acrylonitrite and polyvinylchloride. The incorporation of the nylon web in the Acropor material gives it much greater strength support. Acropor is furnished in a range of pore sizes (0.2 to 5 μ) in punched discs for laboratory use or in continuous rolls up to 21 in. wide. Water-repellent grades are available for certain applications. The material has good chemical stability to a wide range of fluids and reagents. It can be heat-sealed to itself and to certain plastics without affecting the membrane filtering quality. The material is used for filtering pharmaceutical preparations without danger of causing a toxic filtrate. There are several types of Acropor filters.

1. *Hydrophobic Acropor.* Selected grades of Acropor membranes are available with water-repellent characteristics that permit passage of gases and nonaqueous fluids but retain aqueous solutions. Water breakthrough pressures decrease with increasing pore size.

2. *Acropor Ion-Exchange Filters.* Four ion-exchange resins are now available in convenient membrane form as an integral part of a microporous, plastic matrix. These filters are used for easy removal, concentration, and/or recovery of ions in dilute solutions. They are available in strong acid and strong basic form, making possible high recovery of trace quantities of cations and anions. The following ion-exchange materials are used. (*a*) Acropor type SA 6404 (Dowex 50W-X8 resin): strong acid cation-exchange membrane containing sulfonic acid resin groups; capable of removing soluble cobalt, copper, iron, nickel, and zinc. (*b*) Acropor type SB 6407 (Dowex 1-X-8 resin): strong base membrane containing resin composed of quaternary ammonium chloride; used for collection of trace amounts of sulfate and nitrate in water. (*c*) Acropor type SA C10 (Duolite C-10 resin): strong acid membrane. (*d*) Acropor type WA Chelex (Dowex A-1 resin): weak acid membrane, excellent for determination of metallic elements.

Table 7.5 gives the chemical compatibilities of Gelman filter membranes with a number of reagents or solutions, and Table 7.6 summarizes the membrane filter specifications of this manufacturer.

Table 7.5 Chemical Compatibilities of Gelman Filters[a]

Reagent	Filter Types						
	Acropor	Alpha Metricel	DM Metricel	GA Metricel	GN Metricel	TCM Metricel	VM Metr
Acids							
Acetic acid, glacial	Good	Good	Good	No	No	No	Goo
Acetic acid, 90 %	No	No	Good	No	No	No	Goo
Acetic acid, 30 %	Good	No	Good	No	Good	No	Goo
Acetic acid, 10 %	Good	No	Good	Good	Good	Good	Goo
Hydrochloric acid, conc.	No	No	No	No	No	No	Goo
Hydrochloric acid, 6 *N*	No	Good	Good	Good	Good	Good	Goo
Nitric acid, conc.	No	No	No	No	No	No	Goo
Nitric acid, 6 *N*	No	Good	Good	Good	Good	Good	Goo
Sulfuric acid, conc.	No	No	No	No	No	No	Goo
Sulfuric acid, 6 *N*	No	Good	Good	Good	Good	Good	Goo
Alcohols							
Amyl alcohol	Good	Good	Good	Good	Good	Good	Goo
Benzyl alcohol	Good	No	Good	No	No	No	Goo
Butanol	Good	Good	Good	Good	Good	Good	Goo
Ethanol	Good	Good	Good	Good	No	Good	Goo
Isopropanol	Good	Good	Good	Good	Good	Good	Goo
Methanol	No	Good	Good	Good	No	Good	No
Propapol	Good	Good	Good	Good	Good	Good	Goo
Bases							
Ammonium hydroxide, 3 *N*	No	No	No	No	Good	No	No
Ammonium hydroxide, 6 *N*	No	No	No	No	No	No	No
Potassium hydroxide, 3 *N*	No	No	No	No	No	No	No
Sodium hydroxide, 3 *N*	Good	No	Good	No	No	No	Goo
Sodium hydroxide, 6 *N*	No	No	No	No	No	No	No
Esters							
Amyl acetate	Good	Good	Good	No	No	No	Goo
Butyl acetate	No	No	No	No	No	No	No
Cellosolve acetate	Good	Good	Good	No	No	No	Goo
Ethyl acetate	Good	Good	No	No	No	No	No
Isopropyl acetate	No	No	No	No	No	No	No
Methyl acetate	No	No	No	No	No	No	No
Ethers							
Diethyl ether	Good	Good	Good	Good	Good	Good	Goo
Diisopropyl ether	Good	Good	Good	Good	Good	Good	Goo
Dioxane	No	No	No	No	No	No	No
Tetrahydrofuran	No	No	No	No	No	No	No
Glycols							
Ethylene glycol	Good	Good	Good	Good	Good	Good	Goo
Glycerine	Good	Good	Good	Good	Good	Good	Goo
Propylene glycol	Good	Good	Good	Good	No	Good	Goo

ble 7.5 (*Continued*)

Reagent	*Filter Types*						
	Acropor	*Alpha Metricel*	*DM Metricel*	*GA Metricel*	*GN Metricel*	*TCM Metricel*	*VM Metricel*
omatic hydrocarbons							
Benzene	Good	Good	Good	Good	Good	Good	Good
Toluene	Good	Good	Good	Good	Good	Good	Good
Xylene	Good	Good	Good	Good	Good	Good	Good
alogenated hydrocarbons							
Carbon tetrachloride	Good	Good	Good	Good	Good	Good	Good
Chloroform	Good	Good	Good	No	Good	No	Good
Chlorothene NU	Good	Good	Good	Good	Good	Good	Good
Dichloroethane	Good	Good	No	No	Good	No	No
Dowclene WR	Good	Good	Good	Good	Good	Good	Good
Freon TF	Good	Good	Good	Good	Good	Good	Good
Freon TMC	No	No	No	No	Good	No	No
Genosolv D	Good	Good	Good	Good	Good	Good	Good
Methylene chloride	No	No	No	No	No	No	No
Perchloroethylene	Good	Good	Good	Good	Good	Good	Good
Trichloroethylene	Good	Good	Good	Good	Good	Good	Good
etones							
Acetone	No	Good	No	No	No	No	No
Cyclohexanone	No	No	No	No	No	No	No
Methyl ethyl ketone	No	Good	No	No	No	No	No
Methyl isobutyl ketone	No	Good	No	No	No	No	No
ils							
Cottonseed oil	Good	Good	Good	No	Good	No	Good
Lubricating oil, 7808	No	Good	No	No	Good	No	No
Peanut oil	Good	Good	No	No	Good	No	No
Sesame oil	Good	No	No	No	Good	No	No
iscellaneous							
Aniline	No	Good	No	No	No	No	No
Dimethylformamide	No	Good	No	No	No	No	No
Dimethylsulfoxide	No	No	No	No	No	No	No
Formaldehyde	Good	Good	Good	Good	Good	Good	Good
Formalin	Good	Good	No	Good	Good	Good	No
Gasoline	Good	Good	Good	Good	Good	Good	Good
Hexane, dry	Good	Good	Good	Good	Good	Good	Good
JP-4	Good	Good	Good	Good	Good	Good	Good
Kerosene	Good	Good	Good	Good	Good	Good	Good
Phenol, liquified	No	Good	No	No	No	No	No
Pyridine	No	No	No	No	No	No	No
Skydrol 500	Good	Good	Good	Good	No	Good	Good
Turpentine	Good	Good	Good	Good	Good	Good	Good
Water	Good	No	Good	Good	Good	Good	Good

Courtesy of Gelman Instrument Company.

Table 7.6 Gelman Membrane Filter Specifications[a]

Membrane Type and Designation	Composition	Pore Size (μ)	Flow Rate: Air [liters/(min)(cm²) at 70 cm Hg]	Flow Rate: Water [ml/(min)(cm²) at 70 cm Hg]	Bubble Point: Water psi (min) (atm)	Bubble Point: Kerosene psi (min) (atm)	Autoclavable	Sterilizable by Ethylene Oxide	Maximum Operating Temperature
Type GA Triacetate Metricel									
GA-1	Cellulose	5	40	320	8 (0.54)	4 (0.27)	Yes	Yes	135°C (275°
GA-3	triacetate	1.2	35	285	12 (0.8)	6 (0.41)	Yes	Yes	135°C (275°
GA-4		0.8	24	200	16 (1.1)	8 (0.54)	Yes	Yes	135°C (275°
GA-6		0.45	10	70	30 (2)	15 (1.02)	Yes	Yes	135°C (275°
GA-7		0.3	6	35	40 (2.7)	18 (1.2)	Yes	Yes	135°C (275°
GA-8		0.2	4	25	50 (3.4)	24 (1.6)	Yes	Yes	135°C (275°
Type TCM Metricel									
TCM-1200	Cellulose	1.2	35	220 4 psi[b]	14 (1)	6 (0.41)	Yes	Yes	135°C (275°
TCM-450	triacetate	0.45	10	50 8 psi[b]	35 (2.4)	15 (1.02)	Yes	Yes	135°C (275°
TCM-200		0.2	4	20 20 psi[b]	60 (4.1)	24 (1.6)	Yes	Yes	135°C (275°
Type GN Metricel									
GN-6	Mixed esters	0.45	10	70	30 (2)	15 (1.02)	Yes	Yes	110°C (230°
GN-4 Black	of cellulose	0.8	16	120	16 (1.1)	9 (0.61)	Yes	No	135°C (275°
GN-6 Black		0.45	7	50	30 (2)	15 (1.02)	Yes	No	135°C (275°
GN-4 Green		0.8	16	120	16 (1.1)	9 (0.61)	Yes	No	135°C (275°
GN-6 Green		0.45	7	50	32 (2.2)	15 (1.02)	Yes	No	135°C (275°
Type Alpha Metricel									
Alpha-1000	Regenerated	1	26	350[c]		8 (0.54)	Yes	No	149°C (300°
Alpha-6	cellulose	0.45	10	125[c]		17 (1.2)	Yes	No	149°C (300°
Alpha-8		0.2	4	55[c]		26 (1.8)	Yes	No	149°C (300°
Vinyl Metricel									
VM-1	Vinyl	5	100	700	2	1 (0.068)	No	Yes	66°C (150°F

Refractive Index	Extractables (Boiling Water)	Moisture Pick-up 24 hr at 90 % RH/20°C	Recommended Clearing Solution	Description, Applications and Special Properties
1.47	<2 %	3.5 % max.	*Solution:* 70 % *n*-methyl-2-pyrrolidone, 30 % water *Procedure:* wet, but do not soak, membrane in solution; dry 10 min at 93°C (200°F)	For biological filtration; assures high stability and safety in autoclaving; excellent chemical compatibility with alcohols.
1.47	<2 %	3.5 % max.		
1.47	<2 %	3.5 % max.		
1.47	<2 %	3.5 % max.		
1.47	<3 %	3.5 % max.		
1.47	<3 %	3.5 % max.		
1.47	<1 %	3 % max.	Same as above for Type GA Metricel	Cellulose Triacetate membrane specially formulated for tissue culture media filtration; it contains no wetting agent of any kind, yet can be safely autoclaved.
1.47	<1 %	3 % max.		
1.47	<1 %	3 % max.		
1.51	<6 %	6 % max.	*Solution:* 35 % dimethylformamide, 15 % acetic acid, 50 % water *Procedure:* wet, but do not soak, in solution; dry 5–10 min at 82°C (180°F)	Specifically designed for detection of coliforms in water. GN Metricel allows the coliform colony to grow with its characteristic shape and sheen. Autoclavability is excellent, without shrinkage or distortion. GN Metricel is also available in both green and black to provide a contrasting visual background and facilitate detection of light-colored bacterial colonies.
	<9 %	6 % max.		
	<9 %	6 % max.		
	<9 %	6 % max.		
	<9 %	6 % max.		
1.524	<1 %	7.5 % max.		This is the ideal membrane filter for clarification and purification of aggressive organic solvents. The regenerated cellulose pure white membrane is unaffected by acetone, methyl ethyl ketone, methyl isobutyl ketone, and dimethylformamide, and is commonly used for the filtration of photosensitive resists. It can be autoclaved.
1.524	<1 %	7.5 % max.		
1.524	<1 %	7.5 % max.		
	<10 %	<1 %		For air sampling and gravimetric analysis, VM Metricel provides high flow rates and exceptional weight stability.

Table 7.6 *(Continued)*

Membrane Type and Designation	*Composition*	*Pore Size* (μ)	*Flow Rate* *Air* [liters/(min)(cm²) at 70 cm Hg]	*Water* [ml/(min)(cm²) at 70 cm Hg]	*Bubble Point* *Water* psi (min) (atm)	*Kerosene* psi (min) (atm)	*Auto-clavable*	*Steriliz-able by Ethyl-ene Oxide*	*Maxim Opera Tempe ture*
DM Metricel									
DM-800	Copolymer of acrylonitrile and polyvinyl chloride	0.8	24	200	16	8 (0.54)	No	Yes	77°C (1
DM-450		0.45	10	70	30	15 (1.02)	No	Yes	77°C (1
Acropor									
AN-5000	Copolymer of acrylonitrile and polyvinyl chloride on nylon substrate	5	85	400	2 (0.14)	1 (0.068)	No	Yes	121°C (2
AN-3000		3	50	320	4 (0.27)	2 (0.14)	No	Yes	121°C (2
AN-1200		1.2	35	170	8 (0.54)	4 (0.27)	No	Yes	121°C (2
AN-800		0.8	23	120	12 (0.8)	6 (0.41)	No	Yes	121°C (2
AN-450		0.45	7	40	24 (1.6)	8 (0.54)	No	Yes	121°C (2
AN-200		0.2	3	17	45 (3.1)	11 (0.75)	No	Yes	121°C (2
ANH-450		0.45	7	20 psi[b]		13 (0.88)	No	Yes	121°C (2

[a] Courtesy of the Gelman Instrument Company.
[b] Water breakthrough pressure.
[c] Acetone is used for liquid flow rate tests. Alpha-Metricel is not compatible with water.

Woven Filter Media*

Woven filter materials can be used as the primary filter medium, as support or septum material as a backup for finer woven media or loose filter media, or alone—in strainers or roughing filters. Woven filter media can consist of either woven fabric or woven wire. The choice of fabric or wire media is determined in large part by the requirements of the given filtration application. These include chemical resistance to the materials being filtered, the durability and workability of the media, and cost.

In general, there are five basic sources of filter media: the basic material

* Much of the information in this section is based on a private communication with Lawrence G. Loff, Tetko, Inc., Precision Woven Screening Media, 420 Saw Mill Road, Elmsford, New York 10523.

·k- s)	*Refrac- tive Index*	*Extract- ables (Boiling Water)*	*Moisture Pick-up* 24 hr at 90 % RH/20°C	*Recommended Clearing Solution*	*Description, Applications and Special Properties*
0	1.52	<2 %	<1.3 %	*Solution:* 20 % dimethyl-formamide, 80 % ethyl alcohol *Procedure:* wet, but do not soak, in solution; dry 5–10 min at 77°C (170°F)	Excellent for filtration of acid solutions.
0	1.52	<2 %	<1.3 %		
5		<3 %	<1.8 %		Extraordinary strength, flexibility, and ease of handling are characteristics of this remarkably versatile membrane. It is manufactured from acrylonitrile polyvinylchloride copolymer membrane, reinforced with nylon.
5		<3 %	<1.8 %		
5		<3 %	<1.8 %		
5		<3 %	<1.8 %		
5		<3 %	<1.8 %		
5		<3 %	<1.8 %		
5		<3 %	<0.1 %		

manufacturer, the converter, the single-line medium company, the original filter manufacturer, and the filter media company.

The basic material manufacturer may be a producer of woven wire or woven cloth in various weaves and fiber or wire sizes. Such manufacturers supply the medium only and do not convert the material to a particular shape for use in a given filter application. The converter is a company such as a local tent and awning company or tailor shop, which takes the woven medium and converts it to its final form. Converters normally provide fast, effective service in a local situation. However, they do not always have complete technical knowledge about all the available filter fabrics. Since they purchase their woven media from the weaver or his jobber, they may be handicapped in getting one type of medium if it is no longer available and may find it difficult to locate a satisfactory substitute.

Many companies manufacture only one type of filter medium. For example, a company may produce only one type of a synthetic fabric or a range of types. Such specialization in one product can be advantageous if that specific product is required but disadvantageous if one's work calls for a different material.

The original filter manufacturer may not produce any filter media himself, or even convert them to his own apparatus, but he can usually be expected to know which materials are most suitable with his equipment. Also, he usually maintains replacement woven materials as replacements for his equipment.

A filter medium company normally weaves or manufactures many of its own materials, offers full converting services, and often acts as an agent for filter media that it does not itself manufacture. Such companies offer the best opportunity for finding the correct filter medium for a given application.

Both fabric and wire woven media are available in a number of different weaves, each having its own advantages and disadvantages (Fig. 7.3). Fabric media can be made of both natural and synthetic fibers into a variety of weights (expressed in ounces per square yard) and weave patterns. Four basic weave patterns, however, account for the bulk of woven media: plain, twill, plain reverse Dutch, and satin. The woven fabrics are made from three different forms of yarn: *monofilament*, a synthetic fiber made from a single continuous filament; *multifilament*, a yarn made by twisting two or more continuous monofilaments; and *spun-staple*, a yarn made by twisting short lengths of natural or synthetic fiber into a continuous strand.

The monofilament yarn allows minimum blinding with good cleaning and excellent cake discharge characteristics. The multifilament yarn has the greatest tensile strength and allows better cake discharge than spun yarns, which in turn provide best particle retention because of the presence of the hairy filaments, but also are more difficult to handle when cleaning and removing the cake.

Woven wire media can be made using many different metals and in a number of weaves: plain, twilled, plain Dutch, twilled Dutch, plain reverse Dutch, and basket (multibraid). To understand the construction of the different weaves used in woven fabric and wire media, some definitions are necessary. In filter media terminology, "warp wire" or "yarn" refers to the yarn or wire that runs lengthwise in cloth as it is woven on a loom. "Shute" or "fill yarn" or "wires" or "picks," on the other hand, run the short way of the cloth as woven. A *plain* weave medium has each shute or fill wire passing alternately over and under each warp wire. Thus square or rectangular openings are available depending on the relative sizes of the warp and shute wires. The size of the openings formed can be controlled by regulating both wire diameter and the number of wires per inch; openings are straight through the cloth, permitting the filtrate to pass through in a perpendicular path. Finer meshes with smaller openings in plain weaves are lacking in physical strength.

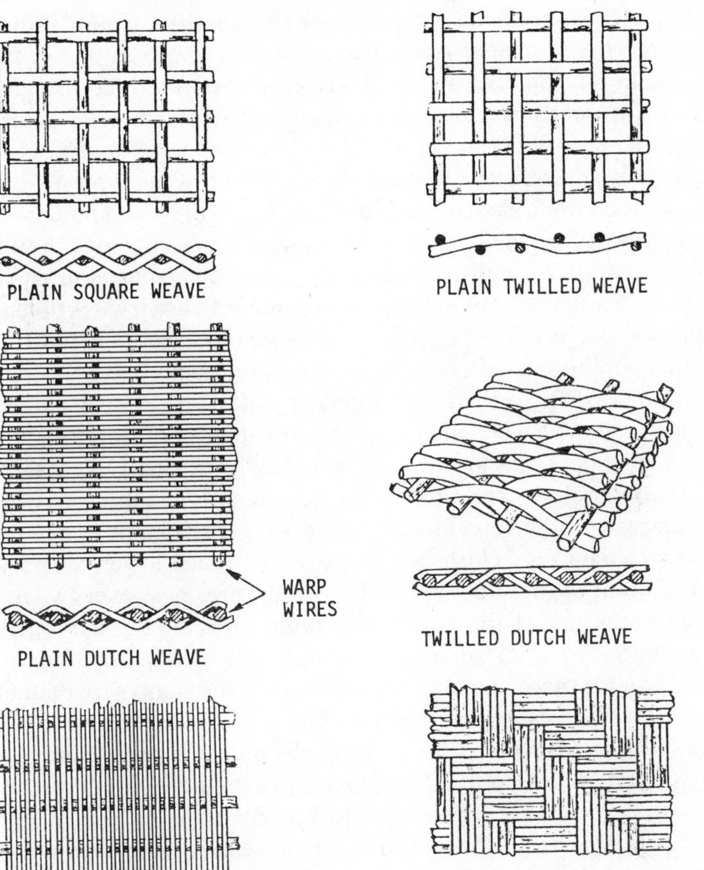

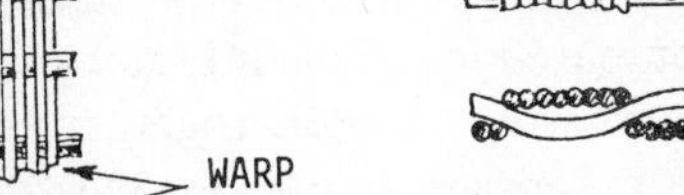

Fig. 7.3. Weaves. From *Tetko Training Manual*, Elmsford, New York, 1972.

Twilled weave filter cloth also has straight-through openings, but each shute or fill wire alternately crosses over two warp wires, then under two warp wires to form a diagonal pattern. Larger wires can be used for a given mesh size to provide proportionally smaller openings and greater physical strength. For example, a 0.0055-$in.^2$ opening can be obtained either with *plain square weave* (100 mesh with 0.0045-in. wire diameter) or with a twilled weave (80 mesh with 0.007-in. wire diameter). "Mesh" is defined as the number of openings or fractions of openings in a linear inch of cloth. Where a fractional part of an inch is specified (i.e., $\frac{1}{2}$ mesh or $\frac{1}{2}$-in. mesh), "mesh" is understood to mean the measurement from the center of one wire to the center of the next wire. In the example above, the twill weave would have longer service life and good wear resistance, but at the expense of a smaller percentage of open area (slower flow rate) in the wire cloth.

Basket weave cloths (or braided or multibraid weave) are twilled weave cloths in which multiple wires are used in both warp and shute, resulting in the formation of a dense, strong fabric. The mesh openings, however, are irregular because the multiple shute wires tend to twist around one another.

Plain Dutch weave cloths are generally made with warp wires that are larger than the shute wires. The warp wires are spread far enough apart to permit each shute wire to pass alternately over one and under one warp wire and to be positioned tightly against the adjacent shute wire. This yields relatively small openings with very high strength, but there are no straight-through openings. The openings are triangular and twist through the material at an angle. The filter cloths are rated on particle retention in microns because the opening sizes and the percentage of open area are difficult to determine.

Twilled Dutch weave cloths are made with a much lower warp wire mesh count than that of the shute wires. The shute wires pass over two and under two warp wires in both directions. The shute wires are pushed together such that one shute wire is on top of the next one, which is directly under the warp wire. This yields twice as many shute wires for the same wire diameter as in the plain Dutch weave, and the result is a very strong and dense weave through which it is impossible to see light on a perpendicular projection. Mesh counts range from 20 × 250 (approximately 100-μ retention) to upward of 375 × 2300 (approximately 4-μ retention).

Reverse Dutch weave cloths (Fig. 7.4) reverse the arrangement of the warp and shute wires as compared to the plain Dutch or twilled Dutch weave. The greater number of wires woven closely together are found in the warp, and the shute wires have the larger diameters. This yields a very strong fabric with high rates of flow compared to twilled Dutch weave having the same micron rating. The reverse Dutch weave cloths are easy to clean and have good resistance to blinding. Mesh counts range from 175 × 50 (approximately 95-μ retention) to about 850 × 155 (14-μ retention).

Filter fabrics woven with natural and synthetic fibers are similar to woven wire in both fabric construction and job capability. Many synthetic fabrics are produced with the objective of replacing as many woven wire or similar competing fabrics in natural or synthetic fiber itself. The plain, twill, and plain reverse Dutch weaves used with natural and synthetic fibers are overlaps from wire cloth weaves, but they have slightly different characteristics when put into the framework of woven, nonmetallic cloth. The satin weave, however, is one in which the shute (or warp) fiber passes over several warp

Fig. 7.4. Plain reverse Dutch weave, PRD 90W-18, 200 × magnification. Courtesy of Tetko, Inc.

(or shute) fibers, then under one, in an alternating pattern. When the pattern involves the shute fiber passing over three warp fibers and under one warp fiber in a satin weave, it is referred to as a three-shaft satin.

Wire cloth can be fabricated using wire made from all the malleable metals, principally including aluminum, brass, bronze, phosphor bronze, copper, monel, "Nichrome," nickel, steel, galvanized steel, tinned steel, and stainless steel. The most frequently used metal is stainless steel in either type 304 or type 316. Both are 18-8 alloys (18 % chromium, 8 % nickel), but the type 316 stainless includes molybdenum for increased corrosion resistance. For special purposes, wire cloths can also be made of gold, platinum, silver, tungsten, molybdenum, tantalum, and titanium. The selection of the proper metal cloth is based on knowledge of all the service conditions to which the cloth will be exposed.

Wire cloths as fine as 120 mesh per inch are usually woven 36 and 48 in. wide, although many meshes are woven in varying widths under 36 in. and in some meshes as wide as 96 in. (Newark Wire Cloth Company, 351 Verona Avenue, Newark, New Jersey 07104). On finer than 120 mesh, a 36-in. width is considered standard. Any width can be produced by welding several standard widths together. Using wire cloth, it is possible to fabricate any size circle, disc, or shape that may be required in a given filtration application.

In ordering wire cloth filter media, it is important to clearly specify the kind of cloth needed—space cloth or mesh cloth. A clear distinction should be made between these two types of cloth. *Space cloth* is woven wire cloth in which the size designation refers to the distance across the open space, in both directions, measured from the inner edge of one wire to the inner edge of the next adjacent wire. A $\frac{1}{8} \times \frac{1}{8}$ space cloth of woven wire would have a clear open space distance of $\frac{1}{8}$ in. between adjacent warp wires and adjacent shute wires. Table 7.7 lists typical ranges of values for several types of space cloth.

Mesh cloth is woven wire cloth in which the size designation refers to the number of openings per linear inch as measured from the center of one wire in both horizontal and vertical directions. With coarser mesh cloth, the count can be made with a ruler. With the finer cloths beginning with about 20 mesh, it is better to use a counting magnifying glass. Tables 7.8 and 7.9 supply characteristics of typical mesh cloth in selected size ranges and for several extrafine mesh cloth weaves, respectively.

Table 7.7 Typical Space Cloth Characteristics of Newark Woven Wire Products[a]

Clear Opening or Space (in.)	*Diameter of Wire*		*Open Area* (%)	*Weight (lb/ft²)*			
	in.	mm		*Plain Steel*	*Pure Copper*	*80–20 Brass*	*Monel*
1 × 1	$\frac{1}{2}$	12.70	44.4	11.25	12.75	12.43	12.62
1 × 1	$\frac{1}{4}$	6.35	64.0	3.26	3.70	3.60	3.66
1 × 1	0.120	3.05	79.7	0.83	0.94	0.92	0.93
$\frac{1}{2} \times \frac{1}{2}$	$\frac{3}{8}$	9.53	32.7	11.19	12.68	12.37	12.56
$\frac{1}{2} \times \frac{1}{2}$	$\frac{1}{4}$	6.35	44.4	5.62	6.37	6.21	6.31
$\frac{1}{2} \times \frac{1}{2}$	0.120	3.05	65.0	1.51	1.71	1.67	1.69
$\frac{1}{8} \times \frac{1}{8}$	0.120	3.05	26.0	4.19	4.75	4.63	4.70
$\frac{1}{8} \times \frac{1}{8}$	0.092	2.34	33.2	2.71	3.07	2.99	3.04
$\frac{1}{8} \times \frac{1}{8}$	0.063	1.60	44.2	1.43	1.62	1.58	1.60
$\frac{1}{8} \times \frac{1}{8}$	0.041	1.04	56.7	0.67	0.76	0.74	0.75

[a] Selected values courtesy of Newark Wire Cloth Company; for example, $\frac{1}{8} \times \frac{1}{8}$ space cloth can be obtained using wire sizes of 0.041, 0.047, 0.054, 0.063, 0.072, 0.080, 0.092, 0.105, and 0.120 in.

Table 7.8 Typical Mesh Cloth Characteristics of Newark Woven Wire Products[a]

Meshes per Linear Inch	*Diameter of Wire*		*Width of Opening*		Open Area	*Weight* (lb/ft²)			
	in.	mm	in.	mm	(%)	*Plain Steel*	*Pure Copper*	80–20 *Brass*	*Monel*
1 × 1	0.250	6.35	0.750	19.05	56.3	4.12	4.68	4.56	4.63
1 × 1	0.072	1.83	0.928	23.57	86.1	0.333	0.377	0.368	0.374
5 × 5	0.120	3.05	0.080	2.03	16.0	5.11	5.79	5.64	5.73
5 × 5	0.025	0.64	0.175	4.45	76.6	0.202	0.229	0.223	0.227
10 × 10	0.063	1.60	0.037	0.94	13.7	2.85	3.24	3.15	3.20
10 × 10	0.017	0.43	0.083	2.11	68.9	0.188	0.213	0.208	0.211
20 × 20	0.032	0.813	0.0180	0.46	13.0	1.48	1.68	1.63	1.66
20 × 20	0.013	0.330	0.0370	0.94	54.8	0.224	0.254	0.248	0.251
30 × 30	0.017	0.432	0.0163	0.41	23.9	0.592	0.671	0.654	0.664
30 × 30	0.0085	0.216	0.0248	0.63	55.4	0.143	0.162	0.158	0.160
40 × 40	0.0135	0.343	0.0115	0.29	21.2	0.53	0.60	0.59	0.60
40 × 40	0.008	0.203	0.0170	0.43	46.2	0.172	0.195	0.190	0.193
50 × 50	0.012	0.305	0.0080	0.2	16.0	0.511	0.579	0.565	0.573
50 × 50	0.0075	0.191	0.0125	0.32	39.0	0.192	0.218	0.212	0.215
100 × 100	0.005	0.127	0.005	0.13	25.0	0.170	0.193	0.188	0.191
100 × 100	0.003	0.076	0.0070	0.18	49.0	0.06	0.068	0.066	0.067

[a] Selected values courtesy of Newark Wire Cloth Company; many wire sizes are available in each mesh range.

In ordering wire cloth, it is necessary to specify the metal wanted, the mesh count or space opening, the wire in decimals of an inch (or millimeters) without regard to any wire gauge number, and the size and shape of the pieces desired. The edges of wire cloth can be supplied with raw cut edges, with selvage edges, or finished off in a number of ways. (A selvage edge is a finished woven edge.) A special sealed edge can be supplied in medium to fine meshes where the edge is made smooth and sealed against fraying. Woven mesh can be supplied with eyelets and grommets, monoclipper seams, zippers and other closures, webbing and hooks, plastic and sewn seams, and special reinforcements. The wire mesh filter media can also be fabricated into products.

Products that can be fabricated using wire mesh filter material include the following:

1. *Wet Filtering:* filter bags, screens, segments, drum covers, tubes, belts, and elements.

Table 7.9 Typical Mesh Cloth Characteristics of Extra Fine Wire Cloth[a]

Meshes per Linear Inch	*Weave*	*Diameter of Wire*		*Width of Opening*		*Open Area*
		in.	mm	in.	mm	(%)
110 × 110	Plain	0.0040	0.1016	0.0051	0.1295	31.4
110 × 110	Twilled	0.0045	0.1143	0.0046	0.1168	25.6
250 × 250	Plain	0.0016	0.0406	0.0024	0.0610	36.0
250 × 250	Twilled	0.0016	0.0406	0.0024	0.0610	36.0
400 × 400	Twilled	0.0010	0.0254	0.0015	0.0370	36.0

Mesh	*Weave*	*Wire Diameter* (in.)		*Approximate Retention* (μ)
		Warp	*Shute*	
12 × 64	Plain Dutch	0.023	0.017	300
14 × 100	Plain Dutch	0.014	0.011	250
24 × 110	Plain Dutch	0.016	0.011	145
30 × 150	Plain Dutch	0.011	0.008	120
20 × 250	Twilled Dutch	0.010	0.0082	84
30 × 250	Twilled Dutch	0.010	0.0082	74
50 × 250	Plain Dutch	0.0055	0.0045	64
28 × 500	Twilled Dutch	0.0085	0.0045	58
50 × 700	Twilled Dutch	0.004	0.003	40

[a] Courtesy of Newark Wire Cloth Company.

2. *Sifting:* sifter screens, sleeves, stockings, and separator bags.
3. *Food and Pharmaceutical Drying Applications:* conveyor belt drying.
4. *Dry Filtering and Dust Control:* air filter bags, bulk bin and loading vent bags, dust collector tubes.

Some typical fabricated mesh parts and some typical construction details used in the manufacture of these parts are presented in Figs. 7.5 and 7.6, respectively. Typically, wire mesh filter media are used in production of standard (dry) testing sieves, baskets, leaf filter elements for pressure and vacuum filters, and line strainers of a mesh size required to remove particles above a given size (Table 7.9).

Filter media come in an endless variety of materials and forms, including fabrics woven with natural and synthetic fibers. The common filter fabrics are woven from natural fibers such as cotton and from such synthetic fibers as nylon, polyester, polypropylene, rayon and acetate, saran and Teflon, and

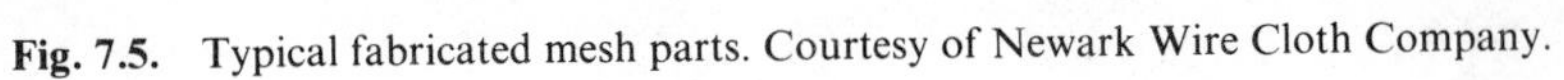
Fig. 7.5. Typical fabricated mesh parts. Courtesy of Newark Wire Cloth Company.

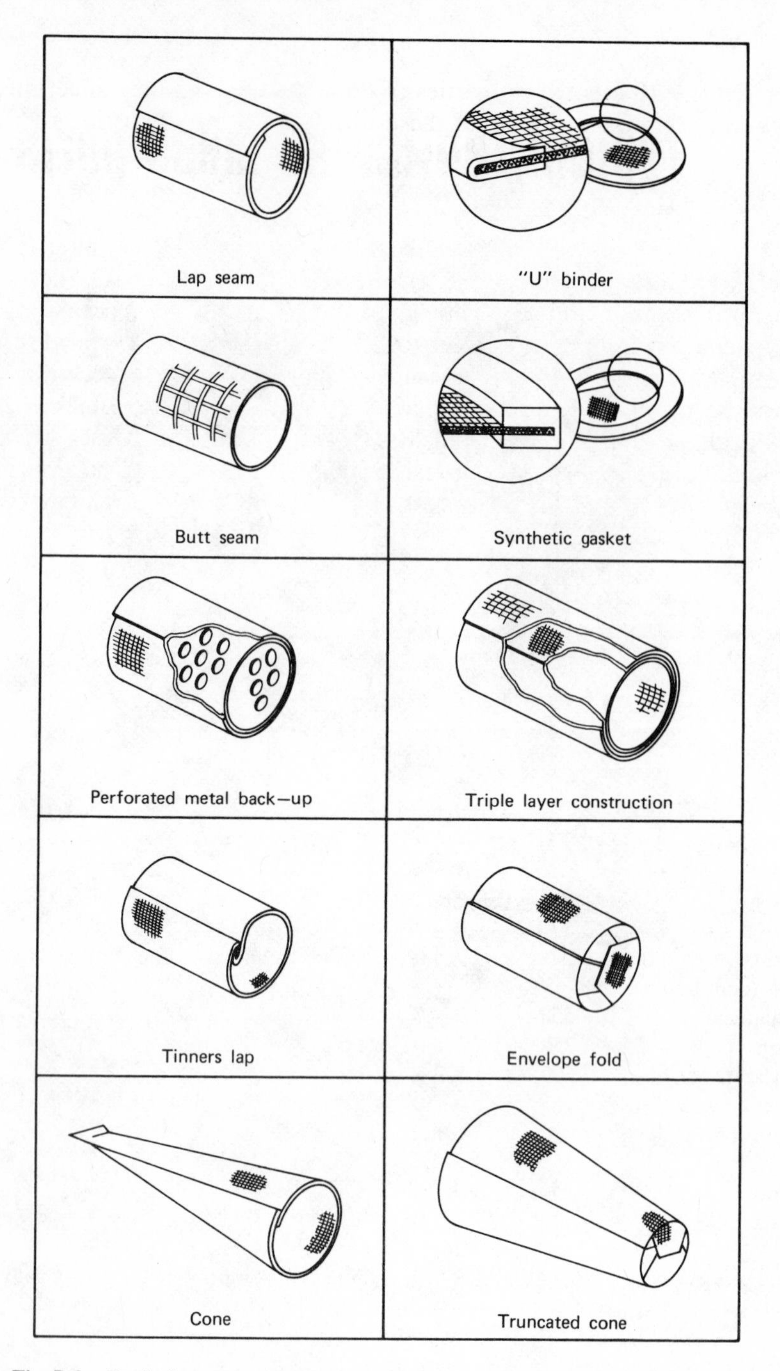

Fig. 7.6. Typical construction details. Courtesy of Newark Wire Cloth Company.

glass. Table 7.10 lists the properties of these materials as they affect media selection. A brief description of these fibers and some discussion of their potential is based on "Filter Media," by Ronald C. French, referenced in Table 7.10.

1. *Cotton.* Cotton has always been the most widely used filter fabric material because of its low price. It provides excellent strength and good abrasion resistance, together with good particle retention, because of its hairy elements. Cotton is resistant to heat degradation. In a typical test it withstood 5 hr of continuous use at 302°F before the cloth decomposed. Cotton is unaffected by cold dilute acids but is vulnerable to attack by hot weak acids or cold concentrated acids. Exposure to caustic causes swelling but no damage to the fiber. Poor resistance of the fiber to mildew and fungi has been offset by various treatments applied to the fabric after weaving.

Table 7.10 Fiber Properties Affecting Media Selection[a]

Fiber	*Temperature Recommended Safe Limit* (°F)	*Wet Breaking Tenacity*[b] (g/denier)	*Acid Resistance*	*Alkali Resistance*	*Price Ratio to Cotton*	*Yarn Forms*[c]
Cotton	210	3.3–6.4	Poor	Fair	1	S
Polyester (Dacron)	300	6.0–8.2	Very good	Good	2.7	F, S
Dynel modacrylic	200	3.0	Excellent	Excellent	3.2	S
Glass (spun)	750	3.0–4.6	Excellent	Fair	6.0	S
Glass (continuous filament)	550	3.9–4.7	Excellent	Fair	2.2	F
Nylon	250	2.1–8.0	Fair	Excellent	2.5	F, S, M
Acrylic (Orlon)	300	1.8–2.1	Excellent	Fair	2.7	S
Polyethylene	165	1.0–3.0	Excellent	Excellent	2	M
Polypropylene	175	3.5–8.0	Excellent	Excellent	1.75	F, S, M
Saran	160	1.2–2.3	Excellent	Excellent	2.5	F, M
Teflon	475	1.9	Excellent	Excellent	25.0	F, M
Polyvinylchloride	165	1.0–3.0	Good	Excellent	2.7	F
Wool	210	0.76–1.6	Very good	Fair	3.7	S
Rayon and acetate	210	1.9–3.9	Poor	Fair	1	F, S

[a] From Ronald C. French, "Filter Media," *Chemical Engineering*, October 14, 1963.
[b] Breaking tenacity is the tensile strength at rupture of a specimen.
[c] F, multifilament; S, staple; M, for monofilament.

2. *Nylon.* Next to cotton, nylon claims the greatest usage in woven filter fabrics. It has excellent abrasion resistance—about 3 times that of wool. Though providing an extremely smooth surface for good cake discharge, it is unaffected by mold, fungus, or bacteria. Nylon has high tensile strength and loses only about 15 % of its strength when wet. It has good elasticity in filter operations where continuous media flexing is present. Nylon is degraded by oxidizing agents and mineral acids, the degree increasing with concentration and temperature. It resists most common alkalies and organic acids of any strength. It is unaffected by most solvents except for *m*-cresol and synasol.

3. *Polyester.* Polyester fibers (Dacron) can be woven into strong, flex-resistant filter cloth capable of withstanding severe abrasion and giving long life. Since Dacron has a high initial shrinkage, the fabrics become quite tight and offer excellent retention of solids. The material has high tolerance of oxidizing agents and good resistance to most acids except for concentrated nitric or sulfuric solutions. Although its alkali resistance is superior in many cases to that of Orlon acrylic fiber, it has only moderate resistance to strong acids at room temperature and totally decomposes in strong boiling acids. With the exception of aromatic phenols, Dacron is very resistant to solvents. It has high resistance also to mold, mildew, and fungus as well as low (0.6 to 0.7 %) moisture adsorption. Hydrolysis, on the other hand, does cause degradation.

4. *Acrylic.* Acrylic fiber (Orlon), formerly available in both filament and staple yarn forms, is now available in staple yarn forms only. Where the combination of good abrasion resistance and high temperature resistance is important, Orlon is unequaled. It offers good to excellent resistance to mineral acids but dissolves in sulfuric acid concentrations above 80 %. It has fair to good resistance to weak alkalies and to strong alkalies at room temperature. It is not harmed by common solvents, oil, greases, neutral salts, and most acid salts. It is unusually resistant to oxidation when exposed to air heated to 300°F. It is excellent for dry filtration and in atmospheres of hot corrosive fumes, and it is virtually impervious to water adsorption.

5. *Dynel.* Dynel modacrylic filter fabrics have good retentivity, gasketing properties, and tensile strength. Dynel also has high dimensional stability, showing shrinkage of only 0.5 % in boiling water. This synthetic is virtually unaffected by most concentrated mineral acids and bases, but it is unsuitable for use with ketones, certain amines, and aromatic hydrocarbons at elevated temperatures. It has poor resistance to most solvents.

6. *Glass.* Glass fiber fabrics have claimed a strong position in the filter media field because of their outstanding resistance to both acids and elevated temperatures. They completely resist acids, except for hydrofluoric and hot phosphoric in their most concentrated forms. They are attacked by strong alkalies at room temperature or weak alkalies at higher temperatures.

Although glass has a high resistance to compression, it is vulnerable to abrasion and flexing. To reduce abrasion, a mineral oil is employed as a lubricant in weaving.

There are two types of glass yarn—chemical and electrical. Chemical glass has slightly higher acid resistance than electrical glass, which is primarily designed for electrical applications. The electrical type is available in both staple and continuous filament yarns, whereas chemical glass fibers are available in staple form only.

7. *Polyethylene.* Polyethylene fabrics have rapidly been replaced by polypropylene fabrics. For a time polyethylene fabrics were woven in both multifilament and monofilament forms of high-density polyethylene, but today they are available only in monofilaments of the conventional low-density form. The material found widespread use because of its low cost and chemical resistance. It withstands both alkalies and inorganic acids at room temperatures as well as many organic chemicals. Although a relatively weak fiber, its elasticity proved to be particularly attractive for use in conventional horizontal filters where air blowback was involved. As a monofilament, polyethylene is recommended for filtering out coarse crystals, fibrous pulps, and other free-filtering materials.

8. *Polypropylene.* Polypropylene has the lowest density of any synthetic filter cloth. This results in greater cloth yield per pound of yarn used, which is reflected in lower initial cloth cost and lower shipping charges. The sleek fiber provides good cake discharge and retardation of blinding. It has a recommended safe operating temperature of 175°F. Chemical properties include good resistance to acids and alkalies as well as to many solvents except for the aromatic and chlorinated hydrocarbons. Since polypropylene has extremely low moisture adsorption (rated at less than 0.03 %), it is possible to use this filter medium in dye production on more than one color or pigment with only a light wash between batches. This synthetic medium promises to become the most universal type, replacing many other synthetics now in use.

9. *Rayon and Acetate.* Although widely used in filtration applications, very little of these fibers appears as woven cloth. Rather, usage is in nonwovens, batting, and roving.

10. *Saran.* Saran filter cloth is available in both staple and monofilament yarns. However its widest application is in open monofilaments and screens. It is one of the heavier fibers and holds up well under abrasion. It is resistant to mineral acids, most alkalies, chlorine, and alcohols. It has excellent durability when exposed to acids, particularly hydrochloric, and to alkalies with the exception of ammonium hydroxide. It may be cleaned with solvents except those bearing oxygen, which cause it to soften.

11. *Teflon.* Only asbestos and quartz media have a higher temperature range of application than Teflon, but the former do not hold together with

sufficient strength for most liquid filtration needs. Teflon is virtually chemically inert, being attacked only by fluorine gas, molten alkali metals, and trichlorine fluoride. It is the smoothest of all synthetics, giving exceptional cake discharge and reduction of blinding.

At 400°F, Teflon releases a toxic gas that is injurious to operators; thus the filter must be enclosed under such conditions. Usage is restricted by high cost, limited retentivity, and poor gasket qualities.

Table 7.11 summarizes properties of Tetko fibers used in filter media, and Table 7.12 furnishes data on their chemical compatibility.

Table 7.11 Properties of Fibers Used in Filter Media[a]

Property	*Nylon (Nitex)*	*Polyester (PeCap)*	*Polypropylen (Propyltex)*
Specific gravity	1.14	1.38	0.91
Tensile strength	(See product manufacturer's listings)		
Elongation at breaking point, dry	20–40 %	15–30 %	20–50 %
Elongation at breaking point, wet	25–45 %	15–30 %	20–50 %
Moisture absorption at 65 % RH, 68°F (20°C)	3.8–4.2 %	0.4 %	0
Melting point, °F (°C)	414–487 (212–253)	482–500 (250–260)	329–347 (165–175)
Softening point, °F (°C)	338–455 (170–235)	428–464 (220–240)	284–329 (140–165)
Resistance to sunlight and weather	Poor to average; loses strength after prolonged exposure to light	Good to very good	Poor to average; strength after prol ed exposure to lig
Abrasion resistance	Very good	Very good	Average to good
Resistance and reaction to acids (see Table 7.12 for detailed listings)	Average; dissolves in cold hydrochloric, sulfuric, and formic acids depending on concentration	Very resistant to most acids	Very resistant to r acids
Resistance to alkalies, reaction with caustics (see Table 7.12 for detailed listings)	Very good at room temperature, partially weakening at higher temperatures	Average to good; resistant to weak lyes; dissolves in strong lyes at high temperatures	Very resistant, exc to some oxidizing agents
Reaction to organic solvents, such as used for dry cleaning (see Table 7.12 for detailed listings)	Generally insoluble; dissolves in some phenol compounds	Generally insoluble; dissolves in some phenol compounds and hot nitrobenzene	Good resistance w few exceptions

[a] Courtesy of Tetko, Inc.

Table 7.12 Chemical Compatibility of Synthetic Fibers Used in Filter Media[a]

Reaction or Resistance to	*Ratings[b] of Fibers at Maximum Temperature Indicated*				
	Nylon (Nitex), 250°F	*Polyester (PeCap),* 300°F	*Poly-ethylene,* 150°F	*Polypro-pylene (Propyl-) tex),* 250°F	*Teflon (Fluortex),* 500°F
Mineral acids					
Aqua regia	C	B	A	A	A
Chromic	C	A	A	B	A
Hydrochloric	C	A	A	A	A
Hydrofluoric	C	B	A	A	A
Nitric	C	A	B	B	A
Phosphoric	B	A	A	A	A
Sulfuric	C	B	A	A	A
Organic acids					
Acetic	B	A	A	A	A
Benzoic	C	A	A	A	A
Carbolic	C	B	B	D	A
Formic	B	A	A	A	A
Lactic	B	A	A	A	A
Oxalic	A	A	A	A	A
Salicylic	B	A	A	A	A
Bases					
Ammonium hydroxide	B	C	A	A	A
Calcium hydroxide	A	A	A	A	A
Potassium hydroxide	B	B	A	B	A
Potassium carbonate	A	B	A	A	A
Sodium hydroxide	B	B	A	B	A
Sodium carbonate	A	A	A	A	A
Salts					
Calcium chloride	C	A	A	A	A
Ferric chloride	B	A	A	A	A
Sodium acetate	A	A	A	A	A
Sodium benzoate	B	A	A	A	A
Sodium bisulfite	B	A	A	A	A
Sodium bromide	A	A	A	A	A
Sodium chloride	A	A	A	A	A
Sodium cyanide	B	A	A	A	A
Sodium nitrate	B	A	A	A	A

Table 7.12 (*Continued*)

Reaction or Resistance to	*Ratings[b] of Fibers at Maximum Temperature Indicated*				
	Nylon (Nitex), 250°F	*Polyester (PeCap)*, 300°F	*Poly-ethylene*, 150°F	*Polypro-pylene (Propyl-) tex)*, 250°F	*Teflon (Fluortex)*, 500°F
Sodium sulfate	A	A	A	A	A
Sodium sulfide	A	A	A	A	A
Zinc chloride	C	C	A	A	A
Oxidizing agents					
Bromine	C	B	C	A	A
Calcium hypochlorite	B	A	A	A	A
Chlorine	C	B	C	A	A
Fluorine	C	B	B	A	A
Hydrogen peroxide	B	B	C	A	A
Iodine	C	A	C	A	A
Ozone	D	A	B	B	A
Peracetic acid	B	D	C	A	A
Potassium chlorite	B	A	A	A	A
Potassium permanganate	B	C	C	C	A
Sodium hypochlorite	B	B	B	A	A
Sodium chlorate	D	D	A	B	A
Organic solvents					
Acetone	A	A	B	B	A
Amyl acetate	A	A	C	B	A
Benzene	A	A	C	A	A
Carbon bisulfide	A	B	A	C	A
Carbon tetrachloride	A	A	C	B	A
Chloroform	A	A	C	B	A
Cyclohexanone	A	A	B	B	A
Diethylene glycol	A	A	A	A	D
Ethyl acetate	A	A	C	B	A
Ethyl alcohol	A	A	A	A	A
Ethyl ether	A	A	B	A	A
Furfural alcohol	D	A	C	D	A
Methanol	A	A	A	A	A
Methyl ethyl ketone	A	A	B	B	A
Naphtha	A	A	B	B	A
Propylene glycol	A	C	B	A	A
Stoddard solvent	A	A	B	B	A

Table 7.12 (*Continued*)

Reaction or Resistance to	Ratings[b] of Fibers at Maximum Temperature Indicated				
	Nylon (Nitex), 250°F	*Polyester (PeCap)*, 300°F	*Poly-ethylene*, 150°F	*Polypro-pylene (Propyl-) tex)*, 250°F	*Teflon (Fluortex)*, 500°F
Trichloroethylene	A	A	B	A	A
Tricresyl phosphate	A	A	C	D	A
Toluene	A	A	C	B	A
Xylene	A	A	C	B	A
Miscellaneous					
Acetaldehyde in water	A	A	D	D	A
Benzaldehyde in water	A	A	D	D	A
Formaldehyde	A	A	A	B	A
Cottonseed oil	A	A	A	A	A
Glycerine	A	A	A	A	A
Glycol	A	C	A	A	A
Lard	A	A	A	A	A
Linseed oil	A	A	A	A	A
Mineral oil	A	A	A	A	A
Nitrobenzene	A	A	A	A	A

[a] Courtesy of Tetko, Inc.
[b] Key: A, Satisfactory; B, conditional (low concentration and temperature); C, unsatisfactory; D, unrated.

The method of weaving filter media controls the size of the pore openings. The plain pattern provides the tightest weave and smallest pore openings, followed by chain, twill, and satin weaves with increasing porosity. Since the plain weave takes fewer picks than the other weaves, it is possible to produce a lighter and more economical material. That is, if a filter user can get by with a 10-oz. plain weave and its lower strength, it would be more economical than a 12 to 14 oz. twill and would give the same results with respect to filtrate clarity.

Even though fibers of the synthetic materials can be of multifilament or monofilament weave, there is now a trend to use monofilament filter fabrics (Fig. 7-7). This has come about mainly because in the past it was very difficult

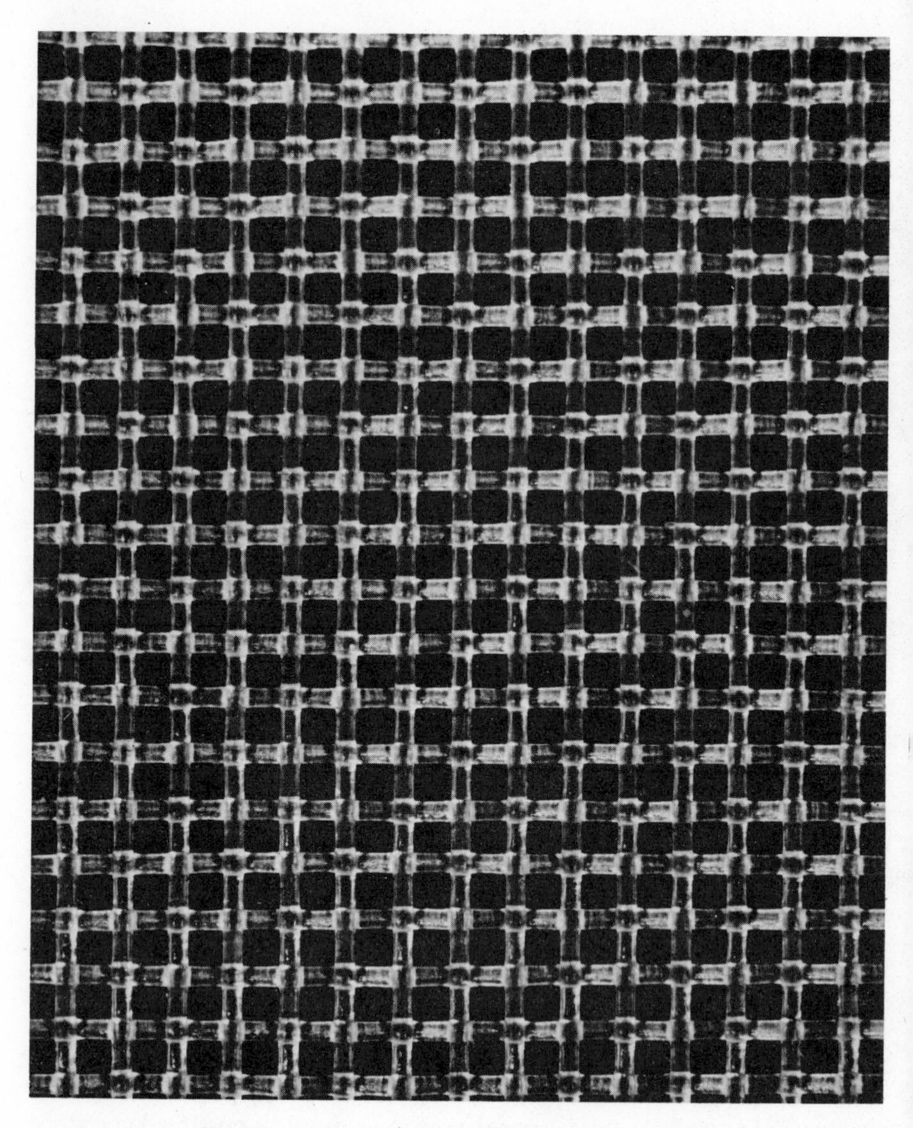

Fig. 7.7. Nylon monofilament in a plain square weave.

to weave monofilaments in the fine, low micron retention areas. Today monofilament fabrics can be woven twilled down to 20 μ and plain reverse Dutch down to 14 μ.

The maximum thread count per inch varies with the ply and yarn dimensions used. The retentivity of a fabric can be increased only by changing the weave pattern. It is possible to put more picks in a satin weave than in either a twill or a chain weave, thus increasing the particle retention of the former. The plain weave takes the fewest picks and has the least porosity and particle retention for a given thread count.

Another method for decreasing filter porosity is through plying. For example, if a fabric made from 800-denier yarn (weight in grams of a single continuous strand of yarn 9000 m long) is replaced with a multifilament yarn composed of four strands of 200-denier yarn plied together, the total multifilament denier remains the same, but the retentivity of the fabric increases significantly. Of course, the cost of the four 200-denier strands is greater. The alternative is to use a lower denier count monofilament fiber and a greater thread count.

The strength and abrasion resistance of filter fabrics can be improved by increasing fiber diameter. However increasing fiber diameter serves to reduce the maximum possible thread count per inch, which in turn lowers both filter retentivity and cake release.

Woven fabrics are also subjected to a variety of finishing processes that affect their filter media characteristics. Calendering uses high-pressure hot rollers, which cause the cloth to shrink and reduce its porosity, giving its surface a smooth polish that improves cake discharge and helps to keep the cloth clean. The process also causes some weakening of the cloth. Napping involves the use of a fine steel comb to produce a soft fuzz on the face of the cloth, thus improving the ability of the cloth to retain fine particles. Heat treatment is sometimes applied to synthetic fibers to stabilize the fabrics and make them more useful at elevated temperatures.

The most important part of the filtration process using synthetic cloth media is the proper selection of the filter fabric. One first determines the material of which the fabric is made and the specific filtration application intended, taking into account the material's resistance to heat, abrasion, and chemicals. The next consideration is the fabric weave and its effects on filtrate clarity, the filter blinding characteristics, and the cake release properties. Multifilaments can yield clearer filtrates but generally have a greater tendency to blinding than monofilaments. When very fine solids must be handled and a clear filtrate is required, one of the high-twist multifilament yarns, twilled-type weave, frequently provides the best compromise between filtrate clarity and cloth-blinding tendencies. This is not to say that a very fine monofilament calendered fabric would not be a good choice. Assumptions are not as effective in selecting a proper fabric as the conduct of bench-scale filtrations or as pilot plant trial of several filter cloths.

The selection of the proper fabric filter media for a specific filtration application can be difficult because of the large number of combinations of materials, thread diameters, mesh openings, and weaves available. The choice must be based on numerous criteria, including flow rate, particle size, environment (shock, vibration, abrasion, temperature, chemicals and their concentration, pressures, pressure drops, etc.), cleanability, cost, and other factors. In ordering materials, it is necessary to specify the type of material and mesh opening in inches or microns, the mesh count (meshes per linear inch), and thread diameter (decimal parts of an inch). Table 7.13 describes some of the characteristics of filter media currently available from Tetko. This table gives typical ranges of filter media that are available, but it is not intended to be complete.

Perforated Metals

Filter media (or filter septa) can also be made in the form of a metal sheet into which holes are punched or etched. It is possible, too, to electrodeposit a metallic sheet around a basic hole pattern by an electrochemical process so precise that the screen formed is all but free of burrs and other defects, greatly reducing hole clogging problems.

Perforated metals can be produced in a wide variety of openings (round, square, slotted, hexagonal, crescent), with hole sizes down to 10 to 12 μ. Smaller openings are likely in the future. The holes themselves can be made cylindrical, conical, or venturi in cross section. Conical holes provide maximum freedom from plugging and permit complete removal of solids during a backwash cleaning cycle. Venturi openings are used when extra metal sheet thickness is required.

Perforated metal can be supplied in which the holes are in straight lines or in staggered or diagonal arrays. A 90° array would produce straight lines of holes in both directions, giving a square metal shape between holes. A 60° array, on the other hand, would produce a triangular metal area between holes, both strengthening the metal sheet and increasing the open area of the perforated sheet.

Perforated metal filter media can be provided of pure nickel, copper, nickel-plated copper, stainless steel, and other metals. Punched holes can be made in almost all metals. Photoetching processes are frequently used in making stainless steel screens. Electrochemical processes find their principal application in the formation of copper or nickel screens. The special processes used in forming the filter media limit the physical size of the perforated sheet that can be prepared without welding sheets together. For example, microetched stainless steel screens normally come in continuous coils with an

Table 7.13 Twill Weave Fabric Characteristics[a]

Material	*Nominal Filter Rating* (μ)	*Bubble Point* (mm wet strength)	*Absolute Filter Rating* (μ)	*Permeability* [liters/(m²)(sec)] *Air*	*Water*	*Weight* (g/m²)
1. Polyester	20[b]	380	24	69	10	280
2. Polyester	27	200	35	215	67	160
3. Polyester	50	125	57	350	75	300
4. Polyester	86	71	98	925	220	370
5. Polyester	103	54	129	2175	625	335
Polypropylene	36	178	40	125	30	220
Polypropylene	79	73	95	555	116	275
Polypropylene	150	41	166	880	150	480

Characteristics	*Polyester Fabric as Above* 1	2	3	4	5
Mesh count per inch	261.6/71	344.2/66	238.8/61	165.1/38.1	179/33
Thread diameter (μ)	120/120	80/140	140/140	170/250	150/250
Cloth thickness (μ)		213	350	460	560
Weight (oz./yd²)		4.72	8.85	10.91	9.88
Bubble point (in. water gauge)		7.87	4.92	2.79	2.13
Tensile strength (lb) (2 × 8 in. section) (warp/weft)		370/304	882/194	847/463	719/410
Elongation (%) (warp/weft)		33/20	30/40	56/45	62/41

[a] Courtesy Tetbo, Inc.
[b] Average pore size of cloth.

18 × 21 in. perforated pattern separated by a $\frac{1}{4}$-in. nominal solid rib occurring between each perforated pattern (Buckbee-Mears Company, 245 East 6th Street, St. Paul, Minnesota 55101). Electroplate perforated metal can be secured in 39.3 × 39.3 in. sheets welded to any larger size required. Some products come in sheets as small as 8 × 8 in.

Electroformed pure nickel or copper perforated metal filter media can be obtained with conical or venturi holes as small as 10 μ. Etched materials, including stainless steel can be produced with openings as small as 100 μ

(0.004 in.). Punched and pierced metals with all hole types and with straight-through, conical, or trapezoidal cross sections can be produced with openings as small as 75 μ (0.003 in.). Table 7.14 presents some typical perforated sheets available from Perforated Products, Inc., 68 Harvard Street, Brookline, Massachusetts 02146.

It is not normally possible to obtain perforated metal filter media thicker than the hole diameter. For example, a perforated metal medium with 0.02-in. straight-through holes will have a maximum plate thickness of 0.02 in. With conical holes, the plate thickness can be twice the diameter of the hole.

Loose Granular Materials

When difficulties are encountered in laboratory filtrations, either in obtaining adequate throughput of filtrate or in the degree of clarity or suspended solids removal in the filtrate, it is customary to add a filter aid to the suspension before the filtration. Frequently used for this purpose are loose filter materials, which may be either granular materials, such as diatomite or perlite filter aids, or fibrous filter materials, such as glass or asbestos fibers or wood pulp derivatives. Selection of the proper loose filter material may substantially increase the rate of filtration or the duration of the effective filtra-

Table 7.14 Typical Perforated Filter Media Available from Perforated Products, Inc.

Material	*Thickness* (in.)	*Holes* (in.)	*Open Area* (%)	*Array*	*Cross Section*
Round holes					
Stainless steel	0.014	0.015	19	60°	Conical
Stainless steel	0.003	0.0138	30	60°	Straight
Stainless steel	0.004	0.007	40	90°	Straight
Square holes					
Copper	0.003	0.080 × 0.080	85	90°	Straight
Mesh sieves[a] (8 × 8 in. pieces)					
Nickel	1270 mesh	0.0002	6,[b] 6.25[c]	60°	Conical
Nickel	806 mesh	0.0004	9,[b] 10[c]	60°	Conical
Nickel	635 mesh	0.0008	22.5,[b] 25[c]	60°	Conical

[a] "Micro Sieves" with holes smaller than 0.0039 in. contain an integral backing to add strength to the material.

[b] Round holes.

[c] Square holes.

tion period or volume of throughput, or it may result in an improved degree of clarification.

The main function of a filter aid is to build up on the supporting septum—which may be a filter paper, a membrane, or a woven fabric—a porous, permeable, and rigid lattice or incompressible structure capable of retaining the solid particles and allowing the fluid to pass through the interstices and channels within the porous structure. If the loose materials added to the suspension are properly related to the concentration of suspended particles the filtration is designed to remove, the impurities are spaced uniformly throughout the cake by the loose material-suspended particles cake that forms, causing the cake to assume the hydraulic characteristics of the loose material instead of the impurity. In general, the loose material needed will be from 2 to 10 (or more) times the concentration of the impurities. The proper concentration can be determined quickly by trial and error after the proper grade of material has been selected. Loose granular materials tend to produce a high-porosity, rigid filter cake matrix that minimizes the compressibility of the collected solids, thus serves to maintain a high filtration rate. Many loose granular materials cannot be used, however, because of their interaction with the fluid, which may add to or change the level of soluble materials present in the fluid. Loose fibrous materials may form a more compressible cake, but their composition may result in their having less contaminating effects on the filtrate.

The characteristics of loose filter materials that affect their use as filter aids include the following:

1. Their structure, which permits formation of a rigid, porous cake: many loose materials, when clean, have porosities that reach 80 to 90 %.
2. A particle size distribution that can be controlled in various ranges (different grades of material) to provide the pore diameter required for the retention of impurities that are to be removed. The coarser the material that can be used to provide the desired retention, the better will be the filtration characteristics of the cake.
3. The material should be of a size and density that permit it to remain suspended fairly uniformly in the suspension while filtration takes place.
4. The loose filter material itself should be free of impurities and should be inert in reaction with the fluid being filtered.
5. The material should be available in a moisture-free condition in cases in which addition of moisture to the fluid would be undesirable.

Diatomaceous earth, diatomite, kieselguhr, fossil silica, and infusorial earth are all terms that refer to hydrated silica made up of the skeletal remains of diatoms, the single-celled plants related to algae that grow in lakes or oceans. Diatomite is the most widely used type of filter aid.

Diatomite is processed from fossil-like skeletons of diatoms, members of the Bacillariophyceae class of algae. In the geological past more than 10,000 species of diatoms fluorished in the waters covering certain of today's coastal areas. When these diatoms died and their skeletons sank to the ocean floor, large deposits of almost pure silica were formed. Later the land rose from the ocean floor, and the deposits are now mined in open quarries. The largest and purest deposit of diatomaceous earth is located near Lompoc, California. Other deposits are mined along the western coast of the United States and Canada and throughout the world. The United States is the world's largest producer and user of diatomite.

Diatomaceous earth has many applications: as a filter aid in the filtration of sugar syrups, beverages, and various chemicals, as well as water; as a mineral filler in lacquers and paints, polishes, plastics, paper, insecticides, and so on; as high-temperature insulation; as an admixture for concrete; as an absorbant; and for countless other industrial applications. Processing the crude diatomaceous earth for use as a filter aid consists of grinding, drying, and flux calcining. When flux calcining takes place, 3 to 10 % by weight of

Table 7.15 Typical Celite[a] Properties

Physical Properties		*Chemical Properties: Average Analysis, Dry Basis* — *Constituent or Property*	*Percentage*
Specific gravity	2.0–2.3		
Refractive index	1.42–1.48		
Specific heat [cal/(g)(°C)]	0.24	Silica (SiO_2)	89.7
Particle charge	Negative	Alumina (Al_2O_3)	3.7
Retained on 325 mesh, maximum	0.5–3.0%	Iron oxide (Fe_2O_3)	1.5
		Titanium oxide (TiO_2)	0.1
Average absorption, Gardner-Coleman method		Lime (CaO)	0.4
Water	150–220%	Magnesia (MgO)	0.7
Linseed oil	120–205%	Alkalies (as Na_2O)	0.8
Bulk density (lb/ft^3)			
Dry, loose	8–10		
In filter cake	15–28	Ignition loss (combined H_2O, CO_2, and organics)	3.1

[a] Registered trademark, diatomite filter aids, Johns-Manville Products Corporation, Manville, New Jersey.

soda ash, sodium chloride, or caustic soda is added to the crude ore. Calcination affects the filtering properties of diatomite by changing the surface texture, agglomerating fines, and converting clay minerals to aluminum silicate slag. The slag particles are largely eliminated in later processing steps. During the processing, the diatomite is separated into different particle size fractions by air classification. Particle sizes of individual diatoms vary from less than 5 to more than 100 μ. Grades of diatomite separated by air classification have mean particle sizes ranging from about 14 to 25 μ.

The performance of diatomite as a filter aid depends on the unique physical structure of the diatom particle. The almost infinite variety of shapes and sizes and the extremely porous framework of the skeletons provide numerous microscopic waterways and microscopic sieves which, when used as filter aids, serve to trap impurities. Since the particles are rigid and strong, contact is limited to their outer points; as a result, packing does not occur and the filter cake formed remains extremely porous. The porosity of a clean filter cake varies from 80 to about 90 % for various grades of diatomite. Table 7.15 gives other typical properties of diatomite filter aids.

Figure 7.8 presents a cross section through a loose material filter cake. The clarity or the quality of the filtered water is a function primarily of the

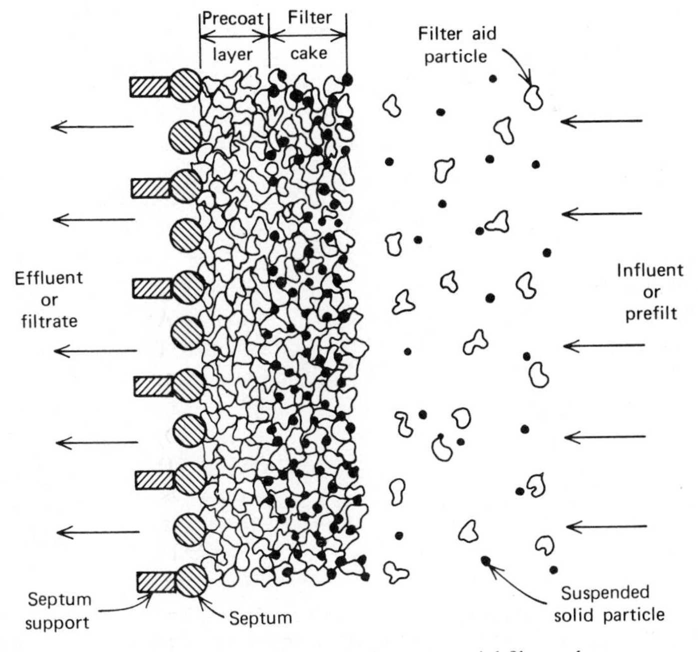

Fig. 7.8. Cross section of a loose material filter cake.

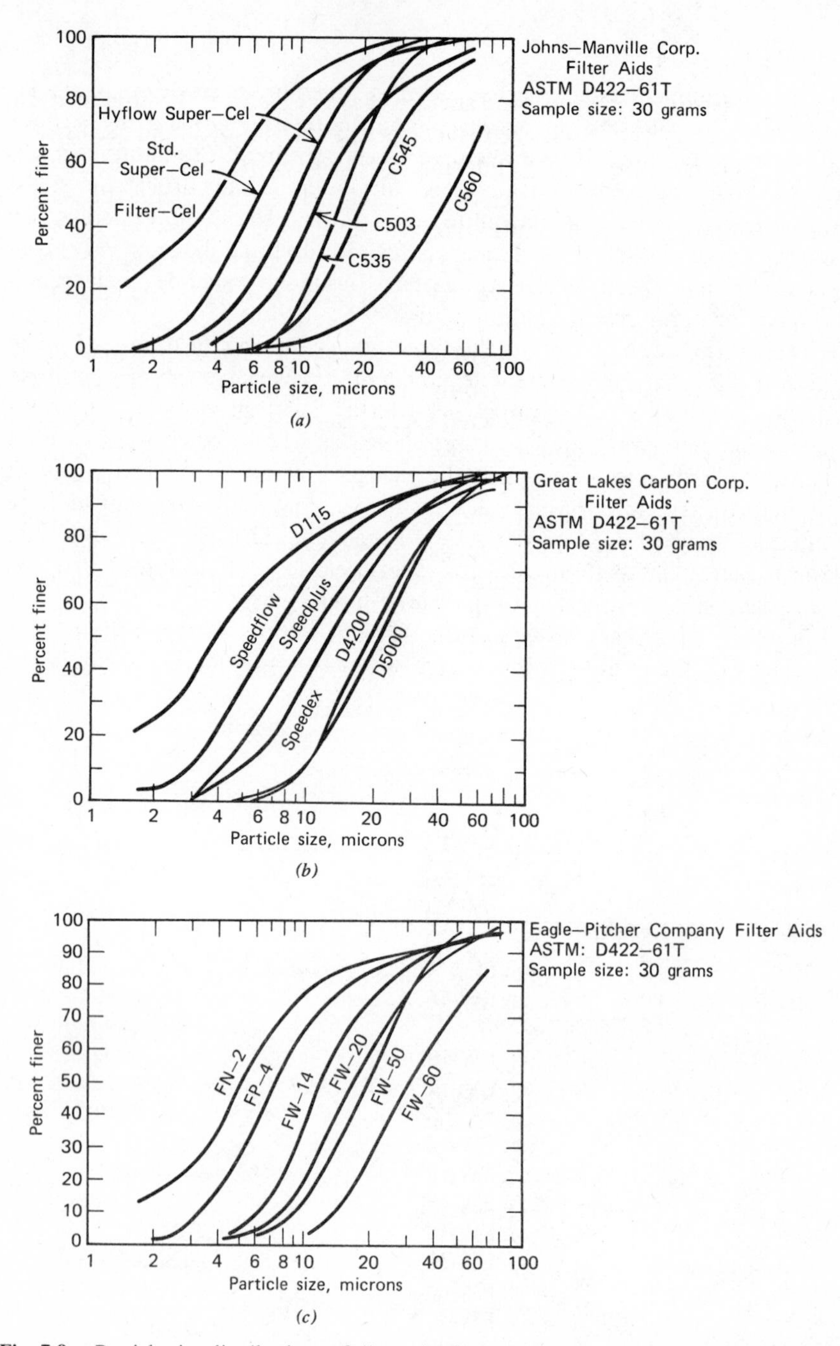

Fig. 7.9. Particle size distributions of diatomite filter aids produced by three manufacturers. From James H. Dillingham and E. Robert Baumann, "Hydraulic and particle size characteristics of some diatomic filter aids," *J. Am. Water Works Assoc.*, **56**, 793–808 (June 1964).

type, grade (size distribution), and surface characteristics of the filter medium used as a precoat. The bench chemist can control the filtered fluid quality and the hydraulic conditions of the filtering operation by regulating the size and surface treatment characteristics of the filter medium used and the rate at which the loose filter material is added to the suspension.

Precoat filter aids are produced by several manufacturers in several grades. Figure 7.9 gives the particle size distributions and grade designations of diatomite filter aids produced by three different companies.

Table 7.16 lists the trade names of several filter aid grades currently on the market that are approximately "equivalent" in performance. Filter aids are considered to be "equivalent" when they produce approximately the same flow rate and filtered solution clarity under the same operating conditions when filtering a standard sugar solution. Grades that produce the highest clarity will also produce the lowest flow rate. In general, the high-clarity filter aids are composed of very small particles of filter aid (mean size in the range of 3 to 6 μ), and high flow rate filter aids are composed of larger particles of media (mean size in the range of 20 to 40 μ). High-clarity diatomite filter aids are cheaper, yield a filtered fluid of better clarity, and produce the highest

Table 7.16 Relative Ratings of Diatomite Filter Aids

Standard Ratios[a]				
Flow Rate	*Clarity*	*Eagle-Picher*	*Johns-Manville*	*Dicalite*
100	1000	Celatom FP-2	Filter Cel	215
125	1000	Celatom FW-2	Celite 505	Superaid, UF
200	995	Celatom FP-4	Standard Super-Cel	Speedflow
300	986	Celatom FW-6	Celite 512	Special Speedflow, 231
400	983	Celatom FW-10	—	341
700	970	Celatom FW-12	Hyflo Super Cel	Speedplus, 689 CP-100
800	965	Celatom FW-14	—	375
950	963	Celatom FW-18	Celite 501	CP-5
1000	960	Celatom FW-20	Celite 503	Speedex, 757
1800	948	Celatom FW-40	—	—
2500	940	Celatom FW-50	Celite 535	4200, CP-8
3000	936	Celatom FW-60	Celite 545	4500
4500	930	Celatom FW-70	Celite 550	5000
5500	927	Celatom FW-80	Celite 560	—

[a] Based on bomb filter tests with 60°Brix raw sugar solution 80°C.

Table 7.17 Grades and Typical Applications of Dicalite Loose Filter Materials.

Dicalite Diatomite Grade	*Relative Flow Ratio*	*Average Filter Cake Density* (lb/ft^3)	*Median Particle Size* (μ)	*Typical Applications*
215 (calcined)	100	23.5	2.5–3.0	Polish filtration of beer and wine and for other liq requiring brilliant clarity
Superaid (calcined)	120	23.5	2.5–5.0	*Industrial:* alcohol, tallow, tar oil (heated); *Food processing:* lard, beet molasses (dilute), citric acid milk sugar, vinegar, wine (polishing)
UF (calcined)	145	23.5	2.5–4.0	*Industrial:* lubricating oil (used Diesel); *food processing:* beer (polishing), wine (polishing), can molasses (dilute); *pharmaceutical:* hormones, pyrogens, vitamins
Speedflow (calcined)	200	23.0	5.5–6.5	*Industrial:* paper mill sulfite liquor, driers (cobalt lead, and manganese tallates), phosphoric acid, p oil, polybutylene, varnish; *pharmaceutical:* agar, hormones, pyrogens, shaving lotion, rochelle salt *food processing:* skim milk (raw), citric acid, cane sugar (washed raw liquor), beer (polishing), bran
231 (calcined)	325	23.0	6.0–7.0	still slop, lemon juice, olive oil, pectin (beet), vinegar, wine (polishing)
Special Speedflow (flux-calcined)	350	21.0	5.5–7.0	*Industrial:* clarification of various fats, lanolin, la removal, sodium chloride; *food processing:* phosphoric acid, lemon juice, citrus pulp, malt syrup, beet (30° Baumé), cane sugar (high raw rem
Speedplus (flux-calcined)	700	21.0	8.8–10.0	*Industrial:* sulfite liquor, soda ash solution, linsee oil, tallow, driers (cobalt and magnesium tallate), magnesium chloride, sodium silicate, tall oil, lubricating oils, Diesel, sodium chloride, sulfuric ac tung oil, soya oil, transformer oil, paper mill liqu (white water, black liquor); *Pharmaceutical:* palm alginate liquors, streptomycin, bitters, shaving lot *food processing:* beer (ruh), citric acid, citrus pulp cane sugar (raw affination syrup), beet sugar (thic juice), cane sugar (raw high remelt, washed raw defecated), grape juice, cottonseed oil, wine (roug malt syrup, beet (30° Baumé), pectin (beet), brand still slop, molasses (dilute cane sugar), dry cleanin
Speedex (flux-calcined)	1030	21.0	10.0–12.5	*Industrial:* pyroxylin solution, hemp oil, dopes, fu oil, mucilage, aluminum hydroxide floc, paints, lacquers, Soln's dye, wax (linoleum emulsion), lubricating oil (used), still residue; *food processing* apple juice, citric acid, cane sugar (raw affination syrup, raw high remelt, washed raw defecated), be sugar (thick juice), coconut oil, beer (wort), drink water

le 7.17 (Continued)

)icalite 'atomite Grade	*Relative Flow Ratio*	*Average Filter Cake Density* (lb/ft^3)	*Median Particle Size* (μ)	*Typical Applications*
0 x-calcined)	1800	23.0	18.0–23.0	*Industrial:* pyroxylin solution, Soln's dye, linseed oil (air-blown), lubricating oil (used), rayon liquor, acetate liquors, dopes, lacquers, varnish, glue and glycerine roller stock, resins, aluminum hydroxide floc, wax (linoleum emulsion), type coating solution, *food processing:* beer (wort), lime juice, grape juice (concentrated), corn gluten, orange juice; *swimming pool water*
0 x-calcined)	1925	23.0	22.0–25.0	Typical uses for Dicalite 4500, 5000, and 6000 are in filtration of the same liquids for which Dicalite 4200 is used. However the higher flow rates of these products are very useful when greater filter output is needed. More specifically: typical liquids filtered are fruit juices, glue, polyethylene, rayon liquor, and varnish.
0 x-calcined)	2050	23.0	25.0–27.0	
0 x-calcined)	2500	25.0	30.0–38.0	

ɔurtesy of Great Lake Carbon Corporation.

rate of pressure drop increase across the filter, therefore shorter filtration cycles. The normal filter aid grades used in potable water filtration without use of media coatings or other chemical pretreatments produce clarities between 936 and 960 as noted in Table 7.16. The performance of diatomite as a filter aid depends on the unique physical structure of the diatom particles available in an almost infinite variety of shapes and sizes, which produce an extremely porous medium with numerous microscopic waterways and sieves that trap impurities. Most diatomite filter aids contain less than 0 to 0.5 % moisture. Specially prepared diatomaceous products for laboratory and/or pilot plant use are now stocked by nearly all laboratory supply houses and are familiar to most laboratory workers. Many of the grades listed are used for laboratory purposes, but more highly refined products (analytical filter aids) are required for many laboratory filtration conditions.

Table 7.17 lists several grades and applications of Dicalite loose filter materials used in industrial and related laboratory filtrations. These grades can normally be used in concentrations up to about 120 g/liter. Figure 7.10 is a precoat chart that displays the screen mesh sizes and screen opening sizes of septa that will be bridged over by various grades of filter aid. The related grades in Table 7.16 may be expected also to form precoats or cakes as indicated in Fig. 7.10. Figure 7.11 shows the approximate size of the suspended

Precoat chart
Ability of Dicalite filteraids to precoat various screens

	Screens				
Mesh	20 × 20	50 × 50	60 × 60	80 × 80	24 × 110
Millimeters – opening	0.89	0.32	0.25	0.14	0.008
Diatomite grades					
215	■	■	■	■	■
Superaid	■	■	■	■	■
UF	■	■	■	■	■
Speedflow	■	■	□	□	□
Special speedflow	■	■	□	□	□
Speedplus	■	■	□	□	□
Speedex	■	■	□	□	□
4200	■	■	□	□	□
4500	■	□	□	□	□
5000	■	□	□	□	□
6000	■	□	□	□	□
Perilite grades					
416	■	■	■	■	■
426	■	■	■	■	■
436	■	■	■	□	□
476	■	□	□	□	□
CP–150	■	□	□	□	□
4106	■	□	□	□	□
CP–175	■	□	□	□	□
4156	■	□	□	□	□

■ No precoat □ Precoat

Fig. 7.10. Retention of Dicalite loose materials on screens. Courtesy of the Great Lakes Carbon Corporation.

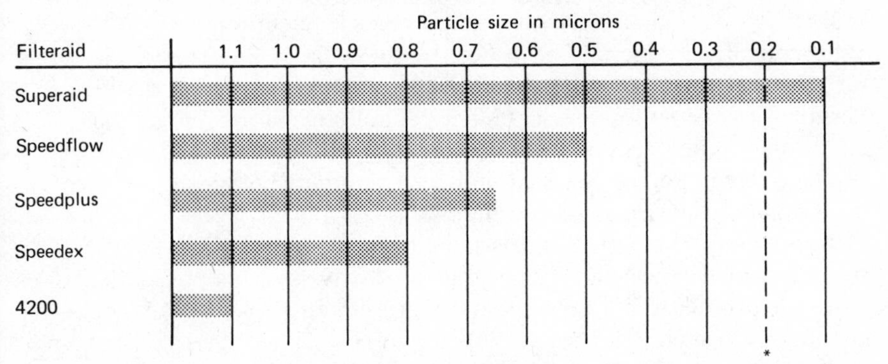

Fig. 7.11. Sizes of particles removed by five grades of diatomite filter aids; the data were established through the use of suspensions of uniform particles of known size. Courtesy of the Great Lakes Carbon Corporation.

Table 7.18 Typical Perlite Properties

Physical Properties[a]		*Chemical Properties: Average Analysis, Dry Basis* — *Property or Constituent*	*Percentage*
Specific gravity	1.70–2.10	Moisture loss at 105°C	0.20
Particle charge	Negative	Total moisture loss after ignition at 800°C	3.38
Bulk density in a filter cake (lb/ft^3)	9.5–13.5	Aluminum oxide (Al_2O_3), including any phosphorus pentoxide or manganese oxide	13.08
		Lime (CaO)	0.72
		Iron oxide (Fe_2O_3)	0.89
		Magnesia (MgO)	0.18
		Potassium oxide (K_2O)	4.44
		Silicon dioxide (SiO_2)	73.20
		Sodium oxide (Na_2O)	3.31
		Sulfur trioxide (SO_3)	0.04
		Titanium dioxide (TiO_2)	0.09

[a] From laboratory tests conducted at Iowa State University.

particles retained by various grades of Dicalite filter aids. Related grades (Table 7.16) should also provide similar particle retention.

Several other materials have been used as loose filter materials. Most successful of these is perlite, which is obtained by processing perlitic rock. Perlitic rock is composed essentially of aluminum silicate and contains 3 to 5 % water. When crushed and heated, the rock expands and fractures to produce a light, porous material similar to diatomite both in appearance and in hydraulic characteristics. Perlite is used in many of the same ways as diatomite. As a filter aid, perlite is available in different grades that vary with respect to particle size and specific gravity. An average analysis of 10 perlites currently produced in 6 different states appears in Table 7.18. A noticeable characteristic of perlite is that its bulk density in a filter cake is about one-half that of diatomite filter aids.

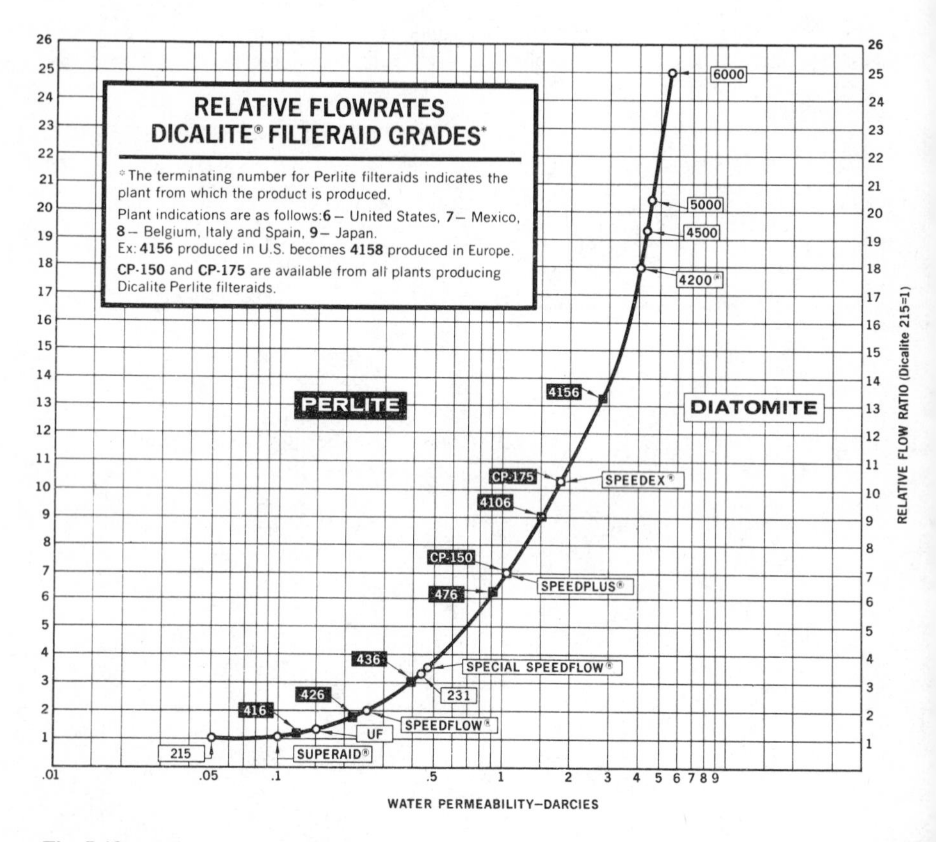

Fig. 7.12. Relative permeability versus flow characteristics of Dicalite diatomite and perlite loose filter materials. Courtesy of the Great Lakes Carbon Corporation.

Like diatomite, perlite filter aids are produced in several grades of differing particle size distributions. It has been found that there may also be differences in the characteristics of filter aid from various production lots of a particular grade and even from various bags of a particular lot. These discrepancies arise from variations between deposits of diatomaceous earth or perlitic rock and the methods of processing the filter aids.

Figure 7.12 plots the permeability versus relative flow characteristics of Dicalite perlite and diatomite filter aids.

SOURCES OF INFORMATION

Up-to-date information on companies that produce filters and filter media for laboratory and commercial use can be found in a filtration engineering catalog obtainable from Filtration Publishing, Inc., 25 West 45th Street, New York, New York 10036.

NAME INDEX

Numbers in parenthesis are reference numbers and indicate that the author's work is referred to although his name is not mentioned in the text. Numbers in *italics* show the pages on which the complete references are listed.

Adams, B. A., 199(5), *251*
Adamson, A. W., 211(19), *251*
Ahrgren, L., 263(19), *291*
Akerstrom, S., 263(18, 19), *291*
Alfa Products, Ventron Corp., 6(9), *22*
Ambler, C. M., 328(13), 330(13), 333(13), *347*
Ambrose, D., 4(4), 7(4), *22*
Amphlett, C. B., 216(31), *251*
Analytical Standards Committee, Report prepared by, 10(25), 17(25), *22*
Anders, M., 240(83), *253*
Anderson, H. M., 169(72), *195*
Anderson, N. G., 244(107, 108), *253*
Anfinsen, C. B., 214(26, 27), *251*
Armington, A. F., 8(15), *22*
Ashani, Y., 276(48), *292*
Asselin, G. F., 134(34), *195*
Atkinson, A., 281(70), *292*
Atwood, G. R., 11(30), *23*
Axen, R., 258(3), 262(3), 263(3, 14, 16, 17), 285(3, 16), *291*

Babb, A. L., 169(75), *195*
Bahr, G., 232(55), *252*
Bailon, P., *292*(45)
Barber, M. L., 244(108), *254*
Baricos, W. H., 283(74), *292*
Barker, S. A., 276(50), *292*
Barnard, A. J., Jr., 8(19), 9(22), 10(22), 17(62, 64, 65), 20(64, 67), 21(64, 73, 74, 75, 78), *22, 24*
Barry, S., 276(36), 283(72), 286(89, 90), *291,* 292(46, 47), *293*
Bartels, C. R., 128(24), *194*
Barth, H., 219(72), 237(72), *253*
Bartling, G. J., 263(20), *291*
Barton, A. F. M., 51(11), 54(11), *74*
Baukenkamp, J., 248(142), *254*
Beeghly, H. F., 10(24), *22*
Befeffy, O., 217(42), *252*
Benenati, R. F., 169(74), *195*
Benson, A. M., 244(111), *254*
Berg, E. W., 211(15), *251*
Biedermann, W., 232(68), *252*
Billiet, H. A. H., 60(22), *75*
Biondi, M. A., 3(2), *22*
BioRad Laboratories, 207(12), 214(12), 216(12), *251*
Blackburn, D. W. J., 219(71), 237(71), *253*
Blatt, W. F., 232(57), *252*
Blattner, F., 244(121), *254*
Blattner, F. R., *253*(105)
Blumberg, P. M., 276(49), *292*
Bond, F. C., 311(4), *347*
Bonn, J. D., 20(67), *24*
Bossinger, C. D., 285(79), *293*
Boyd, G. E., 211(18, 19), *251*
Bowes, W. M., 182(83), *196*
Brandone, A., 216(32), *251*
Braun, T., 217(42, 43), *252*
Breter, H., 244(119), *254*
Brinkworth, R. L., 286(88), *293*
Brodelius, P., 278(58), 279(63), 286(92, 98), *292, 293*
Brooks, J. D., 17(64), 20(64), 21(64, 78), *24*
Brooks, M. S., 8(15, 16), *22*
Broser, W., 232(68), *252*
Brostrom, C., 286(94), *293*
Brown, H. D., 263(20), *291*
Brown, P. R., 244(114, 115, 116, 117, 126), *254*
Buchwald, H., 217(44), *252*

Bulavina, Z. N., 216(33), *252*
Bunger, W. B., 27(1), 28(1), 32(1), 41(1), 67(1), *74*
Burtis, A. C., 244(110), *254*
Burtis, C., 244(120), *254*
Busch, E. W., 244(112), *254*
Bush, H., 241(91), *253*
Bush, M. T., 107(15), *194*
Butler, J. J., 181(82), *196*
Butts, W., 244(120), *254*

Caldwell, I. C., 243(102), *253*
Campbell, A. J. R., 80(10), *194*
Campbell, A. N., 80(10), *194*
Campbell, B. H., 21(79), *24*
Cannon, M. R., 170(76), 171(77), *196*
Contamination Control Handbook, 12(38), *23*
Cardinand, R., 273(35), *291*
Carlsson, B., 241(94), *253*
Carpenter, F. H., 124(22), *194*
Carr, J. J., 167(69, 70), *195*
Casciato, R. J., 316(5), *347*
Cassidy, H. G., 208(13), 232(61, 62, 63, 64, 65, 66, 67), *251, 252*
Chaiken, I. M., 286(84, 85, 86), *293*
Chalmers, R. A., 10(23), *22*
Chambers, R. P., 283(74), *292*
Chan, W. W. C., 286(99, 100, 101), *293*
Cheniae, G. M., 241(85), *253*
Christiansen, E. B., 311(1), *347*
"Clean Room and Work Station Requirements, Controlled Environment," 14(52), *23*
Clegg, J. B., 240(80), *253*
Cline, G. B., 317(8), 324(8), *347*
Clive, R. L., 158(53), *195*
Coas, V., 242(99), *253*
Coetzel, C. J., 247(130), *254*
Cohen, W., 283(74), *292*
Cohn, W., 244(125), *254*
Cohn, W. E., 241(84), 242(100), 243(100), 244(109), 248(136), *253, 254*
Colescott, R. L., 285(79), *293*
Comb, D. P., 243(103), *253*
Comer, M. J., 281(70), *292*
Comings, E. W., 134(34), *195*
Compere, E. L., 134(31), 150(43), *194, 195*
Cook, P. I., 285(79), *293*
Coplan, B. V., 155(47, 48), *195*
Corbin, J. D., 286(94), *293*
Corsuick, J., 240(77), *253*
Cotter, R. L., 11(36), *23*
Cox, J. D., 81(11), *194*
Cozzi, D., 242(99), *253*
Craig, D., 125(23), *194*
Craig, L. C., 78(4,5), 124(22), 125(23), 143(38), 145(39), 146(39), *194, 195*
Crampton, C. F., 244(111), *254*
Craven, D. B., 279(62), 281(65, 70), *292*
Cuatrecases, P., 262(11, 12), 264(21), 265(22, 23), 266(11, 25), 272(32), 273 (21, 34), 285(21), *291*

Dack, M. J. R., 51(10), *74*
Dallmeir, E., 219(72), 237(72), *253*
Darger, B. L., 219(72), 237(72), *253*
Das, K., 283(73), *292*
Davidson, J. K., 155(47, 48), *195*
Dean, P. D. G., 260(6), 262(6), 287(6), 278(57, 59, 60), 279(62), 281(64, 65, 66, 67, 68, 69, 70), *291*(37), *292*
Degeiso, R. C., 248(140), *254*
de Körösy, F., 232(53), *252*
Densen, P. M., 107(15), *194*
Deryaguin, B. V., 7(13), *22*
Desbuquois, B., 266(25), *291*
Desideri, P. G., 242(99), *253*
Determann, H., 214(28), *251*
Diehl, H., 13(44), *23*
Dintzis, H. M., 285(78), *293*
Doherty, D. G., 241(92), *253*
Dolmatov, Y. D., 216(33), *252*
Dolmatova, M. Y., 216(33), *252*
Donaldson, J. D., 247(131), *254*
Dourim, V., 217(45), *252*
Dravid, A., 232(57), *252*
Driscoll, G. L., 20(68), *24*
Dritil, J., 217(45), *252*
Duling, I. N., 20(68), *24*
Dunn, B. M., 286(84, 85), *293*
Dunnill, P., 283(73), *292*
du Vigneand, V., 124(22), *194*
Dybczynski, R., 247(128), 248(137), *254*(138, 139)

Eastman, R. H., 6(10), *22*
Eckschlager, K., 21(80), *24*
Eisner, U., 232(59), *252*

Englin, B. A., *74*(6b)
Enos, C. T., 11(35), *23*
Eon, C., 49(15), 53(15), 59(15), 60(23), 68(23), *75*
Epstein, C. J., 214(26, 27), *251*, 286(91), *293*
Erdey, L., 232(60), *252*
Er-el, Z., 291(38), *292*(39, 43)
Erickson, H., 244(121), *254*
Erickson, H. P., *253*(105)
Erickson, R. P., 286(91), *293*
Ernback, S., 258(3), 262(3), 263(3, 14, 16), 285(3, 16), *291*
Eyring, H., 2(1), 3(1), *22*
Ezrin, M., 232(63), *252*

Failla, D., 285(81), *293*
Farag, A. B., 217(43), *252*
Fedyakin, N. N., 7(13), *22*
Feick, G., 169(72), *195*
Fenske, M. R., 113(17), 136(35), 157(49, 50), *194, 195*
Fenslau, A., 276(51), *292*
Figgins, C. E., 72(26), *75*
Fisch, H. U., 266(24, 27), *291*
Floridi, A., 241(90), *253*
Folley, R. L., *255*(148)
Forrester, L. F., 263(20), *291*
Frankel, F. R., 244(111), *254*
Freeman, D. H., 16(60), *24*
Frenc, M., 78(1), 115(18), *194*
French, W. G., 8(18), *22*
Fuller, M. J., 216(30), 247(131), *251, 254*

Gambill, W. R., 33(6d), *74*
Ganapathi, M., 165(63), *195*
Gans, R., 199(3, 4), *251*
Gapon, T. B., 217(41), *252*
Garn, P. D., 20(71), *24*
Geoffrey, G. L., 11(35), *23*
Gestrelius, S., *293*(76)
Gilbert, H., 14(53), *23*
Girardi, S., 216(32), *251*
Gnichtel, H., 232(68), *252*
Godfrey, N. B., 67(6e), *74*
Gold, M., 240(83), *253*
Goldberger, W. M., 169(74), *195*
Goodman, M., 266(28), *291*
Gordon, R. D., 174(79), *196*
Grande, J. A., 248(142), *254*
Grassetti, D. R., 285(80), *293*
Gray, C. J., 276(50), *292*
Green, G. J., 241(87), *253*
Green, J. C., 244(107, 108), *253*
Griffin, T., 276(36), *291, 292*(46)
Griffith, O. M., 317(9), *347*
Gruden, J. R., 11(34), 13(34), *23*
Grushka, E., 237(70), *252*
Guilford, H., 278(55, 56), *292*
Gurka, D., 42(13), *74*
Gurvich, A. M., 217(41), *252*

Haklits, I., 217(42), *252*
Hallaba, E., 216(34), *252*
Hallquist, L. G., 21(79), *24*
Hamilton, P. B., 240(75), *253*
Hanoune, J., 266(26), *291*
Hansen, C. M., 59(21), *75*
Hanson, H., 285(83), *293*
Hartwick, R. A., 244(114), *254*
Harvey, M. J., 279(62), 281(64, 65, 66, 67, 68, 69, 70), *292*
Hatano, H., 242(97), *253*
Hazen, W. C., 158(53), *195*
Helfferich, F., 205(6), 206(6), 207(11), 211(16, 17), *251*
Herington, E. F. G., 11(28), 20(70), *23, 24*
Herod, J., 244(116), *254*
Hersh, C. K., 14(51), *23*
Hildebrand, J. H., 51(8, 9), *74*
Hofstee, B. H. J., *292*(40)
Hogeboom, G. H., 124(22), *194*
Holgoin, J., 273(35), *291*
Hollis, R. F., *252*(52)
Holmes, E. L., 199(5), *251*
Horvath, C., 36(7), 39(7), 49(7), 50(17a, 17b), 51(7), *74, 75,* 244(113), 254(127)
Horvath, C. G., 218(46), 244(46), *252*
Horvath, J., 14(50), *23*
Houser, J. J., 20(71), *24*
Huber, J. F. K., *254*(146)
Hunter, J. B., 111(16), *194*
Hurlbert, R. B., 241(91), *253*
Hurwitz, J., 240(83), *253*

Imai, Y., 292(42)
Imman, J. K., 285(78), *293*
Inczedy, J., 232(60), *252*
Ireson, J. C., 276(50), *292*

Irving, H. M. N. H., *74*(11a)
Ishii, S., 286(95), *293*

Jacobs, J. J., 216(37, 38), *252*
James, A. T., 11(31), *23*
Jannke, P., 12(39), *23*
Janson, J. C., 261(8), *291*
Jantzen, E., 78(2), 115(19), *194*
Johnson, W. F., 241(88), *253, 255*(148)
Jolley, R. L., 241(86, 88), *253*
Jones, G., Jr., 241(96), *253*
Jones, R. T., 240(77), *253*
Joris, L., 42(13), *74*
Jost, R., *292*(44)
Joy, E. F., 8(19), 17(64, 65), 20(64, 67), 21(64, 73, 74), *22, 24*
J. T. Baker Chemical Co., 21(78, 79), *24*
Junowics, E., 243(106), *253*
Jurs, P. C., 72(26), *75*

Kadar, K., 217(42), *252*
Kagedal, L., 263(18, 19), *291*
Kamogawa, H., 232(67), *252*
Kaplan, N. O., 266(28), *291*
Karasek, F. W., 11(33), 13(33), *23*
Karemaker, H. H., 247(132), *254*
Karger, B. L., 36(7), 39(7), 49(7, 14, 15), 51(7), 53(15), 59(14, 15), 60(23), 68 (23), *74, 75*
Karr, A. E., 79(9), 167(66, 68), 176(80, 81), 182(83), 190(85), *194, 195, 196*
Kasai, K., 286(95), *293*
Katz, S., 241(96), 243(103), *253*
Keith, F. W., Jr., 328(14), 330(14), 333 (14), *347*
Keller, R. A., 49(14), 59(14), *75*
Kennedy, J. K., 8(16), 12(39), *22, 23*
Kershner, N. A., 21(73, 74), *24*
Kesler, R. B., 241(89), *253*
Kesting, R. E., 232(56), *252*
Khym, J. X., 241(84, 85, 86, 92), 243(101), 244(123), *253, 254*
Kiaser, E., 285(79), *293*
Kielland, J., 98(13), *194*
Kirkland, J. J., 11(32), 13(32), *23,* 32(4), 55(20a), 67(4), 68(4), 70(4), *74, 75,* 218(47), 219(47, 50, 51), 244(47, 118), *252, 254*
Kleiman, G., 128(24), *194*
Klines-Sznik, A., 217(43), *252*
Klingenberg, A., 225(73), 237(73), *253*
Klinkenberg, A., 130(27, 28), 131(27), *194*
Knoeck, J., 13(44), *23*
Knox, J. M., 237(70), *252*
Koelsch, R., 285(83), *293*
Koes, M. T., 263(20), *291*
Koh, C. K., 244(109), *254*
Kolf, G., 232(54), *252*
Konkle, T. V., 167(69), *195*
Korenman, I., 4(5), *22*
Kraus, K. A., 217(39, 40), 247(129, 134), *252, 254*
Krishna, M. S., 165(63), *195*
Kristiansen, T., 260(7), 262(7), 266(7), *291*(29)
Krug, F., 266(25), *291*
Kuehner, E. C., 13(43), 16(60), *23, 24*
Kun, K. A., 232(61, 65, 66), *252*
Kunin, R., 206(8), *251*
Kuo, J. Y., 263(13), 285(13), *291*
Kura, G., 248(143), *254*

Laas, T., 261(8), *291*
Lacombe, M. L., 266(26), *291*
Ladd, F. C., Jr., 244(108), *254*
Laland, S. G., 286(96), *293*
Landgraff, L. M., 283(75), *292*
Larsson, P. O., 278(56), 279(61, 63), *292*
Lasch, J., 285(83), *293*
Laub, R. J., 72(27), *75*
Lauer, F. C., 181(82), *196*
Lautsch, W., 232(68), *252*
Lauwerier, N. A., 130(27), 131(27), *194*
Lavanchy, A. C., 328(14), 330(14), 333 (14), *347*
Lawrenson, I. J., 20(69, 70), *24*
Lee, K. S., 241(93), *253*
Lee, T., 331(15), *347*
Lepri, L., 242(99), *253*
Lerman, L. S., 258(2), *291*
Leslie, R. T., 13(43), *23*
Light and Co., England, 8(8), *22*
Lilly, M. D., 283(73), *292*
Lindberg, M., 279(61), *292*
Lindenbaum, S., 248(140), *254*
Lipsky, S. R., 218(46), 244(46, 113), *252, 254*(127)
Little, J. N., 11(36), *23*

Little, K., 17(63, 64), 20(64), 21(64, 78), 24
Littlewood, E. J., 181(82), *196*
Lloyd, W. A., 171(77), *196*
Lo, T. C., 167(68), 176(80), *195, 196*
Lochte, H. L., 162(57), *195*
Long, R. B., 157(49, 50), *195*
Lott, P. F., 21(75), *24*
Lowe, C. R., 260(6), 262(6), 278(57, 59, 60), 281(64, 65, 66, 67, 68, 69), 287(6), *291, 292*
Lowe, M., 286(93), *293*

McArthur, C. K., *252*(52)
McAuliffe, C., *74*(6c)
McCalla, K., 240(77), *253*
McClaren, J. V., 276(50), *292*
MacDonald, F. R., 244(110), *254*
McDonald, P. D., 11(36), *23*
McEwan, C. R., 316(6), *347*
McReynolds, W. O., 72(25), *75*
Mader, W. J., 20(72), *24*
Magnotta, F., 20(68), *24*
Majors, G., 217(42), *252*
Majors, R. E., 6(12), *22,* 219(48, 49), 227 (49), *252*
Malissa, H., 13(45), *23*
Maloney, J. O., 138(36), *195*
Manley, F., *253*(104)
Manley, G. J., *253*(104)
Mansson, M., *293*(76)
Mar, B. W., 169(75), *195*
March, S., 262(11), 266(11), *291*
March, S. C., 262(12), 265(23), *291*
Mark, H. B., 232(59), *252*
Marquardt, I., 285(83), *293*
Martin, A. J. P., 11(31), *23,* 132(30), *194,* 237(69), *252*
Martinsson, E., 241(95), *253*
Masters, C. J., 286(88), *293*
Mather, A. N., 263(20), *291*
May, S. W., 259(4), 260(4, 50), 263(13), 273(4), 283(75), 285(13), *291, 292*
Mears, T. W., 10(24), *22*
Melander, W., 50(17a, 17b), *75*
Melchior, P., 4(6), *22*
Mellan, I., 28(6a), 31(6a), *74*
Meloni, S., 216(32), *251*
Metsch, F. A. V., 35(5), 67(5), *74*
Michaels, A. S., 232(57), *252*
Michaelis, R. E., 10(24), *22*
Michelotti, F. W., 5(7), 17(65), *22, 24*
Mickelson, D. C., 247(129), *254*
Minczewski, J., *254*(138)
Miron, T., 267(31), *291, 292*(44)
Misak, N. Z., 216(34), *252*
Misek, T., 165(62), *195*
Mitchell, J. W., 12(40), 16(40), 20(40), *23*
Moates, G. H., 12(39), *23*
Moffit, E. A., *255*(147)
Molnar, I., 50(17a), *75*
Moore, S., 240(74), *253*
Morie, G. P., 247(135), *254*
Morris, C. J. O. R., 208(14), *251*
Morris, P., 208(14), *251*
Mosbach, K., 278(55, 56, 57, 58), 279(61, 63), 286(92, 98, 100), *292, 293*(76)
Mrochek, J., 244(120), *254*
Munk, M. N., 244(110), *254*
Murakami, F., 242(97), *253*
Murray, J. F., 285(80), *293*
Myers, L. S., 211(19), *251*

Nader, W. K., 311(3), *347*
Nash, A. W., 111(16), *194*
Naughton, M. A., 240(80), *253*
NBS Certificate of Analysis, 10(26), *22*
Neale, G. H., 311(3), *347*
Nelson, F., 247(129, 134), *254*
Nelson, L., 232(57), *252*
Nesher, A., 15(55), *23*
Ney, W. O., Jr., 162(57), *195*
Nichol, L. W., 286(87), *293*
Nickless, G., 247(133), 248(144, 145), *254*
Nishikawa, A. H., *292*(45)
Nunley, C. E., 244(107), *253*
Nystrom, E., *292*(41)

O'Carra, P., 276(36), 283(72), 286(89, 90), *291, 292*(46, 47), *293*
Oglestree, J., 285(77), *293*
Ogston, A. G., 286(87), *293*
Ohaski, S., 248(143), *254*
Ohlsson, R., 278(55, 56, 58), *292*
Oldshue, J. Y., 79(8), 162(58), *194, 195*
Othmer, D. F., 182(83), *196*
Ozsoy, M. W., 167(69), *195*

Palmer, J. H., 14(53), *23*

Pappa, R., 283(71), *292*
Parikh, I., 262(11, 12), 265(22, 23), 266 (11), *291*
Parker, R. C., 276(50), *292*
Parks, R. E., Jr., 244(116, 117), *254*
Parlett, H. W., 190(84), *196*
Pass, L., 286(96), *293*
Patchornik, A., 286(97), *293*
Patterson, C. C., 16(59), *24*
Paulhamus, J. H., 12(37), *23*
Pearson, A. D., 4(3), 8(3, 17, 18), *22*
Pedersen, K. O., 323(10), *347*
Peppard, D. F., 149(42), *195*
Peppard, M. A., 149(42), *195*
Perry, J. H., 311(2), *347*
Peterson, E. A., 214(20, 22, 23), 240(79, 81), *251, 253*
Pettyjohn, E. S., 311(1), *347*
Pfaff, K. T., *255*(147)
Pfann, W. G., 11(27), *23*
Pharmacia Fine Chemicals, Inc., 214(29), *251*
Phillips, H. O., 217(39, 40), 247(129), *252, 254*
Pickels, E. G., 316(7), *347*
Pitt, W. W., 241(88), *253*
Pitt, W. W., Jr., 241(96), *253*
Podbielniak, W. J., 173(78), *196*
Pollard, F. H., 247(133), 248(144, 145), *254*
Pollard, H. B., 273(34), *291*
Porath, J., 214(24), *251*, 258(3), 260(7), 261(8, 9), 262(3, 7, 10), 263(3, 14, 15), 266(7), 285(3), *291*(29)
Post, O., 145(39), 146(39), *195*
Potter, V. R., 241(91), *253*
Preiss, B. A., 218(46), 244(46), *252*
Prendergast, J. A., 11(36), *23*
Preparation Liquid Chromatograph LC-500, 14(47), *23*
Prosperi, G., 283(71), *292*
Purnell, J. H., 72(27), *75*

Rabel, F. M., 14(49), *23*
Rakshys, J. W., 42(13), *74*
Raju, C. J. V. J., 165(63), *195*
Rao, C. V., 165(63), *195*
Re, L., 283(71), *292*
Reddick, J. A., 27(1), 28(1), 32(1), 41(1), 67(1), *74*
Reeve Angel Scientifica Division, 214(21), *251*
Reimann, E. M., 286(94), *293*
Reiney, W., Jr., 244(120), *254*
Reman, G. H., 79(7), 130(27), 131(27), 162(59, 60), 165(61), *194, 195*
Rhower, E. F. C. H., 247(130), *254*
Rieman, W., III, 248(140), *254*
Risby, T. H., 11(35), *23*
Risby, T. N., 72(26), *75*
Robb, W., 216(37, 38), *252*
Robberson, B., *253*(78)
Robert-Gero, M., 276(52), *292*
Robertson, D. E., 16(58), *23*
Robinson, J. W., 21(75), *24*
Robinson, R. A., 98(14), *194*
Rogers, D. E., 248(144, 145), *254*
Rohrschneider, L., 54(18), 55(19), 72(19), *75*
Rokushika, S., 242(97), *253*
Romicon Hollow Fibers, Romicon, Inc., 6(11), *22*
Rosevear, J. W., *255*(147)
Rossini, F. D., 27(2), 32(2), *74*
Rossodivita, A., 283(71), *292*
Rothwell, M. T., 248(144, 145), *254*
Rubin, B., 8(15), *22*
Rubin, R. J., 16(57), *23*
Rubinstein, M., 286(97), *293*
Rushton, J. H., 79(8), 162(58), *194, 195*
Ruvarac, A. L., 216(35), *252*
Ryland, A., 134(31), 150(43), *194, 195*

Sabbioni, F., 216(32), *251*
Saks, B., 266(28), *291*
Salama, H. N., 216(34), *252*
Salmon, J. E., 216(36), *252*
Salomon, Y., 286(93), *293*
Samuelson, O., 206(7), 207(10), 241(93, 94, 95), *251, 253*
Santi, D. V., 285(81), *293*
Sarkstein, E., 257(1), *291*
Sato, R., *292*(42)
Sawyer, W. H., 286(87), *293*
Schaafsma, A., 130(25), *194*
Schachman, H. K., 323(11), *347*
Schechter, Y., 286(97), *293*
Scheibel, E. G., 79(6), 130(29), 134(32, 33), 138(32), 142(37), 148(41), 150(44), 152 (45), 153(46), 160(54, 55), 161(56),

166(64, 65), 176(81), 182(83), 190(85), *194, 195, 196*
Schoenmakers, P. J., 60(22), *75*
Schöffman, E., 13(45), *23*
Schreinemakers, F. A. H., 83(12), *194*
Schroeder, W. A., 240(76, 77), *253*(78)
Schubert, A. E., 138(36), *195*
Schwab, F. W., 20(66), *24*
Schwarz, Klaus, 8(20, 21), *22*
Schwyzer, R., 266(24, 27), *291*
Scott, C. D., 241(86), *253*(148)
Scott, M., 278(55, 56), *292*
Scott, R. L., 51(8, 9), *74*
Scott, R. P. W., 219(71), 237(71), 241(88), *253*
Scouten, W. H., 272(33), *291*
Sela, M., 214(26, 27), *251*
Selinger, Z., 286(93), *293*
Semenza, G., 214(25), *251*
Settle, D. M., 16(59), *24*
Shain, I., 21(76), *24*
Shaltiel, S., *291*(38), *292*(39, 43)
Shorr, J., 232(53), *252*
Silverstein, R. M., 276(53), *292*
Singhal, R., 244(124, 125), *254*
Singhal, R. P., 244(122), *254*
Sjovall, J., *292*(41)
Smit, J., 216(37, 38), *252*
Smith, A. M., 240(82), *253*
Smoot, L. D., 169(75), *195*
Smyth, C. P., 41(12), *74*
Snyder, L. R., 32(4), 36(7), 39(7), 49(7, 14, 15), 51(7), 53(15), 55(20), 59(14, 15), 60(23, 23a), 62(23a), 64(4), 68(4, 23, 24), 69(24), 70(4, 24, 24a), *74, 75,* 237(70), *252*
Sober, H. A., 214(22, 23), 240(79), *251, 253*
Soldano, B., 211(18), *251*
Sotobayashi, T., 13(46), *23*
Speaker, S. S., 171(77), *196*
Speights, R. M., 7(14), 12(14), *22*
Spencer, J. H., 243(106), *253*
Spidmann, H., 286(91), *293*
Spincer, D., 247(133), *254*
Stahman, M. A., 240(82), *253*
Stasin, R. D., 263(20), *291*
Steers, E., Jr., 273(34), *291*
Stein, W. H., 240(74), *253*
Stene, S., 117(21), 130(26), 149(40), *194, 195*
Stokes, R. H., 98(14), *194*
Stopka, P., 21(80), *24*
Strominger, J. L., 276(49), *292*
Strosberg, A. D., 266(26), *291*
Sumizu, K., 242(97), *253*
Sundberg, L., 261(9), *291*(29)
Susuki, T., 13(46), *23*
Svedberg, T., 323(10), *347*
Sweet, T. R., 190(84), *196,* 247(135), *254*
Synge, R. L. M., 132(30), *194,* 237(69), *252*
Szabo, J. J., 171(77), *196*

Taft, R. W., 42(13), *74*
Talaev, M. V., 7(13), *22*
Tanford, C., 50(16), *75*
Taylor, H. C., 286(86), *293*
Tesser, G. I., 266(24, 27), *291*
Thiers, R. E., 16(61), *24*
Thistlewaite, W. P., 217(44), *252*
Thompson, D., 167(69), *195*
Thompson, H. S., 199(1), *251*
Tijssen, R., 60(22), *75*
Todd, D. B., 174(79), *196*
Tompkins, E. R., 248(136), *254*
Trautman, R., 323(12), *347*
Treybal, R. E., 157(51, 52), *195*
Trtanj, M. I., 216(35), *252*
Tustanowski, S., 248(141), *254*
Tuttle, M. H., 78(3), *194*(20)
Tuwiner, S. B., 232(58), *252*

Updegraff, I. H., 232(64), *252*
Uziel, M., 244(109), *254*

Van, R., 216(37, 38), *252*
Van Deemjer, J. J., 225(73), 237(73), *253*
Van Dijck, W. J. C., 130(25), 167(71), *194, 195*
van Raaphorst, J. G., 247(132), *254*
Van Urk-Schoen, A. M., *254*(146)
Vaquelin, G., 266(26), *291*
Varteressian, K. A., 113(17), 136(35), *194, 195*
Venter, J. C., 266(28), *291*
Veprek-Siska, J., 21(80), *24*
Verlander, M. S., 266(28), *291*
Volkin, E., 243(101), *253*

Von Berg, R. L., 169(73), *195*
von R. Schleyer, P., 42(13), *74*
Vretblad, P., 263(17), *291*

Wade, A., 244(111), *254*
Wall, R. A., 242(98), *253*
Waller, J. P., 276(52), *292*
Wallis, K., 276(51), *292*
Walton, H. F., 206(9), *251*
Waters, J. L., 14(48), *23*
Way, J. T., 199(2), *251*
Weast, R. C., 27(3), 32(3), 39(3), *74*
Weatherall, D. J., 240(80), *253*
Weber, C. W., 331(15), *347*
Wellek, R. M., 167(69), *195*
Werber, M. M., 285(82), *293*
Whitfield, W. J., 15(54), *23*
Wickers, E., 20(66), *24*
Wiegandt, H. F., 169(73), *195*
Wilchek, M., 267(30, 31), 286(93), *291*, *292*(44), *293*
Wilcox, W. R., 11(29), 12(29), *23*
Wilkins, T., 219(71), 237(71), *253*
Wilson, I. B., 276(48), *292*
Wilson Pitt, A. B. W., *255*(148)
Winzor, D. J., 286(87, 88), *293*
Wodkiewicz, L., 248(137), *254*(139)

Yatsimirskii, K. B., 21(77), *24*

Zaborsky, O., 285(77), *293*
Zaborsky, O. R., 260(5), *291*
Zahn, R., 244(119), *254*
Zappelli, P., 283(71), *292*
Zebroski, E. L., 155(47, 48), *195*
Ziadenzaig, Y., *291*(38)
Zief, M., 5(7, 14), 11(29, 34), 12(14, 29, 40, 41), 13(34), 14(50), 15(55, 56), 16(40), 20(40), *22*, *23*
Zill, L. P., 241(84, 85), *253*
Zimmer, T. L., 286(96), *293*
Zuiderweg, F. S., 225(73), 237(73), *253*

SUBJECT INDEX

Acetone-water-phenol system, 83
Acid salt exchangers, 216
Activity coefficient, 96
  ion, 98
Adsorbent, group specific, 259
Adsorption, nonspecific, 258
Adsorption processes, in ion-exchange chromatography, 207
Affinity chromatography, 257
  acrylamide supports for, 270
  activation of supports for, 262
  agarose supports for, 260
  commercially available materials for, 287
  coupling to supports, 263
  glass supports for, 270
  group specific ligands in, 276
  hydrophobicity of supports for, 259
  ligand binding in, 275
  multi-point attachment in, 263
  polysaccaride supports for, 260
  spacer arms in, 268, 273
  specific procedures for, 284
  support materials for, 260
  supports for, 259
Agarose supports for affinity chromatography, 260
Agitated columns for extraction, 161
Alternate withdrawal, in countercurrent distribution, 124
Amino acids, by ion exchange chromatography, 240
AMP analogs, in affinity chromatography, 278, 279, 283, 285
Analysis, methods for purity measurements, 18
Analytical applications of extraction procedures, 143
Analytical centrifuges, 321
Analytical standards:
  for purity, 10
    high purity substances for, 10
Anions, by ion exchange chromatography, 248
Applications, in ion exchange chromatography, 226

Balancing in centrifuges, 340
Basket centrifuges, 337
Batch extractions, with fresh solvent, 105
Batch operation, in ion exchange chromatography, 227
Batchwise extraction, simulation of, 107
Binary mixtures, representation of phase data, 79
Binary systems, immiscible, 84
Biologically active compounds, extraction of, 188
Biospecific adsorption, 257
Boiling point of solvents, 32
Bottle centrifuges, 304
Break-through in ion exchange, 204
Bush and Denson procedure for extraction, 107

Cake filtration, 358
Capacity, total, in ion exchange, 204
Carbohydrates, by ion exchange chromatography, 241
Cartridge filters, 365
Catalysts, by ion exchange, 233
Cations, by ion exchange chromatography, 247
Celluloses, in ion exchange chromatography, 211
Centrifugal extractor, 173
Centrifugal force, in centrifuges, 296
Centrifuge:
  definitions, 295
  equipment, 303
Centrifuges:
  balancing in, 340
  basket types, 337

bottle types, 304
care of, 344
conveyor discharge types, 335
disc bowl types, 332
drives for, 343
operation and maintenance for, 340
perforated wall types, 296
preparative types, 314
solid wall types, 296
tubular bowl types, 328
uses for, 340
zonal types, 317
Centrifuging, 295
density gradient in, 317
drainage in, 338
hydrostatic pressure in, 301
isopycnic banding in, 318
"reograd" in, 319
sedimentation in, 298
sedimenting velocity in, 300
separations by, 310
Stokes Law in, 299
stresses in, 302
theory for, 297
Characterization of purity, 16
Chelating, and ligand exchange resins, 233
Chemical properties, of ion-exchange resins, 204
Choice of exchange, in ion-exchange chromatography, 226
Chromatograph, components for, 238
Chromatographic columns, for liquid chromatography, 219
Chromatography:
affinity, 257
HETP in, 237
ion exchange type, 197, 234
principles of, 234
Classification of solvents, schemes for, 51
Clean environments, classes for, 14
Column:
efficiency in chromatography, 237
processes in ion-exchange chromatography, 227
Scheibel, for extraction, 160
Columns, agitated:
for extraction, 161
for liquid chromatography, 219
packing of, 220
pulsed for extraction, 167
unpacked for liquid chromatography, 220
Containment of high purity materials, 16
Contamination control, for pure substances, 14
Continuous column extraction, 109
Continuous counter-current extraction equipment, 175
process for ion-exchange chromatography, 230
Controlled cycling, in extraction, 170
Conveyor Discharge Centrifuges, 335
Countercurrent distribution:
with alternate withdrawal, 124
Craig, 116
equipment, 143
extraction equipment, continuous, 175
extractions in, 107
fractionation in, 115
continuous with two solvents, 125
for small samples, 155
technique, 116
Counter ion, in ion exchange, 205
Coupled column systems, in ion-exchange chromatography, 226
Coupling to affinity chromatographic supports, 263
Covalent bonding, in solvent selection, 47
Craig, apparatus, 143
countercurrent distribution apparatus, 116
fractionation, 115
technique, 116
Critical solution temperature, 80
Cross linked polymers in ion-exchange chromatography, 202
Cupellation, 1
Cyanogen bromide activation in affinity chromatography, 257, 261

Density, effect on filtration, 352
gradient in centrifuging, 317
in solvents, 34
Detection devices, for chromatography, 239
Dextrans, in ion-exchange chromatography, 214
Debye-Huckel Law, 96, 99
Dialysis, solvent selection for, 72
Dielectric constant, effects in filtration, 352

Dipole:
induction, 40
interactions, 39
molecular, permanent, 38
moments, 39
of functional groups, 41
orientation, 40
Disc bowl centriguges, 332
Dispersion:
forces, 38
interactions, 38
Displacement technique in ion-exchange chromatography, 229
Distribution coefficient:
defined, 90
in ionization solutions, 103
effect of pH in ideal systems, 92
apparent, 99
for ionizing solutes, 92
Distribution data for evaluation of ionization constants, 102
Donnan membrane theory in ion-exchange, 209
Drives for centrifuges, 343

Effective size in filter media, 356
Effluents, definition of, 350
Electrostatic interactions, 38
Elution analyses, in ion-exchange chromatography, 229
Elution techniques in ion-exchange chromatography, 228
Enzymes, immobilized, 283, 285, 287
Equilibrium constant, thermodynamic, 96
Exchangers, mixed bed types, 231
Extraction:
in agitated columns, 161
applications of, 180
batchwise, simulation of, 107
Bush and Denson procedure for, 107
centrifugal extractor for, 173
controlled cycling for, 170
mixer-settler devices for, 155
pulse columns for, 167
Scheibel column for, 160
simulated continuous process for, 148
of biologically active compounds, 188
continuous column for, 109
feed state in, 131
fractionation, processes for, 182
using dual solvent with reflux, 139
using single solvent and reflux, 134
graphical method:
of Hunter and Nash for, 111
of Varteressian and Fenske, 113
liquid-liquid type, 78
metal chelates in, 187
optimum reflux ratio in, 142
pilot plant studies in, 179
preparation of small scale samples by, 155
procedure of Maloney and Schubert for, 138
procedures and equipment for, 142, 177
processes, metallurgical, 125
purification of vitamin A by, 190
rejection ratios in, 128
retention ratios in, 128
scale up, 177
separation of heptane and methyl cyclohexane, 136
solvent ratio in, 130
steady state in, 149
operation, 152
simulation in, 153
uranium, 185
Extractions:
analytical applications, 143
batchwise with fresh solvents, 105
countercurrent, 107
nonideal systems, 134
Extractive distillation, solvent selection for, 68

Feed stage location in extraction, 131
Filter aids, chemical types, 364
Filter, apparatus, 365
definition of, 350
media, Diatonite, 411
fibers, 397
filter papers, 350
loose granules, 410
membranes, 350
parameters for, 356
perforated metal, 408
Perlite, 420
weave patterns, 390
woven, 395
fibers, properties of, 407
types, 388

yarn forms, woven, 390
medium, definition of, 350
paper, 368
permeability of, 368, 370
properties of, 369
standards for, 368
Filters:
cartridge types, 365
funnel types, 365
information sources for, 421
leaf types, 365
membrane, particle retention of, 376
pore sizes of, 375
specifications for, 377
types, 365, 374
types of, 365
woven types, 388
Filtrate, definition of, 350
Filtration, 349
cake type, 358
factors affecting, 352
granular bed type, 360
materials to be filtered, 351
particle shape in, 352
porosity in, 355
speed of, 374
suspensions in, 352
theory of, 353
Flow programming, in ion-exchange chromatography, 226
Fractional crystallization, solvent selection for, 67
processes in liquid extraction, 115
Fractionation:
by extraction using dual solvent with reflux, 139
in extraction using single solvent and refllux, 134
processes:
batchwise techniques, 115
in extraction, 182
Free energies of ionization, 98
Frontal development in ion-exchange chromatography, 228

Gas chromatography, solvent selection for, 71
Gibbs-Duhem law, 91
Glass supports, in affinity chromatography, 270
Granular bed filtration, 360
Group specific:
adsorbent, 259, 283
ligands, in affinity chromatography, 276

Halides, by ion-exchange chromatography, 248
Helium, a superfluid, 3
Heptane-methylcyclohexane separation, 136
Heteropoly acid salt exchangers, 216
HETP in chromatography, 237
High purity chemicals, 9
Hildebrand solubility parameter, 51
History of ion-exchange chromatography, 199
HPLC packing materials, 217
Hunter and Nash graphical method for extraction process, 111
Hydrogen bonding, 41
proton-acceptor strengths in, 42
Hydrophilicity of supports, for affinity chromatography, 259
Hydrophobic interactions, 47, 49
Hydrophobicity of supports, for affinity chromatography, 259
Hydrostatic pressure, in centrifuging, 301
Hydrous oxide exchangers, 216

Immiscible binary systems, 84, 88
Immobilized enzymes, 283, 385, 387
Impurity, 3
Injection system, for ion-exchange chromatography, 238
Inorganic ion exchangers, 215
Inorganic subsances, by ion-exchange chromatography, 247
Interactions, intermolecular, 38
Intermolecular interactions, 38
Ion-exchange:
catalysts, 233
celluloses, 211
chromatography, 197, 203, 234
affinity of the resin for counter-ion, 205
applications for organic substances, 240
batch operation, 227
break through in, 204
chelating and ligand exchange resins for, 233

choice of exchanger, 226
column efficiency in, 237
column processes in, 227
components of chromatography, 238
continuous process for, 230
coupled column in, 226
detection devices for, 239
displacement technique for, 229
Donnan membrane theory in, 207
elution analyses in, 229
techniques in, 228
equilibria in, 207
flow programming in, 226
for carbohydrates, 241
cations, 247
halides, 248
inorganic substances, 247
organic acids, 242
rare earth elements, 248
transition metals, 247
frontal development for, 228
in non-aqueous solvents, 234
ion exclusion in, 231
ionophoresis in, 231
HETP in, 237
history of, 199
injection system for, 238
mass action in, 207
mechanisms of reactions in, 207
mixed bed exchangers for, 231
nomenclature in, 207
oxidation-reduction polymers for, 232
packing the column for, 220
polymers for, 203
resolution in, 236
separation factor in, 237
solvent delivery system for, 239
temperature programming in, 226
total capacity in, 204
unpacked columns for, 220
dextrans by, 214
equilibria in, 207
in non-aqueous solvents, 234
kinetics in, 211
membranes in, 231
polyacrylamide gels for, 214
processes in, 227
reactions and adsorption processes in, 207
resins for, 202
chemical properties of, 204
stability of, 207
techniques and applications in, 226
terminology for, 199
Ion-exchangers, acid salts, 216
heteropoly acid salts for, 216
hydrous oxides for, 216
inorganic types, 215
metal sulfides for, 217
organic types, 201
phosphate gels for, 217
Ion-exclusion, in ion-exchange chromatography, 231
Ionic radii, 96, 98
Ionization constant:
defined, 92
evaluation from distribution data, 102
free energy of, 98
Ionizing solutes, distribution data for, 92
Ionophoresis, in ion-exchange chromatography, 231
Isopycnic banding, in centrifuging, 318

Kinetics in ion-exchange chromatography, 211

Leaching process, solvent selection for, 66
Leaf type filters, 365
Ligand binding in affinity chromatography, 275, 284
exchange resins, 233
leakage in affinity chromatography, 261, 265, 266, 267
Liquid chromatography:
columns for, 219
retention in, 236
Liquid extraction:
applications of, 180
for removal, 104
procedures and equipment for, 142
processes, theory of, 103
Liquid-liquid, chromatography:
solvent selection for, 67
extraction by, 78
partition, solvent selection for, 66
Liquid-solid chromatography, solvent selection for, 68
London forces, 36

Maloney and Schubert extraction procedure, 138

Mass Action Law, in ion-exchange chromatography, 207
Materials, standards, classification of, 10
Mechanisms of ion-exchange reactions, 207
Membrane filters, 374
  chemical compatibility of, 376
  particle retention by, 376
  pore sizes in, 375
  specifications for, 377
Membrane type filters, 365
Membranes:
  as filter medium, 350
  for ion-exchange chromatography, 231
Metal chelates in extraction, 187
Metal sulfide exchangers, 217
Metallurgical processes, extraction for, 185
Miscibility of solvents, 34
Mixed bed exchangers, 231
Mixer-settler devices for extraction, 155
Molecular Dipole Moment (debye unit), 40
Molecular species, single, 2
Molecules, identical, 2
Multi-component systems, phase equilibria for, 89
Multi-point attachment in affinity chromatography, 263

Nomenclature in ion-exchange chromatography, 207
Nonideal extraction systems, 134
Nonionizing solutes, phase equilibria for, 90
Nonspecific adsorption, 258
Nucleotide loading in affinity chromatography, 282

Operation of centrifuges, 340
Optimum solvent ratio in extraction, 130
Organic acids, by ion-exchange chromatography, 242
Organic ion exchangers, 201
Oxidation-reduction polymers in ion-exchange chromatography, 232

Packing the column, 220
Packing materials for HPLC, 217
Particle:
  form in ion-exchange chromatography, 202
  shape in filtration, 352
  size in ion-exchange chromatography, 202
  size redistribution in filter media, 356
Peptides, by ion-exchange chromatography, 240
Peripheral properties of solvents, 27
Permeability of filter paper, 368
Phase equilibria, critical solution temperature, 80
  data for multi-component systems, 89
  for acetone-water-phenol system, 83
    nonionizing solutes, 90
    phenol-acetone system, 83
    phenol-water system, 80
    pyridine-water system, 81
    ternary phase, 82
  plait point for, 83
  relationship, 79
Phenol-acetone system, 83
  extraction, 180
  water acetone system, 83
  water system, phase equilibria for, 80
Phosphate gel exchangers, 217
Phosphorus oxyanions, by ion-exchange chromatography, 248
Processes in ion-exchange chromatography, 227
Pilot plant studies for extraction, 179
Plait point, 83, 91
Podbielniak centrifugal extractor, 173
Polarity, of solvents and solutes, 47
  scales for solvents and solutes, 51
  scale of Rohrschneider, 54
Polyacrylamide gels:
  in affinity chromatography, 270
  in ion-exchange chromatography, 214
Polymeric filter aids, 364
Polymers.
  cross linked in ion-exchange chromatography, 202
  oxidation-reduction types for ion-exchange chromatography, 232
Polysaccaride supports, for affinity chromatography, 260
Polywater, 6
Porosity:
  in filtration, 355
  in polymers for ion-exchange chromatography, 203
Preparative centrifuges, 314

Principles of chromatography, 234
Procedures for extraction operation, 177
Properties of solvents, affecting detection, 33
Proteins, by ion-exchange chromatography, 240
Proton-acceptor strengths in compounds, 42
Pulse columns in extraction, 167
Pure compound, 3
Pure materials, containment of, 16
Pure substances, handling and control for, 14
Purity, 3
  at absolute zero, 3
  achievements in, 7
  analytical standards for, 10
  and alchemy, 2
  characterization of , 16
  chemicals, uses for, 9
  concept and reality in, 1
  degree of, 4
  by difference, 4
  evolution of, 1
  history of, 1
  in substances, 10
  methods of analysis for, 18
  need for, 7
  numerical expression for, 4
  spectrographic, 5
Purification, ultra, 10
Pyridine-water system, 81

Rare earth elements, by ion-exchange chromatography, 248
Rate zonal sedimentation, 318
Reflux ratio, optimum in extraction, 142
Rejection ratios in extraction, 128
Reograd in centrifuging, 319
Reorienting gradient, "Reograd," 319
Resolution in liquid chromatography, 236
Retention ratio in extraction, 128
Rohrschneider polarity scale, 54

Scale up in extraction, 177
Scheibel column for extraction, 160
Sedimentation:
  constant of Svedberg, 327
  rate zonal, 318
  theory in centrifuging, 298
Sedimenting velocity in centrifuging, 300
Selectivity parameters for solvents, 59
Separation factor in chromatography, 237
Separation processes:
  solvents for, 26
  solvent selection for, 25, 62
Separations:
  by centrifuging, 310
  solubility factors in, 36
Site of exchange, 200
Solubility:
  factors affecting, 36
  maximizing of, 62
  parameter of Hildebrand, 51, 52
Solutes, polarity of, 47
Solvent:
  classification schemes, 51
  delivery system for chromatography, 239
  extraction process, solvent selection for, 66
  flammability, 34
  fractionation, continuous countercurrent with two solvents, 125
  miscibility, 34
  polarity scales, 51
  properties, 34
    affecting detection, 33
    peripheral, 27
  selection, approach to, in separation processes, 62
    covalent bonding in, 47
    for adsorption and partition, 68
      dialysis, 72
      extractive distillation, 68
      fractional crystallization, 67
      gas chromatography, 71
      leaching process, 66
      liquid-liquid chromatography, 67
        partition, 66
      liquid-solid chromatography, 68
      separation processes, 25
      solvent-adsorption partition, 68
      solvent extraction process, 66
      thermal diffusion, 72
      ultrafiltration, 72
    summary of, 65
Solvent selectivity, 50
  groups, 60
  parameter, 59

Solvent toxicity, 34
Solvent viscosity, 32
Solvents:
  boiling point of, 32
  peripheral properties of, 27, 28
  polarity of, 47
Spacer arms, in affinity chromatography, 268, 273
Spectrographic purity, 5
Stability, in ion-exchange resins, 207
Standard materials, classification of, 10
Steady state:
  in extraction processes, 149, 152
  simulation in extraction processes, 133
Stokes Law, 299
Stresses in centrifuging, 302
Superfluid, helium, 3
Support materials for affinity chromatography, 260
Supports for affinity chromatography, 259
Surface tension effects, in filtration, 352
Suspensions in filtration, 352
Svedberg sedimentation constant, 327
Swelling in polymers, in ion-exchange chromatography, 203

Techniques in ion-exchange chromatography, 226
Temperature programming, in ion-exchange chromatography, 226
Terminology, for ion-exchange chromatography, 199
Ternary mixtures:
  data representation for, 82
  phase equilibria, 82
    for two immiscible binary systems, 84
    one solid component, 85
    three immiscible binary systems, 88
Theory:
  for centrifuging, 297
  of filtration, 353
Thermal diffusion, solvent selection for, 72
Thermodynamic equilibrium constant, 96
Tie lines in phase equilibria, 83
Toxicity of solvents, 34
Transition metals by ion-exchange chromatography, 247
Tubular bowl centrifuges, 328

Ultracentrifuges, 314
  analytical types, 321
  performance index for, 315
Ultrafiltration, solvent selection for, 72
Ultrapurification, 10
Uniformity coefficient in filter media, 356
Unpacked columns, for liquid chromatography, 220
Uranium by extraction, 185
Utrogel support for affinity chromatography, 272

Viscosity effects on filtration, 352
Viscosity in solvents, 32
Vitamin A purification by extraction, 190

Water and polywater, 6
  anomalous, 7
  purification of, 6
Woven fibers, 388

Zonal centrifuges, 317